Advances in Dynamics, Instrumentation and Control Vol. II

Proceedings of the 2006 International Conference (CDIC '06)

Advances in Dynamics, Instrumentation and Control

Vol. II

Queretaro, Mexico 13 – 16 August 2006

Editors

Alejandro Lozano
CONCYTEQ, Mexico

Subhash Rakheja
Concordia University, Canada

Chun-Yi Su
Concordia University, Canada

World Scientific

NEW JERSEY • LONDON • SINGAPORE • BEIJING • SHANGHAI • HONG KONG • TAIPEI • CHENNAI

Published by

World Scientific Publishing Co. Pte. Ltd.
5 Toh Tuck Link, Singapore 596224
USA office: 27 Warren Street, Suite 401-402, Hackensack, NJ 07601
UK office: 57 Shelton Street, Covent Garden, London WC2H 9HE

British Library Cataloguing-in-Publication Data
A catalogue record for this book is available from the British Library.

ADVANCES IN DYNAMICS, INSTRUMENTATION AND CONTROL, VOLUME II
Proceedings of the 2006 International Conference (CDIC '06)

ISBN-13 978-981-270-805-2
ISBN-10 981-270-805-7

Printed in Singapore by World Scientific Printers (S) Pte Ltd

PREFACE

This second volume of *Advances in Dynamics, Instrumentation and Control* is a compilation of 43 articles representing the scientific and technical advances in various aspects of *System Dynamics, Instrumentation, Measurement Techniques, Simulation* and *Controls*. These articles were selected on the basis of rigorous review of over 150 articles submitted for presentation at the Second International Conference of *in Dynamics, Instrumentation and Control* (CDIC'06), held in Querétaro, Mexico, 13-16 August 2006. The conference was organized by *Consejo de Ciencia y Tecnologica del Estado de Querétaro* (CONCYTEQ) jointly with *Concordia University*, Montréal, Canada, with a focus to address the scientific and technological challenges in system dynamics, instrumentation and control, by creating a forum for exchange of knowledge among the experts from the world. This second conference was co-sponsored by Montreal and Querétaro Sections of IEEE, Autonomous University of Querétaro, Canadian Society for Mechanical Engineering (CSME), American Society of Mechanical Engineers (ASME) Québec chapter, Querétaro Institute of Technology and Querétaro Campus of Monterrey Technological Institute. The final technical program featured over 80 stimulating regular technical presentations. In addition, the delegates were very privileged and honored to have the keynote addresses delivered by three leading experts in the field: *Dr. Meyya Meyyappan* from NASA (USA) on Nanostructures and their applications; Professor *Tianyou Chai*, Northeastern University (China) on Hybrid intelligent optimal control methods; and *Dr. Robert Mullins*, Bell-Helicopters-Textron (USA) on the Engineer of 2020. The delegates also participated on a panel discussion on University-Industry Challenge together with representatives from Bombardier Aerospace (Querétaro, Mexico).

The articles included in this volume have undergone a through referring process and represent state-of-the-art contributions in the fields of dynamics and control of nonlinear, hybrid and stochastic systems; nonlinear control theory; vehicular dynamics; adaptive, model predictive and real-time controls; fault diagnostics; and manufacturing systems. The contributions of the technical program committee and the referees are deeply appreciated. Most of all, we would like to express our sincere thanks to the authors for submitting their most recent works and for the patience they have shown during the review process. Special thanks are due to Mr. Steven Patt of the World Scientific Publishing (UK) for enthusiastically supporting this project.

We sincerely hope that this second volume will prove to be an important resource for the scientific community.

Alejandro G. Lozano, CONCYTEQ
Subhash Rakheja, Concordia University
Chun-Yi Su, Concordia University

CONTENTS

COOPERATIVE ROBOTIC SYSTEM USING DISTRIBUTED DECISION MECHANISMS WITH DELIBERATIVE CENTRAL SUPERVISOR*

JULIEN BEAUDRY, RICHARD HURTEAU, RICHARD GOURDEAU

Electrical Engineering Dept., École Polytechnique Montréal, 2900 Édouard-Montpetit Montréal, Québec H3T 1J4, Canada. Email : {julien.beaudry, richard.hurteau, richard.gourdeau}@polymtl.ca

Cooperative multi-robot systems with distributed decision mechanisms and distributed sensing may be the source of decisional conflicts which can lead to severe performance deterioration. A deliberative central supervisor is a simple approach to correct any incoherent decisions in the system. Given an application, the supervisor can be an autonomous software agent or a human-machine interface. Using Hierarchical Decision Machines (HDM) as distributed decision mechanisms, the decision supervision can use simple matrix representations of decisional data. The resulting architecture has been tested on a fully autonomous team of soccer-playing robots and results indicate that it is well adapted for, but not restricted to, the specific needs of autonomous multi-robot systems with real-time distributed sensing and decision taking.

1. Introduction

Following the ascension of behavior-based and hybrid systems [3], many modern robotic system architectures with similarities have been proposed. Architectures such as L-ALLIANCE [12], CAMPOUT [13] or XABSL [9] are good examples, although simpler architectures were used in experimental systems (for example in the Martha project [1]).

Clearly, the use of hybrid systems gives robotic systems a very large spectrum of possibilities: reactive capabilities for deterministic real-time response in highly dynamic environments, but also deliberation capabilities on various timescales for completion of complex tasks. A common way of discretizing level of reactiveness and deliberation in an architecture is by using a hierarchical structure.

The use of communication in a multi-robot system is of particular interest. Depending on the type of system involved, various constraints may apply

* This work is supported by the Institut de recherche d'Hydro-Québec and the Fonds québécois de la recherche sur la nature et les technologies.

concerning availability, bandwidth and reliability of communication protocol. Strict guidelines are therefore considered for the proposed architecture:

- A multi-robot system must be able to implement cooperative behaviors without the use of explicit communication.
- If explicit communication is possible for a given multi-robot system, the system should efficiently benefit of this possibility by augmenting its cooperative capabilities, but minimize the use of bandwidth.

Many other key characteristics have been identified as essential to ensure usefulness of a given architecture: modularity and scalability, multi-platform development, generic level programming and independence to hardware.

2. Multi-robot architecture proposed

It is possible to define a generic control architecture for a single robot as illustrated on figure 1. In a multi-robot system with distributed sensing and decision mechanisms, this architecture is the basic structure of each robot. A multi-robot architecture must then consider this individual level control architecture as its housing for distributed decision mechanisms. Its simple and modular structure allows its implementation with any robotic system and architecture, such as architectures with strong theoretical and experimental background like the 4D/RCS architecture [2]. Additional modules like intermediate reactive controllers or communication modules can be fitted to the generic architecture.

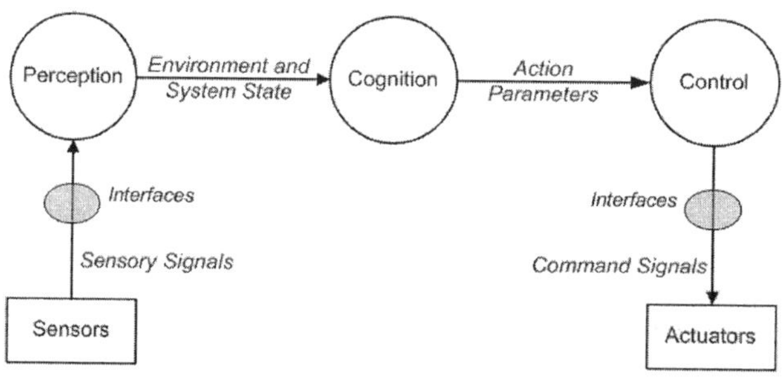

Figure 1. Generic robot level control architecture.

Three major modules are considered: *Perception*, *Cognition* and *Control*. The *Cognition* module is the core of the intelligent system. All the information of the system can be accessed and interpreted by this module, direct link between *Perception* and *Control* are therefore disallowed. The decision mechanisms are enclosed within the *Cognition* module.

2.1. *Distributed decision mechanisms*

The proposed decision mechanism is the *Hierarchical Decision Machine* (*HDM*) [4], a structural representation similarly implemented in other architectures [8], [11]. This decision machine uses a hierarchical structure and works as a succession of sequential decision mechanisms. The graphical representation of the HDM and its decision mechanism are illustrated in figure 2. XABSL [9] is a good example of similar architecture taking profit of this simple representation. The HDM is implemented in object-oriented C++. A given machine may be defined by an arbitrary number of hierarchies, each one containing an arbitrary number of *Decision_node*s. The succession of decisions, called the *Decision Line*, is always terminated by the selection of a *Behaviour_node*, which updates various *Action_parameters* in order to activate specific *Action*s of the set offered by the multi-robot system. Each type of node has access to useful information, like robot internal states and an appropriate model of its dynamic environment.

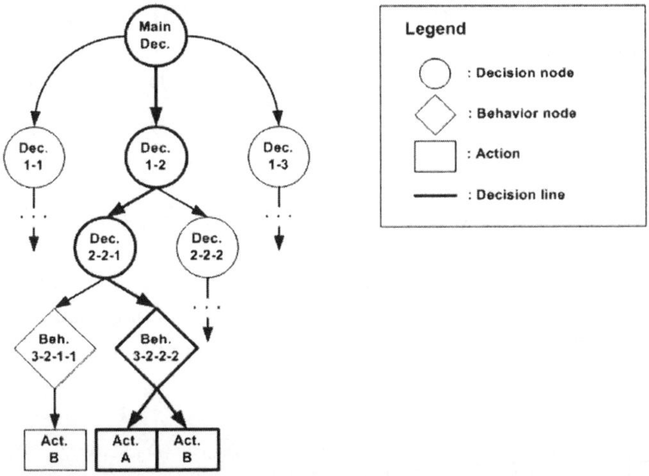

Figure 2. Graphical representation of the Hierarchical Decision Machine.

The HDM is designed with the objective of being distributed on every robot of a multi-robot system. In order to introduce cooperative behaviours between robots, pre-established agreements can be used on each decision machine, similar to locker-room agreement concept [16]. These agreements can implement various cooperative capabilities like resource sharing or dynamic role allocation.

2.2. *Deliberative decision supervision*

As a design characteristic, a multi-robot system must be able to cooperate without the use of external help. It is however possible that such system can be a source of decision conflicts between robots, in particular if perception capabilities are also distributed. In this case, instead of using distributed techniques which can still yield to conflicts, such as negotiation [6] or utility functions [5], the concept of *Decision Supervision* is introduced. A central deliberative process can control the ongoing distributed decision mechanisms on the complete hierarchy of a given decision machine and on every robot of the system.

Technically, the supervisor is a central element connected to every robot of the system. Using a client/server approach, it can centralize and redistribute any pertinent data. With the HDM simple structure, a *Decision Line* for a given robot can be represented as a *Decision Vector* (V_{D_i}) of length equal to the number of hierarchies in the decision machine. The supervisor can retrieve each *Decision Vector* of the system and form the *Decision Matrix* (M_D). These data structures can be defined as follows:

$$V_{D_i} = \begin{bmatrix} D_{i-1} & D_{i-2} & \cdots & D_{i-N_D} \end{bmatrix}, \; M_D = \begin{bmatrix} D_{1-1} & D_{1-2} & \cdots & D_{1-N_D} \\ D_{2-1} & D_{2-2} & \cdots & D_{2-N_D} \\ \vdots & \vdots & \vdots & \vdots \\ D_{N_R-1} & D_{N_R-2} & \cdots & D_{N_R-N_D} \end{bmatrix} \quad (1)$$

With this matrix obtained and updated using communication with robots, the supervisor can define a *Supervision Matrix* (M_S) composed of *Supervision Vectors* (V_{S_i}). These variables can be defined in a similar way:

$$V_{S_i} = \begin{bmatrix} S_{i-1} & S_{i-2} & \cdots & S_{i-N_D} \end{bmatrix}, \; M_S = \begin{bmatrix} S_{1-1} & S_{1-2} & \cdots & S_{1-N_D} \\ S_{2-1} & S_{2-2} & \cdots & S_{2-N_D} \\ \vdots & \vdots & \vdots & \vdots \\ S_{N_R-1} & S_{N_R-2} & \cdots & S_{N_R-N_D} \end{bmatrix} \quad (2)$$

When a robot receives its *Supervision Vector* it can respond, deliberately, to supervision order. Using such simple representation, conflicts can be rapidly addressed and easily solved. Although this scheme requires explicit communication with every robot of the system, the necessary bandwidth is kept minimal with simple numerical values (integers) exchanged. Still, control of decisional data update rate must be possible.

3. Test bench: autonomous soccer-playing robots

A team of fully autonomous and cooperative soccer-playing robots has been used to test and validate the proposed decisional architecture. The environment in which the robots operate is highly dynamic, adversarial and offers unpredictable characteristics.

3.1. *Description of the experimental platform*

The multi-robot system used as a test bench consists of a total of six soccer-playing robots [15]. The team of robots, shown in figure 3, is conforming to the rules of the Middle Size Robot League (MSL) of the RoboCup World Championship [14]. Each robot is fully autonomous, contains an embedded computer, an omnidirectional vision system [10], a wireless LAN connection, and appropriate electromechanical devices.

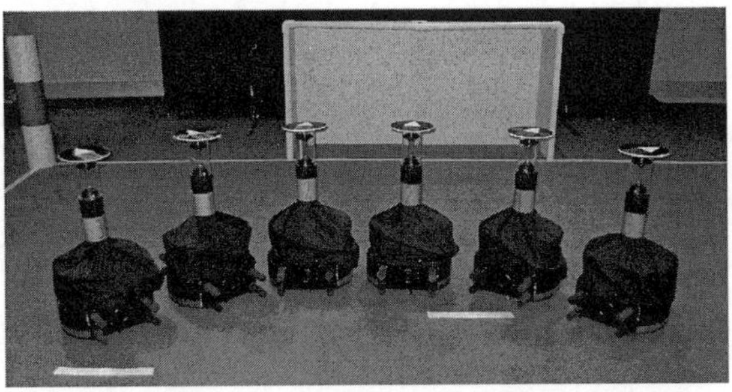

Figure 3. Team of autonomous soccer-playing robots.

A *teamserver* is connected to every robot to act as the decision supervisor. The client/server capabilities are part of MICROB [7], a C++ robotic library used as a valuable toolbox in the software architecture.

3.2. *HDM developed for a team of soccer-playing robots*

The decision machine developed for the soccer-playing robots had to respond to two specific needs, the cooperative team play and individual skills for the game itself, but also conform to all the rules of the MSL. The overall view of the decision machine is shown in figure 4. Even if this graph has been simplified, it can be seen that having a view of a complete decision machine is fastidious. Studying particular segments of a machine is more appropriate.

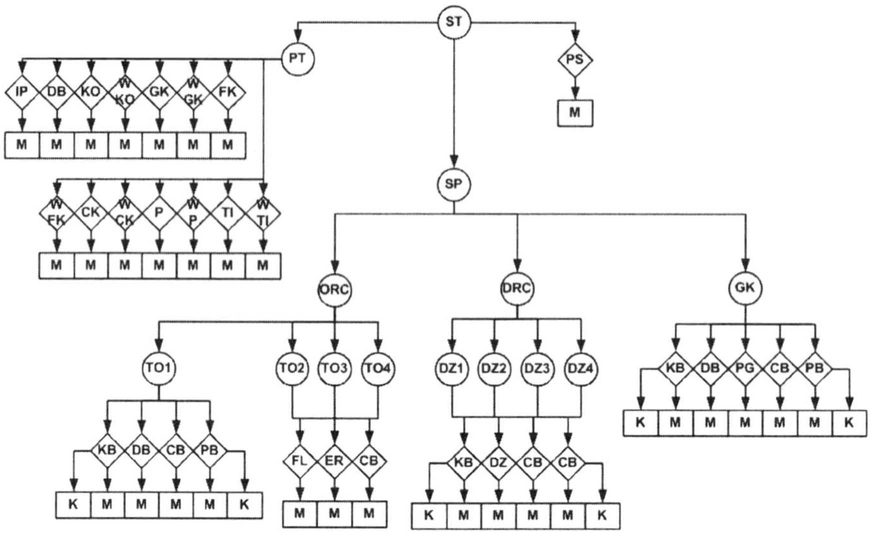

Figure 4. Overall view of a HDM developed for soccer-playing robots.

This decision machine presents a five level hierarchy: *Mode*, *Pattern*, *Role*, *Behaviour* and *Action*. The machine is composed of 128 individual nodes. But, using the object-oriented implementation of the HDM, only 21 different basic nodes have been used, since many are reutilized.

During game play situation, the mode *SoccerPlayer* allows the team of robots to cooperatively play soccer. Basically, there are offensive and defensive patterns, dynamically selected depending on game situation. The pre-established rule can be as simple as:

- if the ball is in defensive zone, defensive pattern is selected
- otherwise, offensive pattern is selected

This is an example of simple pre-established rule for cooperative team play. Such simple rule can be the source of decision conflicts, what happens when the ball is in the center of the field? Similar but more elaborate rules are often needed to minimize source of conflicts.

Offensive and defensive patterns use formations with predefined characteristic. For example, the *OffenseRC* pattern uses a dynamic role allocation procedure to implement an offensive formation where each robot selects an appropriate role. An example of such formation with its corresponding decision mechanism is showed on figure 5. Dynamic role allocation here is based on relative positions of robots to ball and opponents goal.

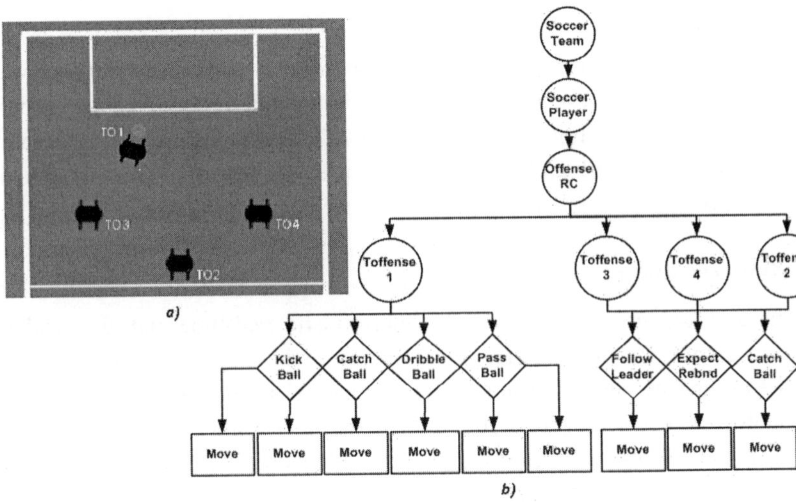

Figure 5. Dynamic offensive formation (a) and corresponding decision machine (b).

3.3. *Decision supervision for this type of multi-robot system*

Decision supervision for this system had to respond to three specific needs:

1. Accurately respond to referee calls
2. Supervise the dynamic selection of the same pattern by each teammate
3. Supervise the selection of an exclusive role by each teammate

Each robot's *Decision Vector* is composed of four elements corresponding to decision machine hierarchy. With a team of five robots a twenty elements *Decision Matrix* is obtained. Consequently, the *Supervision Matrix* contains twenty elements. The dynamic role allocation for offensive pattern (see figure 5) can illustrate the supervision mechanism. When the team is in the offensive pattern, players must select an exclusive role. A *Decision Matrix* showing a conflict for the role selection and a resulting *Supervision Matrix* could be:

$$M_D = \begin{bmatrix} 0 & 0 & X & X \\ 0 & 2 & \mathbf{1} & X \\ 0 & 2 & \mathbf{1} & X \\ 0 & 2 & 3 & X \\ 0 & 2 & 4 & X \end{bmatrix}, \quad M_S = \begin{bmatrix} 0 & -1 & -1 & -1 \\ 0 & 2 & \mathbf{2} & -1 \\ 0 & 2 & \mathbf{1} & -1 \\ 0 & 2 & 3 & -1 \\ 0 & 2 & 4 & -1 \end{bmatrix} \tag{3}$$

Response to supervision in the robots is done by a parallel thread managing communication with the server. In other words, the robot does not count on this communication, but it could receive a useful *Supervision Vector*. It is possible to adjust the update rate of the supervision mechanism and bandwidth of communicated data can therefore be controlled.

4. Results

Using the already existing soccer-playing robots hardware and software modules, every needed functionality for the multi-robot system has been rapidly developed. Enabling intuitive programming of decision machines is a major benefit of the proposed architecture. More precise performance analysis and results are presented in next paragraphs.

4.1. *Resource usage*

The architecture is to be used in deterministic real-time and also scalable robotic systems. Resource usage must therefore be kept minimal. CPU and memory usage has been measured on experimental robots. The CPU usage on the robots has been measured to a mean of 5.2% on Pentium III 800MHz processors and 7.6% on Celeron 566MHz. Plenty of CPU is left for vision algorithms. However, memory usage is important for a relatively small HDM, with 13.36MB used on robots. Duplication of nodes is responsible for this result.

Time of completion for decision mechanisms has also been measured. This time can vary depending on the active decision line. Measurements are given in table 1. Completion time is always kept below 200us which allows for fast and deterministic control loops.

Table 1. Completion time of decision mechanism measured on experimental robots.

Computer configuration	**Min. time (us)**	**Max. time (us)**	**Avg. time (us)**
Pentium III LP 800MHz	107	135	120
Celeron 566MHz	145	172	158

4.2. *Decision supervision relevance*

Measurements have showed that even with meticulously defined cooperative rules, the HDM developed was the source of decision conflicts. Tests have been made considering a two minutes period where dynamic role allocation is measured for the offensive pattern *OffenseRC*. In this pattern, if the same role is used by two robots at the same time a conflict is detected. Figure 6 shows the dynamic role allocation of the four field players without (*a*) and with (*b*) supervision. According to these results, decision conflicts occurring using the cooperation agreements can be reduced from a near 45% to a level of approximately 2.5% of the total working time. This two minutes test considers a short period of time and variability is important, but every other test executed showed similar results.

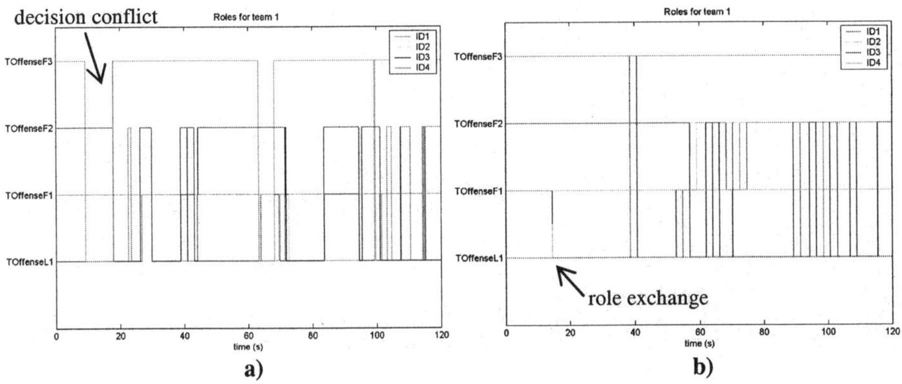

Figure 6. Dynamic role allocation without (*a*) and with (*b*) supervision.

Performance increase in a given system may not be directly measured with decision supervision enabled, but other benefits should also appear. For soccer-playing robots, conflicts can be the source of collisions and other perceptible problems leading to hardware problems. Furthermore, conflicts can increase robots energy consumption and compromise consistent strategic team play.

5. Future Work

The project presented in this paper established some basic elements for multi-robot systems with distributed sensing and decision mechanisms. These elements should serve as building blocks for more elaborate work on cooperative multi-robot systems. A combination of central and distributed deliberation is probably the most powerful approach in terms of global intelligence, robustness and efficiency of a system. More work should be done to demonstrate this point.

An ongoing project that uses the proposed architecture concerns generic learning methods for hierarchical multi-robot systems. Soccer-playing robots are again used as a test bench. Development of graphical possibilities of the HDM and dynamically modified HDMs are examples of future developments.

References

1. R. Alami, S. Fleury, M. Herrb, F. Ingrand, F. Robert. *Multi Robot Cooperation in the Martha Project.* In IEEE Robotics and Automation Magazine, Vol. 5, No. 1. IEEE. 1997, pp.36-45.
2. J.S. Albus et al. *4D/RCS: A Reference Model Architecture for Unmanned Vehicle Systems Version 2.0.* NISTIR 6910, National Institute of Standards and Technology. USA. 2002.
3. R.C. Arkin. *Behavior-Based Robotics.* The MIT Press, Cambridge. 1998.

4. J. Beaudry. *Machine décisionnelle pour systèmes multi-robots à perception distribuée*. M.Sc.A. thesis, Electrical Engineering Department, École Polytechnique de Montréal. 2005.
5. L. Chainowicz, M.F.M. Campos, V. Kumar. *Dynamic Role Assignment for Cooperative Robots*. Proceedings of the 2002 IEEE International Conference on Robotics & Automation. 2002.
6. R. Emery, K. Sikorsky, T. Balch. *Protocols for Collaboration, Coordination and Dynamic Role Assignment in a Robot Team*. Proceedings of the 2002 IEEE International Conference on Robotics & Automation, 2002, pp.3008-3015.
7. R. Houde, M. Blain, J. Côté. *Manuel de l'usager pour Microb*. Internal report IREQ-2000-075, Institut de recherche d'Hydro-Québec, 2000.
8. R. L'Archevêque, E. Dupuis. *Autonomous Robotics and Ground Operations*. Proceeding of the 7th International Symposium on Artificial Intelligence, Robotics and Automation in Space: i-SAIRAS 2003. Japan. 2003.
9. M. Lötzsch, J. Bach, H.-D. Burkhard, M. Jüngel. *Designing Agent Behavior with the Extensible Agent Behavior Specification Language XABSL*. In 7th International Workshop on RoboCup 2003, Lecture Notes in Artificial Intelligence, Padova, Italy. 2003.
10. S. Marleau. *Système embarquée de localisation et de perception pour un robot mobile*. M.Sc.A. thesis, Electrical Engineering Department, École Polytechnique de Montréal. 2005.
11. I.A. Nesnas, A. Wright, M. Bajracharya, R. Simmons, T. Estlin, W.S. Kim. *CLARAty: An Architecture for Reusable Robotic Software*. SPIE Aerosense Conference, 2003, pp.121-132.
12. L.E. Parker. *ALLIANCE: An Architecture for Fault Tolerant Multi-Robot Cooperation*. IEEE Transactions on Robotics and Automation, 1998, pp.220-240.
13. P. Pirjanian, T.L. Huntsberger, A. Trebi-Ollennu, H. Aghazarian, H. Das, S. Joshi, P.S. Schenker. *CAMPOUT: A control architecture for multi-robot planetary outposts*. In Proceedings of the SPIE Symposium on Sensor Fusion and Decentralized Control in Robotic Systems III, Vol. 4196, Boston, MA, Nov. 2000.
14. The RoboCup Federation. *RoboCup Official Site*. The RoboCup Federation, online, April 2006: http://www.robocup.org.
15. Robofoot ÉPM. *Robofoot ÉPM Official Site*. Robofoot ÉPM, online, April 2006: http://robofoot.polymtl.ca.
16. P. Stone. *Layered Learning in Multiagent Systems*. The MIT Press, Cambridge. 2000.

LOCALIZATION OF COMPACT INVARIANT SETS OF THE RIKITAKE SYSTEM AND PIKOVSKY-RABINOVICH-TRAKHTENGERTZ SYSTEM

KONSTANTIN E. STARKOV AND LUIS EDUARDO VARGAS

CITEDI-IPN
Av. del Parque 1310,
Tijuana, B.C. 22510, México
E-mail: konst@citedi.mx

In this paper we describe localization results of all compact invariant sets of the Rikitake system and the Pikovski, Rabinovich and Trakhtengerts (PRT) system. We derive localizing sets corresponding different quadratic surfaces for the Rikitake system and ellipsoidal and conic localizing sets for the PRT system. Our approach is based on the solution of the first order extremum problem.

1. Introduction

The problem of localizing compact invariant sets is studied in many papers due to increasing interest to the long-time behaviour of a chaotical system. During the last decade the interest of many researchers has been attracted to the idea of finding some geometrical bounds for attractors, periodical orbits and chaotic dynamics of a nonlinear autonomous differentiable right-side system

$$\dot{x} = f(x), \quad x = (x_1, \dots x_n)^T \in \mathbf{R}^n, \tag{1}$$

$f(x) = (f_1(x), \dots f_n(x))^T \in C^\infty(\mathbf{R}^n)$. Three main approaches to the solution of this problem should be mentioned. One of them is the method based on using Lyapunov–type functions, see [1], [6], [7] and others. The second method is based on finding families of semipermeable surfaces, see [8]. In this paper we apply the third method proposed originally in [3] for the localization of periodic orbits. Later this method has been successfully applied for the Lanford system in [4] and for the Lorenz system in [5]. Here when we talk about a localization we have in mind the following problem: find the set $\Omega \subset \mathbf{R}^n$ (a localization set) that contains all compact invariant sets of the system (1).

The structure of the paper is as follows. In Sec. 2 we formulate basic definitions and present main assertions applied in the localization process. In Sec. 3 we study the Rikitake system introduced in [10]. In the next section we consider the Pikovsky-Rabinovich-Trakhtengertz (PRT) system of plasma dynamics obtained in [9]. In Sec. 5 we give conclusions.

2. Some preliminaries

We remind two standard well-known concepts of qualitative theory of ordinary differential equations. By $\varphi(x,t)$ we denote a solution of (1), with $\varphi(x,0) = x$ for any $x \in \mathbf{R}^n$.

A set $G \subset \mathbf{R}^n$ is called invariant for (1) if for any $x \in G$ we have that $\varphi(x,t) \in G$ for all t from the interval of existence of this solution.

The union of equlibrium points with trajectories connecting them is referred to as heteroclinic orbits when they connect disctinct points and homoclinic orbits when they connect a point to itself.

Compact invariant sets can contain equlibrium points, periodic orbits, heteroclinic orbits, homoclinic orbits and trajectories of more complex structure. We define a maximal (with respect to inclusions) compact invariant set of (1) as a compact invariant set containing any compact invariant set of (1). A maximal compact invariant set may not exist.

In this section we describe localization sets which contain all compact invariant sets of the system (1). The localization of invariant subsets such as periodic orbits, homoclinic orbits, heteroclinic orbits, invariant tori inside an invariant set claims to apply additional ideas which is beyond the scope of this paper.

Let $f = \sum_{i=1}^{n} f_i(x)\partial/\partial x_i$ be a vector field on $\mathbf{R}^n$ corresponding to the system (1). Let $L_f h(x) = \sum_{i=1}^{n} f_i(x)\partial h(x)/\partial x_i$ be a Lie derivative of the function $h \in C^\infty(\mathbf{R}^n)$ with respect to the vector field f. We define a set

$$S(h) = \{x : L_f h(x) = 0\}.$$

Below we shall use notations:

$$\begin{aligned} h_{sup} &= \sup_{S(h)} h(x), \\ h_{inf} &= \inf_{S(h)} h(x). \end{aligned} \tag{2}$$

For any function $h \in C^\infty(\mathbf{R}^n)$ the following assertions are valid, see e.g in [4], [5].

Proposition 2.1. *Let Q be a set in $\mathbf{R}^n$. If $S(h) \cap Q = \emptyset$ then the system (1) has no compact invariant sets (totally) contained in Q. Now let $S(h) \cap Q \neq \emptyset$. If $L_f^2 h \leq 0$ on $S(h) \cap Q$ or $L_f^2 h \geq 0$ on $S(h) \cap Q$ then the system (1) has no compact invariant sets (totally) contained in Q.*

Theorem 2.1. *Any compact invariant set G of the system (1) is contained in the set*

$$K_h = \{x : h_{inf} \leq h(x) \leq h_{sup}\}.$$

Proposition 2.2. *If the set K_h is compact then the system (1) has a maximal compact invariant set which is contained in K_h.*

Corollary 2.1. *Any compact invariant set of the system (1) is contained in the set $K = \{\cap K_h, h \in C^\infty(\mathbf{R}^n)\}$. If the set K is compact then the system (1) has a maximal compact invariant set which is contained in Ω.*

The generalization of Theorem 4 leads to

Theorem 2.2. *Let $h_m(x), m = 0, 1, 2, \ldots$ be a sequence of functions from $C^\infty(\mathbf{R}^n)$. Sets*

$$K_0 = K_{h_0}, \quad K_m = K_{m-1} \cap K_{m-1,m}, \quad m > 0,$$

with

$$\begin{aligned} K_{m-1,m} &= \{x : h_{m,inf} \leq h_m(x) \leq h_{m,sup}\}, \\ h_{m,sup} &= \sup_{S(h_m) \cap K_{m-1}} h_m(x), \\ h_{m,inf} &= \inf_{S(h_m) \cap K_{m-1}} h_m(x), \end{aligned}$$

contain any compact invariant set of the system (1) and $K_0 \supseteq K_1 \supseteq \cdots \supseteq K_m \supseteq \ldots$. Here we talk that this is $(h_0, h_1, ..., h_m, ...)$- localization.

3. Localization of compact invariant sets of the Rikitake system

We examine the Rikitake system, [10],

$$\begin{aligned} \dot{x} &= -\mu x + yz, \\ \dot{y} &= -\alpha x - \mu y + xz, \\ \dot{z} &= 1 - xy, \end{aligned} \tag{3}$$

describing the Earth's geomagnetic field. Here all parameters are some positive constants. Dynamics of this system was studied in [7]. Some preliminary results concerning the localization problem of compact invariant sets of the Rikitake system are contained in the paper [11].

Firstly, we note that these equations have a noncompact invariant set. Namely, it is the straight line $x = y = 0$. Our main results concerning the Rikitake system are presented in two theorems. Let f be the corresponding vector field of (3).

Theorem 3.1. *All compact invariant sets of (3) are located in the set defined by* $\{p_1x^2 + p_2y^2 + (p_1 + p_2)z^2 - 2p_2\alpha z \geq \xi(p_1, p_2);\ p_1 + p_2 > 0\}$ *and in the set* $\{p_1x^2 + p_2y^2 + (p_1 + p_2)z^2 - 2p_2\alpha z \leq \xi(p_1, p_2);\ p_1 + p_2 < 0\}$*, where*

$$\xi(p_1, p_2) := -\frac{(p_1 + p_2 - 2p_2\alpha\mu)^2}{4\mu^2(p_1 + p_2)} - \frac{p_2\alpha}{\mu}$$

and $p_1; p_2$ are real parameters.

Proof.

We consider $h(x, y, z) = p_1x^2 + p_2y^2 + (p_1 + p_2)z^2 + c_3z$, with $c_3 = -2p_2\alpha$. The surface $L_fh(x) = 0$ is the set given by $2\mu(p_1x^2 + p_2y^2) - 2(p_1 + p_2)z + 2p_2\alpha = 0$ or by $p_1x^2 + p_2y^2 = \mu^{-1}(p_1 + p_2)z - \mu^{-1}p_2\alpha$. Thus

$$h(x, y, z) \mid_{S(h)} = \mu^{-1}(p_1 + p_2)z - \mu^{-1}p_2\alpha + (p_1 + p_2)z^2 - 2p_2\alpha z =$$
$$(p_1 + p_2)(z + \frac{p_1 + p_2 - 2p_2\alpha\mu^2}{2\mu(p_1 + p_2)})^2 - \frac{(p_1 + p_2 - 2p_2\alpha\mu^2)^2}{4\mu^2(p_1 + p_2)} - \frac{p_2\alpha}{\mu}.$$

Hence, by using Theorem 2.1, all compact invariant sets are located in the set $K(p_1, p_2)$ defined by

$$\{p_1x^2 + p_2y^2 + (p_1 + p_2)z^2 - 2p_2\alpha z \geq \xi(p_1, p_2);\ p_1 + p_2 > 0\}$$

and in

$$\{p_1x^2 + p_2y^2 + (p_1 + p_2)z^2 - 2p_2\alpha z \leq \xi(p_1, p_2);\ p_1 + p_2 < 0\},$$

with

$$\xi(p_1, p_2) := -\frac{(p_1 + p_2 - 2p_2\alpha\mu)^2}{4\mu^2(p_1 + p_2)} - \frac{p_2\alpha}{\mu}$$

and we come to the desirable conclusion.■

So by using Corollary 2.1, all compact invariant sets are located in the set $K = \cap_{p_1,p_2}K(p_1, p_2)$. Since signs of $p_1; p_2; p_1 + p_2$ and $\xi(p_1, p_2)$ can be different we obtain K as a infinite intersection of sets bounded by quadratic

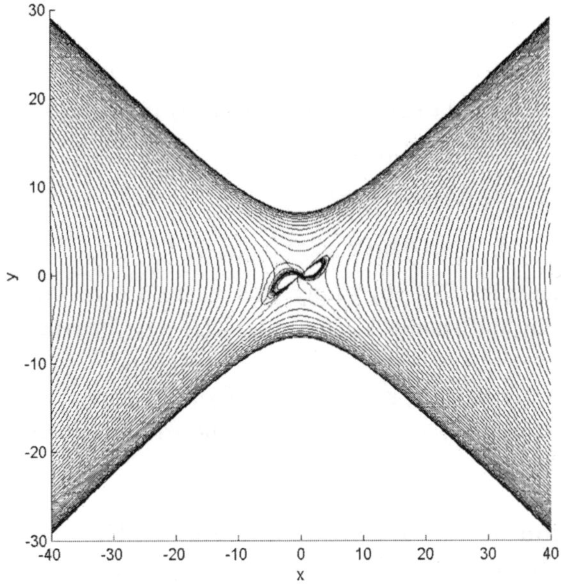

Figure 1. Localization of the Chaotic Attractor of system (3) with parameters $\mu = 2$, $a = 5$ and values of $p_1 < 0$ and $p_2 > 0$; here we find that the attractor is localized inside a one sheet hyperboloid.

surfaces of different topological types depended on signs of p_1; p_2; $p_1 + p_2$; $\xi(p_1, p_2)$.

In what follows in this section , we examine the location of compact invariant sets respecting half-spaces $z < q$; $z > q$, where q is real.

We introduce $V_<(q) := \{z < q\}; V_>(q) := \{z > q\}$.

Theorem 3.2. *1. There are no compact invariant sets in* $\mathbf{R}^3 - \{0\}$ *contained in half-spaces* $V_<(\mu+\alpha), V_>(\mu+\alpha)$. *2. There are no compact invariant sets without common points with the plane* $y = 0$ *which are contained in the frustrum* $\{0 < z < \alpha\}$.

Proof. 1. Since the straight line $x = y = 0$ is a noncompact invariant set we can consider the localization problem on $G = \{x^2 + y^2 > 0\}$.

We take $h_1(x, y, z) = z$. Then $L_f h_1(x, y, z) = 1 - xy$ and $L_f^2 h_1(x, y, z) =$

$(-z+\mu+\alpha)(x^2+y^2)$. Now we derive that

$$L_f^2 h_1(x,y,z)\mid_{S(h_1)\cap G}=\{(\mu+\alpha-z)(x\mp y)^2 \pm 2(\mu+\alpha-z)\}\mid_{S(h_1)\cap G}. \quad (4)$$

By applying the formula (4), we obtain that a) if $\mu+\alpha-z>0$ then

$$L_f^2 h_1(x,y,z)\mid_{S(h_1)\cap G}>0; \quad (5)$$

b) if $\mu+\alpha-z<0$ then

$$L_f^2 h_1(x,y,z)\mid_{S(h_1)\cap G}<0. \quad (6)$$

Thus inequalitiies (5)- (6) imply that there are no compact invariant sets contained in half spaces $V_<(\mu+\alpha), V_>(\mu+\alpha)$ because of Proposition 2.1.

2. Let us apply $h_2(x,y,z)=\frac{x}{y}$. Then $L_f h_2(x,y,z)=z-h_2^2(z-\alpha)>0$ on the set $\{0<z<\alpha\}$ which entails the necessary conclusion in view of Proposition 3.■

4. Localization of compact invariant sets of the Pikovsky-Rabinovich-Trakhtengertz system

In this section we shall examine the localization problem of all compact invariant sets for the Pikovsky-Rabinovich-Trakhtengertz (PRT) system of plasma dynamics, see [9]:

$$\begin{aligned}\dot{x}&=-\nu_1 x+\beta y-yz\\ \dot{y}&=\beta x-\nu_2 y+xz\\ \dot{z}&=-\nu_3 z+xy\end{aligned} \quad (7)$$

These equations for the amplitudes of the three waves are obtained from the hydrodynamic equation for the radio-frequency oscillations of an electron gas and from the kinetic equation for the ion acoustic wave. Here all parameters are positive numbers. Let f be the corresponding vector field of the PRT system.

We introduce numbers depended on real parameters $a_1; a_2$

$$\xi_1:=\frac{\beta(2a_1+a_2)\nu_3^{1/2}}{2a_2^{1/2}(a_1+a_2)^{1/2}\nu_1^{1/2}};\ \xi_2:=\frac{\beta(2a_1+a_2)\nu_3^{1/2}}{2a_1^{1/2}a_2{}^{1/2}\nu_2^{1/2}};$$

$$\xi_3:=\frac{\beta(2a_1+a_2)}{a_2};\ \eta:=(a_1+a_2)\xi_1^2+a_1\min(\xi_2^2;\xi_1^2+2\xi_3^2).$$

Theorem 4.1. *All compact invariant sets of the PRT system (7) are located in the ellipsoid defined by*

$$\{(a_1+a_2)x^2+a_1y^2+a_2z^2-2\beta(2a_1+a_2)z\le\frac{\beta^2(2a_1+a_2)^2\nu_3}{4a_2}(\frac{1}{\nu_1}+\frac{1}{\nu_2});a_1;a_2>0\}.$$

Proof.

Let us take

$$h_0(x,y,z) = (a_2+a_3)x^2 + a_2y^2 + a_3z^2 - 2\beta(2a_2+a_3)z.$$

Then we compute that

$$L_f h_0(x,y,z) = -2(a_2+a_3)\nu_1 x^2 - 2a_2\nu_2 y^2 - 2a_3\nu_3 z^2 + 2\beta(2a_2+a_3)\nu_3 z.$$

The latter gives the formula for $S(h_0)$

$$(a_2+a_3)\nu_1 x^2 + a_2\nu_2 y^2 + a_3\nu_3(z - \frac{\beta(2a_2+a_3)}{2a_3})^2 = \frac{\beta^2(2a_2+a_3)^2\nu_3}{4a_3}. \quad (8)$$

Suppose that $a_2 > 0$ and $a_3 > 0$. In this case the set (8) defines an ellipsoid. Now we deduce that $S(h_0)$ is located inside the polytope Π defined by

$$\begin{aligned} | x | &\le \xi_1 := \frac{\beta(2a_2+a_3)\nu_3^{1/2}}{2a_3^{1/2}(a_2+a_3)^{1/2}\nu_1^{1/2}}; \quad (9)\\ | y | &\le \xi_2 := \frac{\beta(2a_2+a_3)\nu_3^{1/2}}{2a_2^{1/2}{a_3}^{1/2}\nu_2^{1/2}};\\ 0 \le z &\le \xi_3 := \frac{\beta(2a_2+a_3)}{a_3}. \end{aligned}$$

Let $\psi(z) = a_3z^2 - 2\beta(2a_2+a_3)z$. Then

$$\inf_{\Pi} \psi = \psi(\frac{\beta(2a_2+a_3)}{a_3}) = -\frac{\beta^2(2a_2+a_3)^2}{a_3}$$

which entails

$$\sup_{\Pi} \psi = 0.$$

Since

$$h_{0\,\sup} \le \sup_{\Pi} h_0; h_{0\,\inf} \ge \inf_{\Pi} h_0$$

and in view of inequalities (9) we obtain that

$$h_{0\,\sup} \le \eta_1 := \frac{\beta^2(2a_2+a_3)^2\nu_3}{4a_3}(\frac{1}{\nu_1} + \frac{1}{\nu_2}); h_{0\,\inf} = 0.$$

Thus, by Theorem 2.1, we deduce that all compact invariant sets are contained in the interior of the ellipsoid $Ell = \{h_0(x) \le \eta_1\}$.■

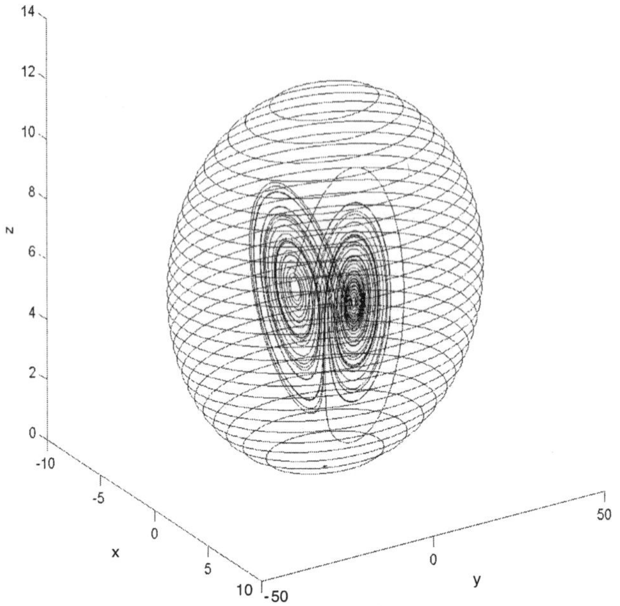

Figure 2. Localization of Chaotic Attractor of System (7) with parameters $\nu_1 = \nu_3 = 1$, $\nu_2 = 4$, $\beta = 6$ and parameters $a_1 = 2.4955$ and $a_2 = 112.03$; here we find that the chaotic attractor is located inside an ellipsoid described by (8).

Theorem 4.2. *Let $\nu_1 < \nu_2$, $\nu_3 < \nu_2$ and let us choose positive $a_1;a_2$ satisfying the inequality*

$$\frac{1}{\nu_2} > \frac{a_1}{(a_1 + a_2)\nu_1} + \frac{8a_1}{a_2\nu_3}.$$

Then all compact invariant sets are located in the set $\{(a_1+a_2)x^2+a_1y^2+a_2z^2 - 2\beta(2a_1+a_2)z \leq \eta;\ y^2 \leq x^2+2z^2\}$.

Proof. Let us take $h_2(x, y, z) = -x^2 + y^2 - 2z^2$ and then obtain

$$S(h_2) = \{-\nu_1 x^2 + \nu_2 y^2 - 2\nu_3 z^2 = 0\}$$

We shall apply Theorem 2.2 to the $(h_2, h_1)-$ localization. Therefore we obtain that

$$h_2(x, y, z) \mid_{S(h_2)} = (-1 + \nu_1 \nu_2^{-1})x^2 + 2(\nu_3 \nu_2^{-1} - 1)z^2.$$

Now let $\nu_1 < \nu_2$ and $\nu_3 < \nu_2$. This case corresponds some of values of parameters for which the system (7) exhibits chaos. For example, one can choose

$$\nu_1 = \nu_3 = 1; \nu_2 = 4; \beta = 6, \tag{10}$$

see [8].

In this case $h_{2\,\sup} = 0$ and the localization set $K_2 = \{y^2 \leq x^2 + 2z^2\}$ which is a solid elliptic cone. Now we deduce that in $\Pi \cap K_2$ we have: $y^2 \leq \xi_1^2 + 2\xi_3^2$. Thus

$$h_{1\,\sup} = \sup_{S(h_1)\cap K_2} h_1 \leq \sup_{\Pi\cap K_2} h_1 \leq \eta_2 := (a_2 + a_3)\xi_1^2 + a_2 \min(\xi_2^2; \xi_1^2 + 2\xi_3^2).$$

Therefore $K_{21} = \{(a_2+a_3)x^2+a_2y^2+a_3z^2-2\beta(2a_2+a_3)z \leq \eta_2\}$. Comparing η_1 with η_2 we conclude that if positive coefficients a_2 and a_3 are such that

$$\frac{1}{\nu_2} > \frac{a_2}{(a_2 + a_3)\nu_1} + \frac{8a_2}{a_3\nu_3} \tag{11}$$

then $\eta_1 \geq \eta_2$ and we get the improved localization set $K_1 = K_2 \cap K_{21}$. In case of values of parameters (10) the inequality (11) has the form $a_3^2 - 35a_2a_3 - 32a_2^2 > 0$.■

5. Conclusions

In this paper we have computed localizing sets for the Rikitake system and for the Pikovsky-Rabinovich-Trakhtengertz system. Localizing sets constructed for the Rikitake system are not compact whereas we find a compact localizing set of the ellipsoidal type for the PRT system. In case $\nu_1 < \nu_2$ and $\nu_3 < \nu_2$ the ellipsoidal localization for the PRT system can be improved with help of crossing with the set K_2. Some sufficient conditions of nonexistence of compact invariant sets in some half spaces were described for the Rikitake system.

References

1. C.R. Doering and J.D. Gibbon *Dynam. Stability Systems.* **10**, 255 (1995).
2. H. Giacomini and S. Neukirch, S. *Phys. Lett. A.* **240**, 157 (1997).
3. A.P. Krishchenko, *Computers Math. Applic.* **34**, 325 (1997).
4. A.P. Krishchenko and K.E. Starkov *International Journal of Bifurcations and Chaos* **N9** (2006),
5. A.P. Krishchenko, K.E. Starkov *Physics Letters A.* **353**, 383 (2006).
6. A. Leonov, A.I. Bunin, N. Koksch, Z. *Angew Math Mech.* **67**, 649 (1987).

7. T. McMillen, *The Nonlinear Journal.* **1** , 1 (1998).
8. S. Neukirch, *Physical Review E.* **63.** 036202-1 (2001)
9. A.S. Pikovski, M.I. Rabinovich and V.Y. Trakhtengerts, *Sov. Phys. JETP.* **47** , 715 (1978).
10. T. Rikitake, *Proc. Cambridge Philos. Soc.* **54** , 89 (1958).
11. K. Starkov and A.P. Krishchenko *Chaos, Solitons & Fractals.* **3**, 981 (2005).

STABILIZATION OF LINEAR SYSTEMS: A POLYNOMIAL APPROACH

BALTAZAR AGUIRRE, JULIO SOLIS-DAUN AND RODOLFO SUÁREZ

Departamento de Matemáticas
Universidad Autónoma Metropolitana - Iztapalapa
Apdo. Postal 55-534, 09340, México, D.F., México
E-mail: bahe@xanum.uam.mx

In this paper, given a stable open-loop system we design a control $u = -kc^T x$ which is a high-gain feedback. The importance of our design is that in general the origin is not necessarily asymptotically stable for all $k > 0$, even when $c \in \mathbb{R}^n$ is chosen in such a way that $u = -kc^T x$ is a high-gain control. In this paper by means a matrix inequality we find a cone of gains c such that $u = -kc^T x$ is a stabilizing control for all $k > 0$.

1. Introduction

The aim of this paper is designing stabilizing control feedbacks for stable open-loop systems. Consider the controllable system $\dot{x} = Ax + bu$, which is written in the canonical form (5) below. Define a feedback control u given by

$$u = -kc^T x \tag{1}$$

where $c \in \mathbb{R}^n$, and $k > 0$. Suppose that c is such that the closed-loop polynomial $p_c(t)$ is a Hurwitz polynomial, that is $u = -kc^T x$ is a stabilizing control. If $k \gg 1$ the control feedback (1) is known as *high-gain feedback*, since high control gains kc^T are induced. In the last thirty years, different approaches have been used to study the high-gain controls (see for instance Refs. 8, 10, 12, 13, 14). In practical applications, high-gain feedback is commonly used to reduce the effects of bounded disturbance and nonlinearities. [7] It is well known (see Refs. 10, 14) that when $k \to \infty$ a closed-loop eigenvalue, say λ_1, has the property that $\frac{\lambda_1}{k} \to -c_1$ and, the other eigenvalues converge to the roots of the polynomial $c_1 t^{n-1} + c_2 t^{n-2} + \ldots + c_n$. Consequently, the closed-loop system is asymptotically stable at the origin when k is sufficiently large. Nevertheless, the origin is not necessarily asymptotically stable for all $k > 0$, even when $c \in \mathbb{R}^n$ is chosen in such a way that the polynomial $c_1 t^{n-1} + c_2 t^{n-2} + ... + c_n$ is Hurwitz. Therefore, it is important have techniques to find vectors c such that $u = -kc^T x$ is a stabilizing control. In this paper we present a sufficient condition on c such that $u = -kc^T x$ is a stabilizing control. The sufficient condition is the matrix inequality (6) below which is a very simple algebraic test. In terms of polynomials, we obtain a sufficient condition for a conic set $p_0 + K$ to consist only of stable polynomials. Here p_0 is an $n-$degree stable polynomial and K is a convex cone of $(n-1)-$degree polynomials. In the framework of Ref. 4 this correspond to have infinite robustness of the polynomial p_0 with respect to perturbations in the directions contained in K. We will illustrate the above ideas by the following example.

Consider the system

$$\dot{x} = \begin{pmatrix} 0 & 1 & 0 \\ 0 & 0 & 1 \\ -5 & -11 & -7 \end{pmatrix} x + \begin{pmatrix} 0 \\ 0 \\ 1 \end{pmatrix} (-65k \ , -3k \ , -k \ ) \, x \tag{2}$$

whose open-loop polynomial $p_0(t) = t^3 + 7t^2 + 11t + 5 = (t+1)^2(t+5)$ is Hurwitz and its closed-loop polynomial $p_c(t)$ is given by $p_c(t) = t^3 + (7 + k)t^2 + (11 + 3k)t + 5 + 65k$. One of the eigenvalues, say λ_1, has the property that $\frac{\lambda_1}{k} \to -1$ and the other two eigenvalues converge to the roots of the polynomial $p_c^*(t) = t^2 + 3t + 65$. Thus, the origin is locally asymptotically stable for large values of k. In fact, the origin is locally asymptotically stable for $k \in [0, 3) \cup (8, \infty)$. The closed-loop system is not asymptotically stable at the origin for $k \in [3, 8]$ since the corresponding Hurwitz condition $(7 + k)(11 + 3k) - (5 + 65k) > 0$ is not satisfied.

The rest of the paper is organized as follows: in Sec. 2, sufficient conditions on $c = (c_1, c_2, ..., c_n)^T$ assuring that the corresponding polynomial $c_1 t^{n-1} + c_2 t^{n-2} + ... + c_n$ is Hurwitz are established (inequality (6)). Moreover, for c satisfying inequality (6) it is proved that the corresponding closed-loop system is asymptotically stable for every value of the high-gain parameter k. In Sec. 3 we use the Kuhn-Fourier method to study the problem of the solutions of the linear inequalities. Finally in Sec. 4 an example illustrating the results of the paper is presented.

2. Main results

The aim of this section two fold. First, we obtain algebraic conditions for the stability of rays of polynomials (the matrix inequality (6) below).

Second, we will use this to find a conic set of gains c such that the control $u = -kc^T x$ is a stabilizing control for all values of $k > 0$.

Given a real polynomial $p_0(t) = t^n + a_1 t^{n-1} + ... + a_n$ define the matrix

$$E = \begin{pmatrix} a_1 & -1 & 0 & 0 & \dots & 0 & 0 \\ -a_3 & a_2 & -a_1 & 1 & \dots & 0 & 0 \\ a_5 & -a_4 & a_3 & -a_2 & \dots & 0 & 0 \\ \dots & \dots & \dots & \dots & \dots & \dots & \dots \\ 0 & 0 & 0 & 0 & \dots & a_{n-1} & -a_{n-2} \\ 0 & 0 & 0 & 0 & \dots & 0 & a_n \end{pmatrix} \tag{3}$$

and let $E_{i.}, E_{.j}$ denote the i-th row and the j-th column of the matrix E, respectively.

Theorem 2.1. *Let us consider the following system*

$$\dot{x} = Ax + bu \tag{4}$$

where $u = -kc^T x$ is a feedback control, $x, b \in \mathbb{R}^n$ and the controllable pair (A, b) is given in canonical form (see Ref. 2)

$$A = \begin{bmatrix} 0 & 1 & 0 & \cdots & 0 \\ 0 & 0 & 1 & \cdots & 0 \\ \vdots & \vdots & \vdots & & \vdots \\ 0 & 0 & 0 & \cdots & 1 \\ -a_n & -a_{n-1} & -a_{n-2} & \cdots & -a_1 \end{bmatrix} \quad b = \begin{bmatrix} 0 \\ 0 \\ \vdots \\ 0 \\ 1 \end{bmatrix} \tag{5}$$

Suppose that the open-loop polynomial $p_0(t) = t^n + a_1 t^{n-1} + ... + a_n$ is a Hurwitz polynomial. Let E be the corresponding matrix defined by (3). If the vector c is a solution to the system of linear inequalities

$$\boxed{E_{i.}c > 0, i = 1, ..., n,} \tag{6}$$

then the control $u = -kc^T x$ is a high-gain feedback.

Proof: We present the proof for n even (set $n = 2m$), being the case when n is odd analogous. Let $F(t) = p_0(-t)$ and $f(t) = c_1 t^{n-1} + c_2 t^{n-2} + ... + c_n$. Then, $F(t)$ is a real polynomial of degree n with all its roots in $\mathbb{C}^+$. Consider the polynomial $F(t)f(t)$, which has degree $2n - 1$. Notice that $p_0(i\omega)$ and $f(i\omega)$ can be written as

$$p_0(i\omega) = P(\omega^2) + i\omega Q(\omega^2) \quad \text{and} \quad f(i\omega) = p(\omega^2) + i\omega q(\omega^2),$$

where P, Q, p, and q are real polynomials. We have

$$\begin{aligned} F(i\omega)f(i\omega) &= [P(\omega^2) - i\omega Q(\omega^2)][p(\omega^2) + i\omega q(\omega^2)] \\ &= [Pp + \omega^2 Qq] + i\omega(Pq - Qp). \end{aligned}$$

After some calculations we get

$$(Pp + \omega^2 Qq) = -\sum_{i=1}^{n}(E_{i.}c)\omega^{2(n-i)}$$

Note the correspondence between the coefficients of the last polynomial and the linear inequalities (6). Consequently since $E_{i.}c > 0$, $i = 1, ..., n$, it follows that $F(i\omega)f(i\omega)$ does not intersect the imaginary axis for all $\omega > 0$. Let l and r be the number of roots of $F(t)f(t)$ contained in $\mathbb{C}^-$ and $\mathbb{C}^+$, respectively. Since $F(i\omega)f(i\omega)$ does not intersect the imaginary axis for $\omega > 0$, then $F(t)f(t)$ does not have roots on the imaginary axis.

Let $\theta(\omega)$ be the argument of $F(i\omega)f(i\omega)$. Denote by $\Delta_0^\infty\theta = \theta(\infty) - \theta(0)$ the net change in the argument. It is known that $\Delta_0^\infty\theta = \frac{\pi}{2}(l - r)$.(Ref. 6, p. 406; Ref. 3, p. 174). The fact that $F(i\omega)f(i\omega)$ does not intersect the imaginary axis for $\omega > 0$ implies $|\Delta_0^\infty\theta| \le \pi$.

On the other hand, we know that at least n roots of $F(t)f(t)$ are in $\mathbb{C}^+$, then it follows that $r \ge n$ and $l \le n-1$. Hence, $l - r < 0$. Additionally, $l - r$ is an odd number since $l + r = 2n - 1$. Thus, the equality $\Delta_0^\infty\theta = \frac{\pi}{2}(l - r)$ implies that $\Delta_0^\infty\theta = -\frac{\pi}{2}$. Consequently, $l - r = -1$, from where it follows that $r = n$ and $l = n - 1$. Finally, the $n - 1$ roots of $f(t)$ are contained in $\mathbb{C}^-$, as we wanted to prove. Then, the polynomial $f(t) = \sum_{i=1}^{n} c_i t^{n-i}$ is Hurwitz. It is well known that one of the eigenvalues of the closed-loop system $\dot{x} = Ax - kbc^T x$, say λ_1, has the property that $\frac{\lambda_1}{k} \to -c_1$ when $k \to \infty$ and, the other ones converge to the roots of the polynomial $f(t) = c_1 t^{n-1} + c_2 t^{n-2} + ... + c_n$, [9,14] which is a Hurwitz polynomial. Consequently, the closed-loop system is asymptotically stable at the origin for k sufficiently large. This shows that the control $u(t) = -kc^T x$ is a high-gain feedback.

Remark 2.1. In the introduction it was pointed out that in general, a high-gain feedback is not necessarily a stabilizing control for every value of the parameter k and we illustrated this fact with an example. This enhances the importance of the following stronger result which is useful for the design of high-gain stabilizing feedbacks.

Theorem 2.2. *Consider the linear system (4) written in the canonical form (5). Suppose A is Hurwitz, that is, the open-loop polynomial $p_0(t) = t^n + a_1t^{n-1} + ... + a_n$ is Hurwitz. If $c \succ 0$ is a solution to (6), then, for all $k > 0$, the control $u(t) = -kc^Tx$ is a stabilizing control feedback.*

Proof: Suppose n is even (the case n odd is analogous). Let $n = 2m$ and $k > 0$. To prove this item, it is enough to see that the closed-loop polynomial is Hurwitz. Let $p_c(t)$ and $p_0(t)$ denote the closed-loop and the open-loop polynomials, respectively. Consider the polynomial $p_c(t)p_0(-t)$ and let $\theta_1(\omega)$ be the argument of $p_c(i\omega)p_0(-i\omega)$ and $\Delta_0^\infty\theta_1 = \theta_1(\infty) - \theta_1(0)$ denotes the net change in the argument of $p_c(i\omega)p_0(-i\omega)$.

Following similar ideas as in the proof of Theorem 2.1 we get that $|\Delta_0^\infty\theta_1| \le \pi$. On the other hand, $p_c(0)p_0(0) = a_{2m}(a_{2m} + kc_{2m})$, which is a positive real number. Hence $\theta_1(0) = 0$.

Now we will analyze $\theta_1(\omega)$ when ω is large. First, we have for large ω that $p_c(i\omega)p_0(-i\omega) \approx \omega^{4m} - ic_1\delta\omega^{4m-1}$. Therefore, $p_c(i\omega)p_0(-i\omega)$ lies in the 4th. quadrant when ω is large and $\dfrac{\mathrm{Im}[p_c(i\omega)p_0(-i\omega)]}{\mathrm{Re}[p_c(i\omega)p_0(-i\omega)]} \to 0$ when $\omega \to \infty$. Hence, $\theta_1(\infty) = 2s\pi$, where s is an integer. Then, since $\Delta_0^\infty\theta_1 = \theta_1(\infty) - \theta_1(0) = 2s\pi$ and $|\Delta_0^\infty\theta_1| \le \pi$, we get that $\Delta_0^\infty\theta_1(\omega) = 0$.

Consequently, the polynomial $p_c(t)p_0(-t)$ has as many roots in $\mathbb{C}^-$ as in $\mathbb{C}^+$. Since such polynomial has degree $2n$, then there are n roots in $\mathbb{C}^+$. In fact the roots in $\mathbb{C}^+$ correspond to the roots of $p_0(-t)$ because the open-loop polynomial $p_0(t)$ is a Hurwitz polynomial. Finally, it follows that the n roots in $\mathbb{C}^-$ correspond to the roots of $p_c(t)$, which means that $p_c(t)$ is Hurwitz.

Theorem 2.1 can be rewritten in terms of polynomials in the following way.

Corollary 2.1. *Given a Hurwitz polynomial $p_0(t)$ let G be the family of polynomials $p_1(t) = c_1t^{n-1} + c_2t^{n-2} + ... + c_n$ such that $c^T = (c_1, c_2, ..., c_n)^T \succ 0$ satisfies the inequality (6). We have that for each $p_1 \in G$, the ray of polynomials $p_0(t) + kp_1(t)$, $k \ge 0$ is Hurwitz.*

Remark 2.2. A similar inequality was obtained in Ref. 1 and the results were gotten in terms of rays of polynomials. Given a real polynomial $p_0(t) = t^n + a_1t^{n-1} + ... + a_n$ define the matrix $D \in \mathcal{M}_{nxn}$ by

$$D = \begin{pmatrix} 1 & 0 & 0 & 0 & \dots & 0 & 0 \\ -a_2 & a_1 & -1 & 0 & \dots & 0 & 0 \\ a_4 & -a_3 & a_2 & -a_1 & \dots & 0 & 0 \\ \dots & \dots & \dots & \dots & \dots & \dots & \dots \\ 0 & 0 & 0 & 0 & \dots & a_{n-2} & -a_{n-3} \\ 0 & 0 & 0 & 0 & \dots & -a_n & a_{n-1} \end{pmatrix}$$

and let $D_{i.}$ denote the i-th row of matrix D.

If $c = (c_1, c_2, ..., c_n)^T \succ 0$ is a solution to $Dc > 0$, then, the polynomial $p_1(t) = c_1 t^n + c_2 t^{n-1} + ... + c_n$ is Hurwitz and besides $p_0(t) + kp_1(t)$ is Hurwitz for all $k > 0$.

3. The solutions of the inequalities

Given a real polynomial $p_0(t) = t^n + a_1 t^{n-1} + ... + a_n$, consider the matrix E as defined in (3). Let H be the set of solutions to (4), that is

$$H = \{c \in \mathbb{R}^n : c \succ 0 \ \text{ and } \ E^i c > 0, \ i = 1, ..., n\}$$

It can be seen that for $n = 3$ and 4, the set H is not empty and the solutions can be explicitly given as follows:

For $n = 3$, H is the set of three-dimensional vectors $c = (c_1, c_2, c_3)^T$, whose coordinates satisfy:

$$\frac{a_3 c_1 + a_1 c_3}{a_2} < c_2 < a_1 c_1$$

$$0 < c_3 < \frac{a_1 a_2 - a_3}{a_1} c_1$$

$$0 < c_1$$

For $n = 4$, H is the set of four-dimensional vectors $c = (c_1, c_2, c_3, c_4)^T$, whose coordinates satisfy:

$$a_3 c_1 - a_2 c_2 + a_1 c_3 < c_4 < \frac{a_3 c_3 - a_4 c_2}{a_2}$$

$$\frac{c_2}{a_1} < c_1 < \frac{(a_2^2 - a_4) c_2 + (a_3 - a_1 a_2) c_3}{a_2 a_3}$$

$$c_3 < \frac{a_1 (a_2^2 - a_4) - a_2 a_3)}{a_1 (a_1 a_2 - a_3)} c_2$$

$$0 < c_2$$

In both cases, the nonemptiness of the set H turns out to be a consequence of the stability conditions of the closed-loop system.

3.1. *Kuhn-Fourier Method*

For higher dimensions the algebraic problem becomes more complicated. One of the methods that could be applied to address the higher-dimensional case is the *elimination procedure of Kuhn-Fourier* (see Refs. 5, 11).

The algorithm of Fourier generalizes the elimination method for systems of linear equations.

Denote $\Gamma : Ec > 0$ our system of inequalities. The algorithm consists in obtain a new system of inequalities Γ' where one of the variables (say c_n) does not appear such that:

$\begin{pmatrix} c_1 \\ \dots \\ c_{n-1} \end{pmatrix}$ is solution of Γ' if and only if

$\exists\ c_n$ which satisfies that $\begin{pmatrix} c_1 \\ \dots \\ c_{n-1} \\ c_n \end{pmatrix}$ is solution of Γ.

With this general procedure every solution of Γ can be find for successive eliminations.

4. An example

Consider the following system

$$\dot{x} = \begin{pmatrix} 0 & 1 & 0 \\ 0 & 0 & 1 \\ -1 & -\frac{3}{2} & -2 \end{pmatrix} x + \begin{pmatrix} 0 \\ 0 \\ 1 \end{pmatrix} (-2k, -4k, -2k)\, x. \tag{7}$$

The matrix E and the corresponding product with $(2, 4, 2)^T$ are given by

$$E = \begin{pmatrix} 2 & -1 & 0 \\ -1 & \frac{3}{2} & -2 \\ 0 & 0 & 1 \end{pmatrix} \text{ and } \begin{pmatrix} 2 & -1 & 0 \\ -1 & \frac{3}{2} & -2 \\ 0 & 0 & 1 \end{pmatrix} \begin{pmatrix} 2 \\ 4 \\ 2 \end{pmatrix} = \begin{pmatrix} 0 \\ 0 \\ 2 \end{pmatrix}$$

Consequently the system (7) is stable $\forall k > 0$. But this fact can not be verified by the results in Ref. 1 since

$$D = \begin{pmatrix} 1 & 0 & 0 \\ -\frac{3}{2} & 2 & -1 \\ 0 & -1 & \frac{3}{2} \end{pmatrix} \text{ and } \begin{pmatrix} 1 & 0 & 0 \\ -\frac{3}{2} & 2 & -1 \\ 0 & -1 & \frac{3}{2} \end{pmatrix} \begin{pmatrix} 2 \\ 4 \\ 2 \end{pmatrix} = \begin{pmatrix} 2 \\ 3 \\ -1 \end{pmatrix}.$$

References

1. Aguirre, B.; Ibarra, C. and Suarez, R. Sufficient algebraic conditions for stability of cones of polynomials. *Systems & Control Letters.* **46**, (2002), 255-263.
2. Barnett, S. and Cameron, R.G. *Introduction to Mathematical Control Theory* (Clarendon Press, Oxford), 1985.
3. Gille, J.C.; Pelegrin, M.J. and Decaulne, P. *Feedback Control Systems: Analysis, Syntesis and Design.* New York: McGraw-Hill, 1959.
4. Hinrichsen, D. and Kharitonov, V. L. Stability of polynomial with conic uncertainty. *Math. Control Signal Systems* **8**, (1995) 97-117.
5. Kuhn, H.W. [1956] Solvability and consistency for linear equations and inequalities. *Amer. Math. Monthly* **63**, 217-232.
6. Ming-tzu, Ho; Datta, A. and Bhattacharyya, S.P. An elementary derivation of the Routh-Hurwitz criterion. *IEEE Transactions on Automatic Control*,. **43**, No. 3 (1998) 405-409.
7. Morari, M. and Zafiriou, E. *Robust Process Control* (Prentice-Hall, Englewood Cliffs, N.J.), 1989.
8. Schumacher, J.M. Almost stabilizability subspaces and high-gain feedback. *IEEE Trans. Automat. Contr.*, **AC-29**, (1984) 620-628.
9. Shaked, V. and Kouvaritakis, B. Asymptotic behavior of root loci of linear multivariable systems. *Int. J. Contr.* **23**, (1976) 297-340.
10. Shaked, V. and Kouvaritakis, B. The zeroes of linear optimal control systems and their role in high feedback gain stability design. *IEEE Trans. Automat. Contr.*, **AC-22**, (1977) 597-599.
11. Stoer, J. and Witzgall Ch. [1970] *Convexity and Optimization in Finite Dimensions I* (Springer-Verlag, N. Y.)
12. Willems, J. L. Disturbance isolation in linear feedback systems. *Int. J. Syst. Sci.,* **6**, No. 3, (1975) 233-238.
13. Willems, J. L. Almost invariant subspaces: An approach to high-gain feedback design - Part I and Part II. *IEEE Trans. Automat. Contr.,* **AC-26**, 235-252 and **AC-27**, (1981) 1071-1085.
14. Young, K. D.; Kokotovic, P. V. and Utkin, V. I. A singular perturbation analysis of high gain feedback systems. *IEEE Trans.,* **AC-22**, (1977) 931-939.

MODELLING OF AN ELECTRICALLY POWERED HELICOPTER PROTOTYPE

JOSÉ GERARDO BENITEZ-MORALES

Engineering Program in Mechatronics, Polithecnic University of Pachuca,
Pachuca, Hidalgo, Mexico
E-mail: jose_gerardobm@hotmail.com

RAFAEL CASTRO-LINARES*

Department of Electrical Engineering, CINVESTAV,
Av. IPN No. 2508, Col. San Pedro Zacatenco, Mexico City, 07360, Mexico
**E-mail: rcastro@cinvestav.mx*

EDUARDO LICEAGA-CASTRO

ESIME.IPN, Ticomán Unit,
Mexico City, Mexico
E-mail: eliceagac@uk2.net

The dynamics modelling of an electrically powered helicopter prototype is presented. The modelling includes the dynamics of the rotor thrust given by a DC electrical motor. Some experimental results are shown for a vertical flight in order to evaluate the model obtained.

Keywords: Modelling; Electrically powered helicopters; Helicopter dynamics; Rotor speed control.

1. Introduction

In recent years, the automation control of helicopters has been investigated. Since the helicopter dynamics have a nonlinear behavior and are strongly coupled, the models obtained are usually simplified in order to have a minimum number of state and/or inputs so that the control design can be carried out more easily. For example, Morris, van Niewwstandt and Bendotti[8] propose a four degrees of freedom discrete time model where the lateral and vertical motions are eliminated; the helicopter dynamics model is obtained in hover using a linear estimation technique. Following the same line of research, some continuous time models with three or four degrees of freedom have been proposed by Sira-Ramirez, Zribi and Ahmad,[10] Liceaga-Castro *et al*[6] and van Nieuwstand and Murray.[9] A more complete model is presented by Mahony, Hamel and Dzul,[7] where the rotor motion is included. Fantoni and Lozano[4] include the aerodynamics equation in the model, but only the vertical, yaw, and angular rotor motion are considered.

The present paper presents the dynamics modelling of an electrically powered helicopter prototype that includes the aerodynamics given by the rotor thrust. The performance of the model is evaluated in an experimental setup based on an electrically powered helicopter built by the company Kiosho.[3]

2. Experimental Setup

The experimental setup is an electric powered helicopter built by the company Kiosho.[3] The helicopter has a 571 mm diameter main rotor and a 122 mm diameter tail rotor. The total length and weight of helicopter is 550 mm and 461 $grams$, respectively. The main and tail helicopter rotors are driven by a *Le Mans AP29* high power electric DC motor through a set of gears. A photography of the electrically powered helicopter is shown in Fig. 1. Since the main rotor does not have a collective command pitch, its speed has to be handled in order to generate the thrust that allows the helicopter to fly. Also, the tail rotor is geared to the main rotor with a gear ratio of about 81 : 11.

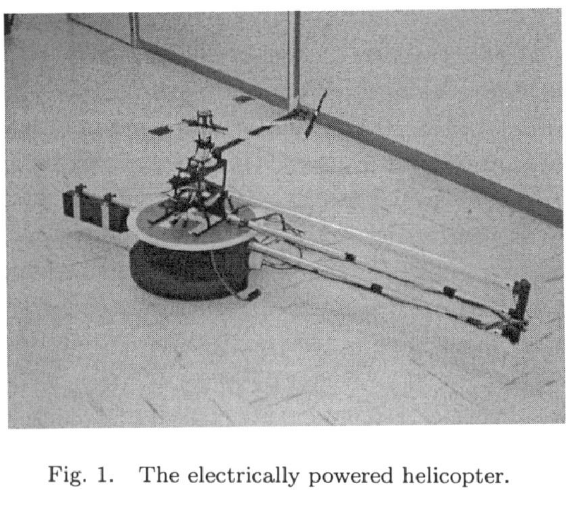

Fig. 1. The electrically powered helicopter.

Two servomotors were implemented to manipulate the cyclic command of the main rotor while a third servomotor was implemented to manipulate the collective command of the tail rotor. The angular position of each servomotor is handled by means of a PWM signal which is genereated using a data acquisition board *Lab-PC+* installed in a personal computer. The data generated by the computer is sent to each microcontroller of the type*PIC16F84A* so that the PWM signal generates an angular position. The block diagram of the electronic interface for the servomotor is shown in Fig. 2.

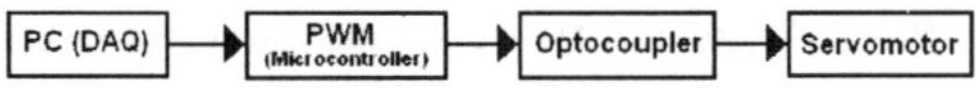

Fig. 2. Block diagram of the servomotor electronics.

The lift is handled by the variation of the rotor speed that is manipulated through an interface power. This interface consists of a power amplifier circuit connected to the DC motor. The input to this circuit comes from a data acquisition card *Lab-PC+* installed in a personal computer. In this computer a main rotor speed control is implemented; the control signal generated by the computer is sent the power interface. The block diagram of the powered interface is shown in Fig 3.

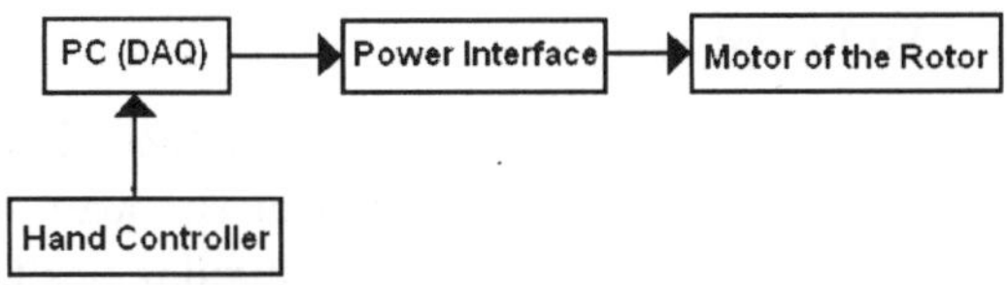

Fig. 3. Block diagram of the rotor power interface.

For security reasons the helicopter is attached to a mechanical arm with five links. This arm also allows to measure the helicopter position in the space via precision potentiometers installed in each link that measure their angular displacement. Such measurements permit the reconstruction of the helicopter position in the space using the direct kinematics associated to it.

The helicopter model proposed in this work considers the dynamics of the rotor actuator which is a DC motor with a permanent magnet stator. It was thus necessary to measure the main variables associated with this kind of actuators such as the armature voltage, the armature current and the rotor speed. Since there is no room left for a sensor to directly measure the main rotor speed, an optical shaft encoder is mounted on the tail rotor shaft; such an implementation gives a better resolution of the rotor speed since the tail rotor rotates faster than the main rotor. The block diagram associated to these electronics is shown in Fig. 4

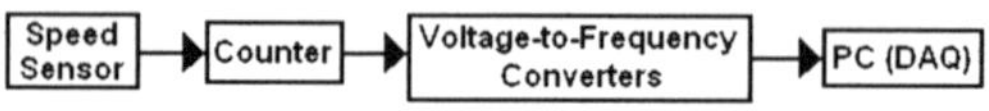

Fig. 4. Block diagram of the speed measurement electronics.

The armature voltage measurement is made by means of a set of operational amplifiers which are connected to the terminals of the motor. The current is measured using a Hall Effect sensor *LEM HX 15-P*. In Benitez[1] a detailed description of the hardware and software tools developed for the electrically powered helicopter is presented. Also, a control strategy was proposed for the main rotor speed so that it can follow a prescribed trajectory allowing to validate the helicopter model obtained.[2] The control strategy is based on the theory of differential flat systems recently introduced in the literature.[11]

3. Model of the Helicopter

The mechanical arm to which the helicopter is attached has five degrees of freedom (see Fig. 5). For each articulation of the arm, a precision potentiometer has been installed to measure the angular position so that the Denavit-Hartenberg method can be used to get the position of the helicopter in the space.

It is considered that the helicopter is described by means of the reference systems given in Fig. 6 and the body diagram shown in Fig. 7. The XYZ, $X_FY_FZ_F$ and $X_rY_rZ_r$ reference systems are the inertial fixed frame, the body fixed frame and fixed frame at the center of the rotor, respectively. The degrees of freedom are the horizontal displacement (x), the vertical displacement (z), the pitch (θ) and the rotation angle of the main rotor (Ψ).

The helicopter dynamics model was obtained using the Euler Lagrange method. The model obtained is given by (see Benitez[1] for a detailed computation)

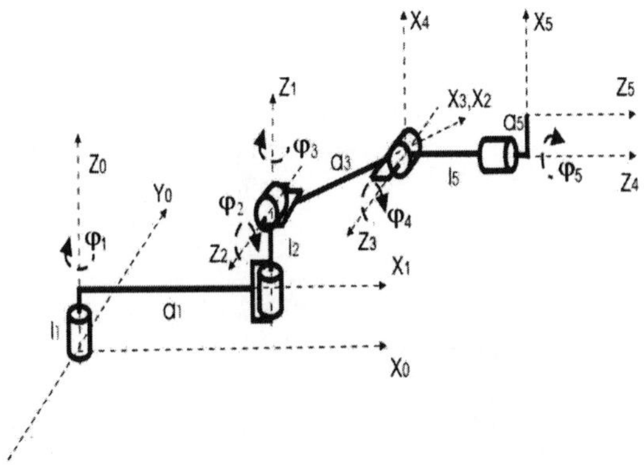

Fig. 5. Schematic of the mechanical arm.

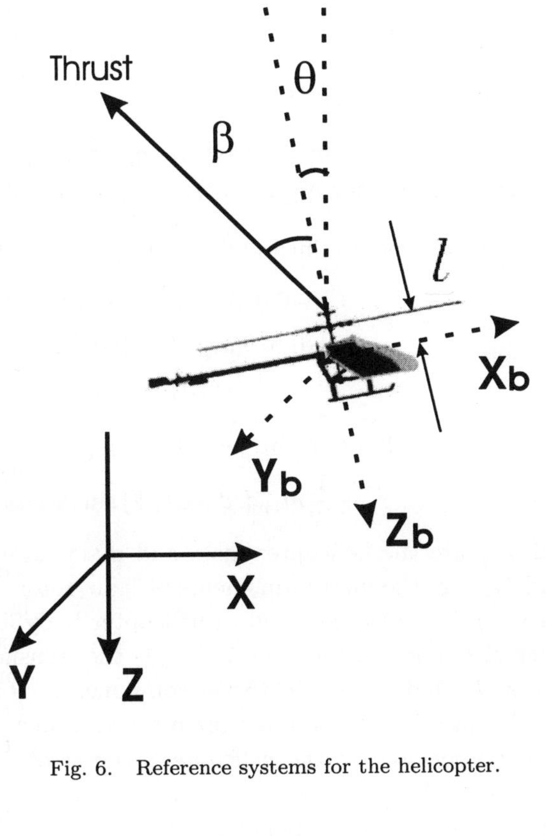

Fig. 6. Reference systems for the helicopter.

$$F_x = (m_1 + m_2)\ddot{x}_1 - (m_1 l_1 + m_{rp} l_2)\cos(\theta)\ddot{\theta} + (m_1 l_1 + m_{rp} l_2)\sin(\theta)\dot{\theta}^2$$

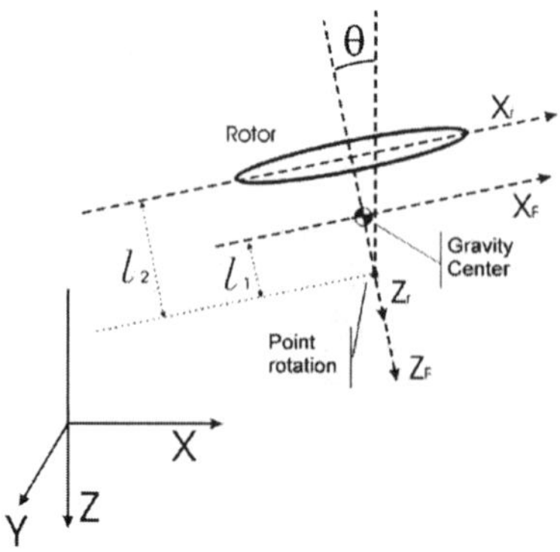

Fig. 7. Body diagram of the helicopter in the space.

$$F_z = (m_1 + m_{rp})\ddot{x}_1 + (m_1 l_1 + m_{rp} l_2)\sin(\theta)\ddot{\theta} + (m_1 l_1 + m_{rp} l_2)\cos(\theta)\dot{\theta}^2 - (m_1 + m_{rp})\, g$$

$$\begin{aligned}\tau_\theta = &\left[m_1 l_1^2 + J_F + m_{rp} l_2^2 + 0.0133 m_{rp} c^2 \sin^2(\alpha)\right]\ddot{\theta} \\ &+ (m_1 l_1 + m_{rp} l_2)\sin(\theta)\ddot{z}_1 - (m_1 l_1 + m_{rp} l_2)\,\ddot{x}_1 \cos(\theta) \\ &- \frac{1}{6} m_{rp} d^2 \cos(\Psi)\sin(\Psi)\dot{\Psi}\dot{\theta} + \frac{1}{12} m_{rp} d^2 \cos^2(\Psi)\,\ddot{\theta} \\ &- 0.075 m_{rp} c d \sin(\alpha)\sin(\Psi)\,\ddot{\Psi} \\ &- 0.075 m_{rp} c d \sin(\alpha)\cos(\Psi)\,\dot{\Psi}^2 \\ &+ (m_1 l_1 + m_{rp} l_2)\, g \sin(\theta)\end{aligned}$$

$$\tau_\Psi = I_{zz1}\ddot{\Psi} - 0.075 m_{rp} c d \sin(\alpha)\sin(\Psi)\,\ddot{\theta} + \frac{1}{12} m_{rp} d^2 \dot{\theta}^2 \cos(\Psi)\sin(\Psi)$$

where m_1 and m_{rp} are the helicopter mass and main rotor mass, respectively. I_{yy1} and I_{zz1} are the inertial moments of main rotor, and I_{yz1} is the inertial product. J_F is inertial moment of helicopter body. v_1 is the translational velocity of the body helicopter and v_2 is the translational velocity of the main rotor. F_x and F_y are the thrust components, τ_θ and τ_Ψ are the torque in the helicopter body and in the main rotor, respectively. l_1, l_2 and d are parameters associated to the helicopter's center of gravity and the DC motor position in the helicopter structure.

On other hand, the thrust components are expressed as $F_x = T\sin(\beta + \theta)$, $F_z = T\cos(\beta+\theta)$ and $\tau_\theta = T\sin(\beta) l_2$, where $T = -C_T \pi r^2 \rho \left(\dot{\Psi} r\right)^2$ is the equation of the so called *aerodynamics force.* Precisely, the rotor torque τ_θ is given by the DC motor attached to the main rotor through a gear and it's model is

$$JN\ddot{\Psi} + fN\dot{\Psi} + \hat{Q}\dot{\Psi}^2 = k_m y$$
$$L\dot{y} + Ry = V - k_e N\dot{\Psi}$$

where $\dot{\Psi}$ is the motor shaft rotation speed amd N is the gear ratio in the main rotor. J and f are the inertia moment and the friction coefficient of the motor shaft, respectively, while L and R are the inductance and resistance of the armature windings, k_e is the back *emf* coefficient and k_m is the motor torque coefficient. V is the armature voltage and y is the current through the armature windings.

4. Experimental Evaluation of the Helicopter Model

Since it is difficult to drive the helicopter in a three dimensional space, due to its six degrees of freedom, a vertical flight was considered. For this kind of flight, the helicopter was first driven at a height of 0.385 *cm* for 4 *s*, then it was set to height of 8.344 *cm* for 13 *s*, and, finally, it was set to a height of 26.476 *cm*. When this experiment was carried out, it was noticed that the Y displacement was considerable for the last part of the flight schedule. This is due to the mechanical arm to which the helicopter is attached and that induces a circular trajectory for the helicopter. For this reason, only the first two parts of the flight schedule were considered so that the helicopter was really moving in a plane following a vertical flight. Fig. 8 shows a comparison between the experimental data obtained from the prototype and the ones obtained from the model (solid line) for the vertical displacement variable when the flight was restricted to such a plane. From these results one can notice the good performance of the model.

5. Conclusions

The dynamics modelling of an electrically powered helicopter prototype has been presented. The model includes the aerodynamics introduced by the thrust of the DC motor that moves the main and tail rotors of the helicopter. An experiment for a vertical flight condition was presented where the performance of the model obtained was compared with the data obtained from direct measurements made in the experimental platform. The model showed a good performance in this kind of flight and it is now under validation for flight conditions in the trhee dimensional space.

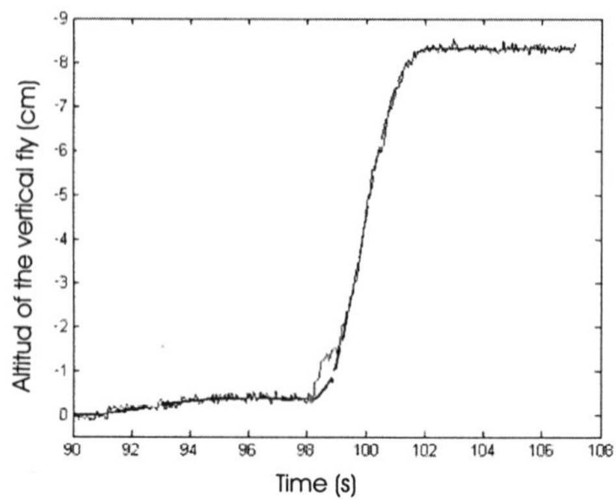

Fig. 8. Comparison between data obtained from the model and the prototype for a vertical displacement.

References

1. J. G. Benitez-Morales, Design, Construction and Modelling of an Experimental Helicopter Setup, Master's thesis (in Spanish), CINVESTAV, Department of Electrical Engineering, (Mexico City, 2005).
2. R. Castro-Linares, E. Liceaga-Castro, and J. G. Benitez-Morales, Trajectory tracking of the main rotor speed in an electrically powered helicopter prototype, in *Proc.* 2005 *IEEE Conference on Control Applications*, (Toronto, Canada, 2005).
3. *Convert, Kyosho Electrically Powered Helicopter, Instruction Manual*, (Kyosho Company, Japan, 1996).
4. I. Fantoni and R. Lozano, in *Non-linear Control for Underactuated Mechanical Systems*, Springer, (London, 2002).
5. D. M. Layton, *Helicopter Performance*, (Matrix Publishers Inc., 1984).
6. J. Liceaga-Castro, C. Verde, J. OReilly and W. E. Leithead, *IEE Proc. Control Theory Appl.*, **141**, No. 1 (1995).
7. R. Mahony, T. Hamel, and A. Dzul, Hover control via Lyapunov control for an autonomous model helicopter, in *Proc.* 38th *Conference on Decision Control*, (Phoenix, Arizona, 1999).
8. J. Morris, M. van Nieuwstandt, and Pascale Bendotti, Identification and control of a model helicopter in hover, in *Proc. of the ASEE Annual Conference*, (Baltimore, Maryland, 1994).
9. Van Nieuwstand and Richard M. Murray, Outer flatness: trayectory generation for a model helicopter, in *Proc. European Control Conference* (1997).
10. H. Sira-Ramirez, M. Zribi and S. Ahmad, Dynamical sliding mode control approach for vertical flight regulation in helicopters *IEE Proc. Control Theory Appl.*, **142**, No. 1 (1995).
11. H. Sira-Ramirez, and S. K. Agrawal, *Differentially Flat Systems*, (Marcel Dekker, Inc., 2004).

TIMOSHENKO BEAM THEORY BASED MATHEMATICAL MODELLING OF A LIGHTWEIGHT FLEXIBLE LINK ROBOT MANIPULATOR

MALIK LOUDINI

Institut National d'Informatique, BP 68M
16270 Oued Smar, El-Harrach, Algiers, Algeria

DJAMEL BOUKHETALA, MOHAMED TADJINE

Laboratoire de Commande des Processus, Ecole Nationale Polytechnique, 10, avenue Hassan Badi, BP 182
16200 El-Harrach, Algiers, Algeria

This paper deals with the mathematical modelling of a planar one-link lightweight flexible robot manipulator clamped at its actuated base and carrying a payload at its free end-point. Taking into account the effects of shearing and rotational inertia of cross-section of the elastic link considered as a beam and neglecting gravity, torsion, and longitudinal elongation, the Timoshenko beam theory has been used to characterize the structural link elasticity with taking into consideration of two damping mechanisms: internal structural viscoelasticity effect (Kelvin-Voigt) and external viscous air damping. Then a closed form dynamic model is derived on the basis of the combined Lagrange-assumed modes method.

1. Introduction

Modelling and control of flexible manipulators have been an active research field in recent years. Traditional robots have been designed for "rigidity" with short arms and a heavy structure, which significantly restrict their range of applications. Lightweight manipulators with lower arm cost, higher motion speed, better energy efficiency, safer operation and improved mobility are highly desirable. A major problem with such flexible robots, however, is that the end-point accuracy is severely impaired due to structural deformation of the flexible links. This elasticity is the most complicating parameter in the task of deriving a mathematical model of such systems. Detailed discussions about modelling and control problems for flexible link robot manipulators can be found in [1-3].

In this paper, we aim to present the details of our investigations concerned with deriving accurate governing equations of motion of a flexible single link robot arm clamped at its base and carrying a payload at its end-point by the use of the Timoshenko beam (TB) theory.

Earlier works, in the same context of the present one, can be found in [4-8]. The considerations that characterize our contribution are summarized below:

- Inclusion of damping effects, i.e. internal viscoelastic effect and external viscous air damping, clearly pointed out as prominent physical characteristics in [9-10] in addition to rotary inertia and shear deformation.
- Clamped-mass type of the constrained mode shapes rather than the pinned-pinned one and thus, the link inertia is small compared to the hub inertia [11]. Indeed, the clamped assumption is more adequate and especially when closing a feedback control loop around the joint where the link is clamped.
- The robot is supposed be actuated by a high gear motor rather than a direct drive one.

The outline of this paper is as fellows. In section 2, the different stages followed in obtaining the flexible manipulator motion governing equation, based on TB theory concepts, are detailed. Section 3 is devoted to the deriving procedure to obtain a dynamic model of the studied robot on the basis of the combined Lagrange-assumed modes approach. Section 4 reports simulation results. The paper is concluded with section 5.

2. Mathematical Modelling of the Flexible Robot Using TB Theory

The flexible robot physical system under consideration is shown in Figure 1. It consists of a clamped-free with tip payload planar moving flexible arm which can bend freely in the horizontal plane. The deflection which is the transverse displacement of the link from the X-axis is denoted by $w(x,t)$.

This figure is a “top” view of the manipulator in deflection and the axis of rotation of the rigid hub (Z_0) is perpendicular to robot evolution plane. The $X_0 - Y_0$ coordinate frame is the inertial frame of reference. The one indicated by $X - Y$ is a frame of reference that rotates with the overall structure. The X-axis is tangent to the clamped beam at the base [11].

As usual, the flexible link can be considered as a beam. The height of the beam cross-section is assumed to be larger than the base. This constrains deflections to occur only in the horizontal plane. Thus, those due to gravity are assumed negligible.

As depicted in Figure 1, the robot manipulator is essentially composed of a rigid hub, a flexible link and a payload. These three parts are characterized by different physical and mechanical parameters (see the list of symbols at the end of the paper).

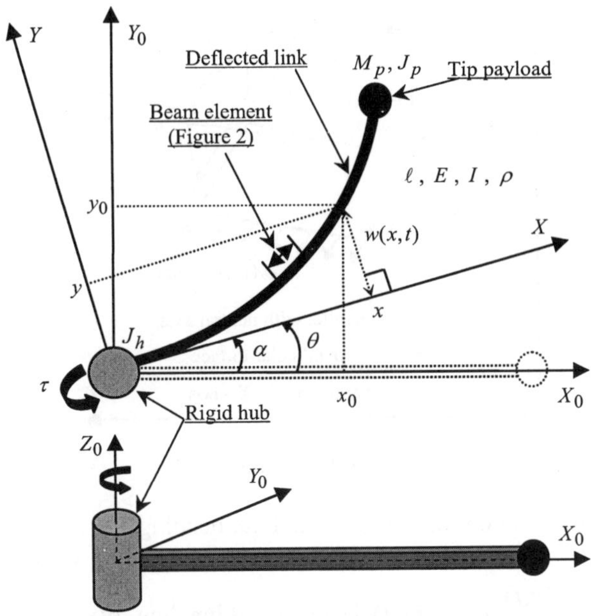

Figure 1. Single link flexible manipulator arm.

The TB theory [12] is widely used in the beams transverse vibration analysis. It has been shown in the literature that the predictions of the TB beam model are in excellent agreement with the results obtained from the exact elasticity equations and experimental results [13-15].

This theory accounts for both the effect of rotary inertia and shear deformation, which are neglected when applied to Euler-Bernoulli beam theory [16]. The transverse vibration of the beam depends on its geometrical and material properties as well as the external applied torque.

Inspect an element of the deflected link with width dx, at position x (Figure 2). It is subjected to a shearing force $S(x,t)$, and a bending moment $M(x,t)$. On the opposite side of this segment, which corresponds to a position $x+dx$, the shearing force ($S+dS$) is

$$S(x+dx,t) = S(x,t) + \frac{\partial S(x,t)}{\partial x}dx. \tag{1}$$

Likewise the moment force ($M+dM$) at the position $x+dx$ is

$$M(x+dx,t) = M(x,t) + \frac{\partial M(x,t)}{\partial x}dx. \tag{2}$$

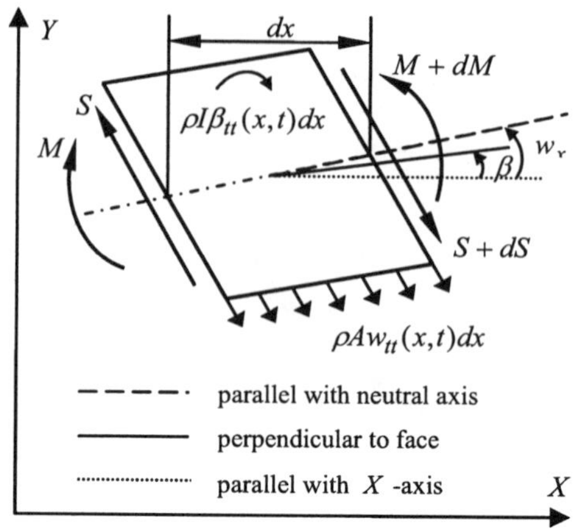

Figure 2. Moments and forces acting on a bending element.

Note that the total deflection is due to both bending and shear forces, so that the shear angle $\sigma(x,t)$ (or loss of slope) is equal to the slope of centreline (neutral axis) $\frac{\partial w(x,t)}{\partial x}$ or $w_x(x,t)$ less slope of bending $\beta(\mathrm{x},\mathrm{t})$ in the form:

$$\sigma(x,t) = w_x(x,t) - \beta(x,t). \tag{3}$$

The shear force S can be written as

$$S(x,t) = kAG\sigma(x,t) = kAG[w_x(x,t) - \beta(x,t)] \tag{4}$$

The total internal moment (bending and damping) M is given by [9-10]

$$M(x,t) = EI\beta_x(x,t) + K_V I\beta_{xt}(x,t). \tag{5}$$

where K_V is the Kelvin-Voigt damping coefficient.

The equation of motion of the studied single link flexible robot arm can be derived by considering both the equilibrium of the moments and the forces. Taking moments as positive in the counter-clockwise direction, their summation with disregarding the second order term of dx, yields this relation

$$M_x(x,t) = -S(x,t) + \rho I\beta_{tt}(x,t). \tag{6}$$

where the term $\rho I\beta_{tt}(x,t)$ stands for the distributed rotational inertia.

The relation that fellows balancing forces is

$$S_x(x,t) - \gamma w_t(x,t) = \rho Aw_{tt}(x,t). \tag{7}$$

where the terms $\gamma w_t(x,t)$, $\rho A w_{tt}(x,t)$ represent, respectively, the air resistance force and the distributed transverse inertial force.

Substitution of Eq. (4) and Eq. (5) into Eq. (6) and likewise Eq. (4) into Eq. (7) yields the two coupled equations of the damped Timoshenko beam motion:

$$\begin{cases} (K_V I \beta_{xt})_x + (EI\beta_x)_x + kAG(w_x - \beta) - \rho I \beta_{tt} = 0 \\ [kAG(w_x - \beta)]_x - \rho A w_{tt} - \gamma w_t = 0. \end{cases} \tag{8}$$

If the damping effects terms are suppressed, the classical set of two coupled PDE developed by Timoshenko arises [12]

$$\begin{cases} (EI\beta_x)_x + kAG(w_x - \beta) - \rho I \beta_{tt} = 0 \\ [kAG(w_x - \beta)]_x - \rho A w_{tt} = 0. \end{cases} \tag{9}$$

The modelled beam cross-sectional area and density being uniform, the two equations given in (8) can be easily decoupled as follows:

$$\begin{aligned} K_V I w_{xxxxt} - \frac{K_V I \rho}{KG} w_{xxttt} + EI w_{xxxx} - \rho I\left(1 + \frac{E}{KG} + \frac{K_V \gamma}{\rho KAG}\right) w_{xxtt} \\ + \frac{\rho^2 I}{KG} w_{tttt} - \frac{EI\gamma}{kAG} w_{xxt} + \frac{\rho I \gamma}{kAG} w_{ttt} + \rho A w_{tt} + \gamma w_t = 0. \end{aligned} \tag{10}$$

$$\begin{aligned} K_V I \beta_{xxxxt} - \frac{K_V I \rho}{KG} \beta_{xxttt} + EI \beta_{xxxx} - \rho I\left(1 + \frac{E}{KG} + \frac{K_V \gamma}{\rho KAG}\right) \beta_{xxtt} \\ + \frac{\rho^2 I}{KG} \beta_{tttt} - \frac{EI\gamma}{kAG} \beta_{xxt} + \frac{\rho I \gamma}{kAG} \beta_{ttt} + \rho A \beta_{tt} + \gamma \beta_t = 0. \end{aligned} \tag{11}$$

Eq. (10) is the fifth order TB homogeneous linear PDE with internal and external damping effects expressing the deflection $w(x,t)$ with the following initial and boundary conditions (BC):

$$\begin{cases} w(x,0) = w_0, \ w_t(x,0) = \dot{w}_0; \quad w(0,t) = w_x(0,t) = 0; \\ M_x(\ell,t) = M_p w_{tt}(\ell,t), \ M(\ell,t) = -J_p w_{xtt}(\ell,t). \end{cases} \tag{12}$$

The classical fourth order Timoshenko beam PDE is retrieved if the damping effects terms are suppressed:

$$EI w_{xxxx} - \rho I\left(1 + \frac{E}{KG}\right) w_{xxtt} + \frac{\rho^2 I}{KG} w_{tttt} + \rho A w_{tt} = 0. \tag{13}$$

To solve the PDE with mixed derivative terms (Eq. (10)), we have opted for the eigenfunction expansion method [17]. Thus, $w(x,t)$ can take the following expanded separated form which consists of an infinite sum of products between the chosen eigenfunctions or mode shapes $W_n(x)$, that must satisfy the clamped-mass BC and the time-dependant modal generalized coordinates $\delta_n(t)$:

$$w(x,t)=\sum_{n=1}^{\infty}W_n(x)\delta_n(t)=\sum_{n=1}^{\infty}\left[1-\cos\frac{m\pi x}{2L}\right].\delta_n(t);\quad m=2n-1 \tag{14}$$

By substituting Eq. (14) into Eq. (10), we find this fourth order temporal ODE:

$$\begin{aligned}cf_1\overset{(4)}{\delta}_n(t)+\left(cf_2m^2+cf_3\right)\dddot{\delta}_n(t)+\left(cf_4m^2+cf_5\right)\ddot{\delta}_n(t)\\+\left(cf_6m^4+cf_7\,m^2+cf_8\right)\dot{\delta}_n(t)+cf_9m^4\delta_n(t)=0\end{aligned} \tag{15}$$

where

$$cf_1=\frac{\rho^2I}{KG};cf_2=\frac{K_VI\rho\pi^2}{KGL^2};cf_3=\frac{\rho I\gamma}{KAG};cf_4=\frac{\rho I\pi^2}{L^2}\left(1+\frac{E}{KG}+\frac{K_V\gamma}{\rho KAG}\right);$$

$$cf_5=\rho A;cf_6=\frac{K_VI\pi^4}{L^4};cf_7=\frac{EI\gamma\pi^2}{KAGL^2};cf_8=\gamma;cf_9=\frac{EI\pi^4}{L^4}.$$

Considering, for example, an aluminum beam, as in [7], we found, after numerical calculations, that the coefficients of $\overset{(4)}{\delta}_n(t)$ and $\dddot{\delta}_n(t)$ are very small compared to those of the lower order terms of Eq. (15). This latter is, then, approximately reduced to the following second order ODE:

$$cof_1\ddot{\delta}_n(t)+cof_2\dot{\delta}_n(t)+cof_3\delta_n(t)=0. \tag{16}$$

where $cof_1=cf_4m^2+cf_5$; $cof_2=cf_6m^4+cf_7\,m^2+cf_8$; $cof_3=cf_9m^4$.

Eq. (16) has the general form of a second order ODE characterizing a linear elastic system with damping. Considering the underdamping mode case, with a set of initial conditions ($\delta(0)$, $\dot{\delta}(0)$), its solution is given by

$$\delta(t)=C_ne^{-\xi\omega_nt}\cos\left(\omega_dt-\psi_n\right) \tag{17}$$

where $\omega_n=\sqrt{\dfrac{cof_3}{cof_1}}=\sqrt{\dfrac{cf_9m^4}{cf_4m^2+cf_5}}$; $\xi_n=\dfrac{cof_2}{2cof_1\omega_n}=\dfrac{cf_6m^4+cf_7\,m^2+cf_8}{2(cf_4m^2+cf_5)\omega_n}$;

$$\omega_d = \omega_n \sqrt{1-\xi_n^2}\ ;\ C_n = \left\{ \left[\frac{\dot{\delta}(0) + \delta(0)\xi_n \omega_n}{\omega_d} \right]^2 + [\delta(0)]^2 \right\}^{1/2} ;$$

$$\psi_n = \arctan\left[\frac{\dot{\delta}(0) + \delta(0)\xi_n \omega_n}{\omega_d \delta(0)} \right];\ n = 1, 2, \cdots;\ m = 1, 3, 5, \cdots.$$

The robot link transverse displacement, approximately obtained, as a solution of the damped Timoshenko equation (Eq. 10) is finally expressed by

$$w(x,t) = \sum_{n=1}^{\infty} C_n e^{-\xi\omega_n t} \cos(\omega_d t - \psi_n)\left(1 - \cos\frac{m\pi x}{2L}\right). \tag{18}$$

3. Robot Dynamic Model

In order to obtain a set of ordinary differential equations of motion to adequately describe the dynamics of the flexible link robot, the Lagrange's approach is used.

A dynamic system completely located by n generalized coordinates q_i must satisfy n differential equations of the form:

$$\frac{d}{dt}\left(\frac{\partial L}{\partial \dot{q}_i}\right) - \frac{\partial L}{\partial q_i} + \frac{\partial D}{\partial \dot{q}_i} = Q_i\ ,\ i = 0, 1, 2, \cdots \tag{19}$$

where L is the so called Lagrangian which is given by $L = T - U$.
T represents the kinetic energy of the modelled system, U its potential energy and. D is the Rayleigh's dissipation function. Q_i is the generalized external force acting on the corresponding coordinate q_i.

Theoretically there are infinite number of ODE, but for practical considerations, this number is truncated at a finite one n [4, 18].

The total kinetic energy of the robot elastic link and its potential energy due to the internal bending moment and the shear force are, respectively, given by

$$T = \frac{1}{2}\int_0^{\ell} \rho A[w_t(x,t)]^2 dx + \frac{1}{2}\int_0^{\ell} \rho I[\beta_t(x,t)]^2 dx. \tag{20}$$

$$U = \frac{1}{2}\int_0^{\ell} EI[\beta_x(x,t)]^2 dx + \frac{1}{2}\int_0^{\ell} KAG[\sigma(x,t)]^2 dx. \tag{21}$$

The dissipated energy due to the damping effects is:

$$D = \frac{1}{2}\int_0^{\ell} \gamma\left[w_t(x,t)\right]^2 dx + \frac{1}{2}\int_0^{\ell} K_V I\left[w_{xxt}(x,t)\right]^2 dx. \tag{22}$$

Substituting these energies expressions into Eq. (19) accordingly and using the transverse deflection separated form (Eq. (14)) we can, after tedious manipulations, derive the desired dynamic equations of motion in the mass (B), damping (H) and stiffness (K) matrix form:

$$B.\ddot{q}(t) + H.\dot{q}(t) + K.q(t) = Q(t) \tag{23}$$

with $q(t) = \left[\theta(t) \quad \delta_1(t) \quad \delta_2(t) \quad \cdots \quad \delta_n(t)\right]^T$; $Q(t) = \left[\tau \quad 0 \quad 0 \quad \cdots \quad 0\right]^T$.

The matrix differential equation in Eq. (23) can be easily represented in a state-space form as

$$\begin{cases} \dot{z}(t) = A_z z(t) + B_z u(t) \\ y(t) = C_z z(t) \end{cases} \tag{24}$$

with $u(t) = \left[\tau \quad 0 \quad \cdots \quad 0\right]^T$; $z(t) = \left[\theta(t) \ \delta_1(t) \ \cdots \ \delta_n(t) \ \dot{\theta}(t) \ \dot{\delta}_1(t) \ \cdots \ \dot{\delta}_n(t)\right]^T$
Solving the state-space matrices gives the vector of states $z(t)$, that is, the angular displacement, the modal amplitudes and their velocities.
Details of the obtained dynamic model are not included here.

4. Simulation Results

To simulate the vibrational behaviour of the modelled flexible manipulator, its physical parameters numerical values are taken from [7]. Only the hub and the payload inertias are given, respectively, by $J_h = 0.4\,Kg.m^2$, $J_p = 0.005\,Kg.m^2$.

To formulate a simple, physically correct finite dimensional dynamic model for behaviour analysis, the general governing equation is truncated to only the two lower (dominant) modes of vibration ($n = 2$). The corresponding two mode shape functions $W_1(x)$ and $W_2(x)$ are shown in Figure 3 where the third is also shown for illustration.

Comparing the two responses, we note, as expected, that the frequency of vibration is reduced but the amplitude of vibration is increased as a result of adding a payload. Indeed, the effect of the added payload is the increasing of the whole structure (link + attached payload) mass and inertia. Thus, when the end-loaded link vibrates, the oscillations amplitudes are larger and their frequency slower than those of the free link vibrational evolution.

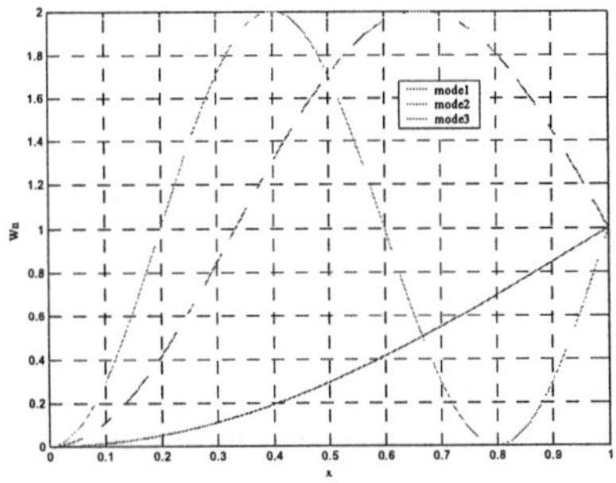

Figure 3. The first three mode shape functions.

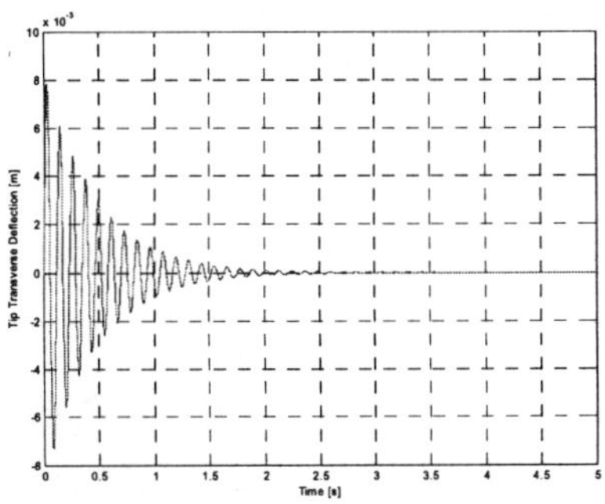

Figure 4. Tip deflection without payload

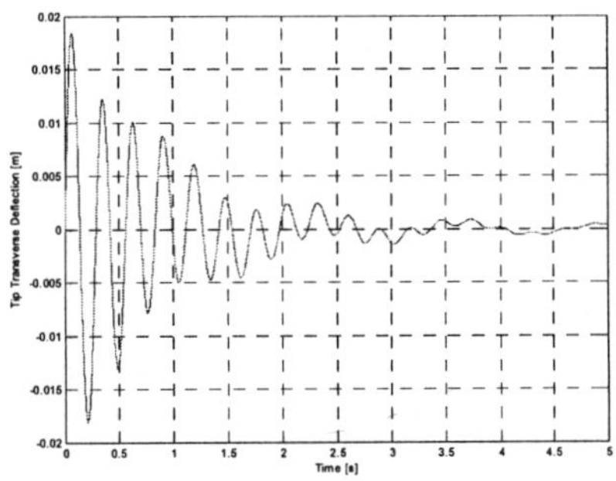

Figure 5. Tip deflection with payload.

5. Conclusions

In this paper, the development of a flexible link robot manipulator mathematical model using Timoshenko beam theory has been reported.

The emphasis has been, essentially, set on obtaining accurate and complete equations of motion that display the most relevant aspects of structural properties inherent to the modelled lightweight flexible link. To derive a finite-dimensional dynamic model, the main steps of the Lagrangian approach combined with the assumed modes method have been presented.

In order to reveal the vibrational behavior of the modelled system, the system has been digitally simulated by solving the corresponding motion governing PDE. As expected, it has been found that, with the addition of a payload, the frequency of vibration of the manipulator end-point is reduced but its amplitude of vibration increases significantly.

Nomenclature

A : link cross-section area; B : inertia matrix; D : dissipated energy; E : Young's modulus of elasticity; G : shear modulus; H : damping matrix; I : link moment of inertia; J_h : hub + actuator total inertia; J_p : payload inertia; k : shear correction factor; K : stiffness matrix; K_V : Kelvin-Voigt damping coefficient; ℓ : link length; L : Lagrangian; M : bending moment; M_p : payload mass; n : mode number; q : vector of generalized coordinates; Q : vector of external forces; t : time; S : shear force; T : kinetic energy; U : potential energy; w : transverse deflection; W_n : n th mode shape function; x : coordinate along the beam; α : angular position of a point of the deflected link; β : rotation of cross-section; δ_n : n th modal amplitude; γ : viscous air damping coefficient; θ : hub angular position; ρ : link mass density; σ : shear angle; τ : actuator torque; ω_n : n th natural frequency of vibration; ξ_n : n th damping ratio.

References

1. M. Benosman, F. Boyer, G. L. Vey and D. Primautt, *J. of Intelligent and Robotic Systems*, **34**, 381-414 (2002).
2. M. Benosman and G. L. Vey, *Robotica*, **22**, 533-545 (2004).
3. S. K. Dwivedy and P. Eberhard, *Mechanism and Machine Theory*, **41**, 749-777 (2006).
4. X. Qi and G. Chen, *Proc. IEEE Conf. on Contr. Appl.*, **1**, 288-293 (1992).
5. M. W. D. White and G. R. Heppler, *Proc. the ACC*, **4**, 2815-2819 (1995).
6. F. -Y. Wang, P. Zhou and P. Lever, *Proc. IEEE Int. Conf. on Systems, Man and Cybernetics*, **3**, 2315-2320 (1996).
7. P. Sooraksa and G. Chen, *Mathl. Comput. Modelling*, **27**, 73-93 (1998).
8. M. Karkoub and K. Tamma, *Computers and Structures*, **79**, 543-551 (2001).
9. H. T. Banks, Y. Wang, and D. J. Inman, *J. of Applied Mechanics*, **116**, 188-198 (1994).
10. H. T. Banks, R. C. Smith and Y. Wang, *Wiley-Masson, New York* (1996).
11. F. Bellezza, L. Lanari, and G. Ulivi, *Proc. IEEE Conf. Rob. Autom.* 734-739 (1990).
12. S. Timoshenko, D. H. Young, W. Jr. Weaver, *Wiley, New York* (1974).
13. T. C. Huang, *J. of Applied Mechanics*, **28**, 579-584 (1961).
14. N. G. Stephen, *J. of Sound and Vibration,* **80**, 578-582 (1982).
15. S. M. Han, H. Benaroya and T. Wei, *J. of Sound and Vibration,* **225**, 935-988 (1999).
16. L. Meirovitch, *Prentice-Hall, Upper Saddle River, NJ, USA*, (1997).
17. S. Ekwaro-Osire, D. H. S. Maithripala and J. M. Berg, *J. of sound and vibration,* **240**, 667-678 (2001).
18. H. Kanoh and H. G. Lee, *Proc. of the 24th CDC*, 1172-1177 (1985).

A NEW APPROACH FOR MODELING, SIMULATION AND CONTROL OF COMPLEX ELECTROMECHANICAL SYSTEMS: THE COMPUTATIONAL MECHATRONICS SCHEME

†LUIS IVÁN LUGO VILLEDA*, ‡VICENTE PARRA VEGA and †GUSTAVO NUÑEZ ESQUER

†Mechatronics Research, Universidad Politécnica de Pachuca
Carr. Pachuca-Cd.Sahagún Km 20,Ex-Hacienda de Santa Bárbara, Zempoala Hgo. 43080, México.
*E-mail: *lilugo@upp.edu.mx, gustavone@upp.edu.mx*
‡Robotics and Advanced Manufacturing Division Research Center for Advanced Studies
Carretera Saltillo-Monterrey Km 13.5 Ramos Arizpe, Coah, 25900 México.
E-mail: vparra@cinvestav.mx

Major analysis and synthesis of dynamic models and their controllers imply deeper understanding of closed-loop response. Nevertheless, the numerical solution is useful in practice as long as it reflects a real behavior, that is the solution is as close as possible to reality as long as the model embodies the real complexity of the system and a number of factors are considered for the closed-loop simulation, for instance: state variables, mass distributions, unmodeled disturbances, friction, input and output noise, quantization, actuator saturation, bandwidths, or any signals that excite the dynamical system beyond the modeling hypotheses. However, the complexity of the simulator may increase exponentially if all these factors are included. Conventional techniques of closed-loop models is routinely made by manipulation of ordinary differential equations, which obtains results nearly to the reality, but with lack of validity if each factor is considered isolated, because each factor imposes diverse constraints to other elements of the closed-loop system. In this article, a novel close-loop approach for the modeling, simulation and control of dynamic systems, specifically applied to complex electromechanical systems, is presented based on the so-called Computational Mechatronics scheme (CMk) to obtain results of similar realness but with less complex models in comparison to conventional modeling and simulation procedures. We obtained these tools by using the synergetic mechatronics approach for design and advanced computational CAE tools. The main idea of the CMk is to provide a simulator/modeler that can be executed on PC in the continuous and discreet time under different conditions of operation, with application of parametric uncertainties, kinematics and dynamics constraints, electronics noise, and quantization or filtering.

Keywords: Mechatronics, planar biped robot, pendulum, CAE.

1. Computational Mechatronics (CMk) Scheme

Consider the computational mechatronics scheme as follows:

The synergetic and dynamical integration at all steeps and at all stages of modeling, simulation, control design and engineering specifications of dynamical systems —using detailed CAD models as the dynamical system, not differential equations— is coined as a Computational Mechatronics Scheme (CMk)

This virtual system, which models the input-output characteristic response of all elements of a real system, including the information processing; it is considered as mechatronic system because the mechanical and informatic systems prevail, and it is modeled through CAD with dynamic properties, rather only geometric modeling, as conventional modeling approach does. Computational mechatronics can be used at bachelor or graduate schools because CMk platform is easy to program and is more *cheap* and does not require dedicated and expensive training with respect to another CAE-based simulator like CATIA, SABER, ADAMS. This new scheme comprises the following intrinsically related steeps:

(1) Mathematical modeling of dynamic system and mechanical geometric conceptualization and assembly parts (CAD), under hardware integration "philosophy" of mechatronics approach.
(2) Integration to the CAD model of the mechanical properties as well as electric/electronic, kinematics and dynamics constraints all inside CAE dynamic based simulator
(3) Analysis and synthesis of control techniques applied to constrained CAD
(4) Integration of information processing tools to the mechanical design constrained, including all final constraints and dynamic response of sensors and actuators, as well as unmodeled parametric perturbations.

The CMk stands a powerful simulator at hand of bachelor and graduate school level because it includes a mechatronics scheme in open architecture CAE simulator, see Figure 1. Next, we further discuss in detail this idea with two case study: Novel Biped Robot [4], and Simple Pendulum, which exemplifies the use and advantages of this simulator over standard simulations, for instance like Matlab®-based simulators.

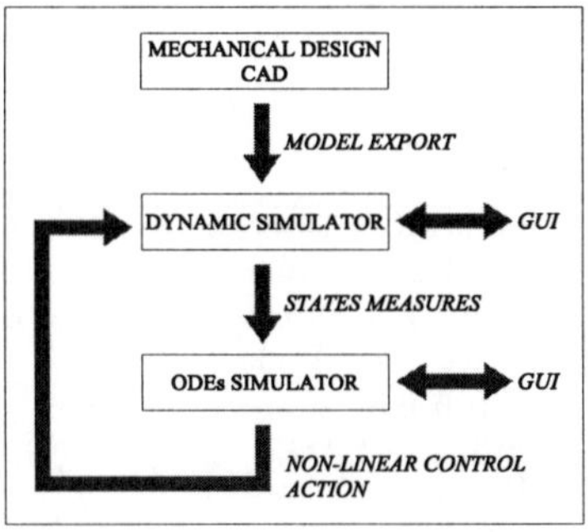

Fig. 1. CMk simulator flow.

2. Analysis, Synthesis and Control of Novel Biped Robot using CMk

Considering the novel planar biped robot [3,6,8], see Figure 2, it is composed of rotary central body interconnected without any friction, two links or legs in the plane, called stance leg and swing leg, respectively. Therefore, there

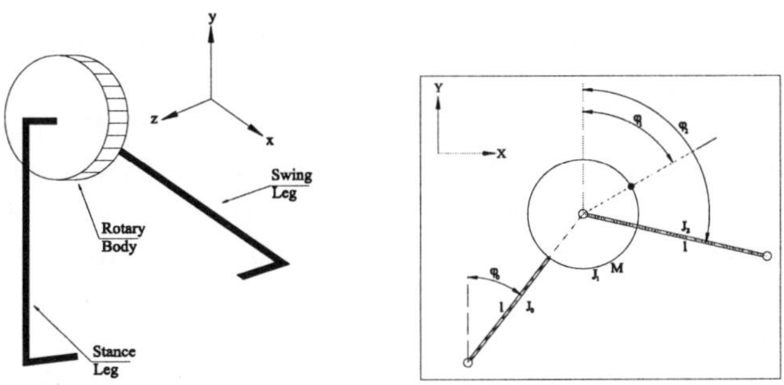

(a) Planar biped robot architecture

(b) Absolute coordinates.

Fig. 2. 3 DOF planar biped robot : The simplest mechanical system that may walk

are 3 DOF, 2 of them associated to the legs and the third one associated to the frictionless rotary body. Applying conventional modeling techniques[6], we obtain

$$J_0\ddot{\varphi}_0 = Mgl\sin\varphi_0 - \tau_1 + \Gamma(t) \quad (1)$$

$$J_1\ddot{\varphi}_1 = \tau_1 - \tau_2 \quad (2)$$

$$J_2\ddot{\varphi}_2 = \tau_2 \quad (3)$$

whose variables are described in Table 1.

Table 1. Parameters Notation

Symbol	Description	Units
J_0	Stance Leg Discrete Inertia	$kg-m^2$
J_1	Hip Discrete Inertia	$kg-m^2$
J_2	Swing Leg Discrete Inertia	$kg-m^2$
φ_0	Stance Leg Angle	rad
φ_1	Hip Angle	rad
φ_2	Swing Leg Angle	rad
M	Pendulum Mass	kg
l	Leg Length	m
g	gravity constant	m/s^2
τ_1	Stance Leg Torque	Nm
τ_2	Swing Leg Torque	Nm
$\Gamma(t)$	$\leq 10Hz$ White Noise	Nm

Since closed-loop parameters design delivers a more dedicated and specific design, an advanced controller must be considered, in particular a model-free controller because, in principle, the model is uncertain. Using a second order sliding mode control with the flatness of the system[4], and with nominal parameters of the systems, the biped robot may walk using a conventional simulation of ODEs, considering a theoretical model and neglecting all types of constraints. Now applying the CMk scheme to the planar biped robot with similar physical and control parameters, we have a different response in the control system in comparison to the conventional modeling approach. Figure 3 shows snapshots with a not surprising match between real behavior and CMk-based simulations, under harsh and extreme conditions.

3. Case Study: Analysis, Synthesis and Control of Simple Pendulum

Consider all steeps in CMk approach given in section 1, we apply CMk on the pendulum system, every steep is developed in the next subsections.

3.1. *Mathematical Modeling and Mechanical Design Conceptualization*

Consider governing dynamics equations of simple pendulum as follows

$$J_p\ddot{q} + b_0\dot{q} - Mgl\sin q = \tau \tag{4}$$

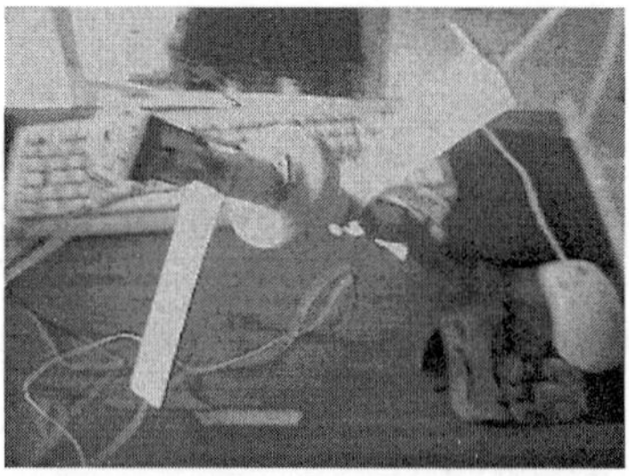

(a) Instability due to torsional torques along Z axis.

(b) The CAE simulator reproduces the seemingly the above fall thanks to implementation of mechatronics approach.

Fig. 3. Real time and CMk simulator results. Notice that very good matching results are obtained. The CMk simulator uses the real parameters obtained through real constraints and quantization of the real system.

where $J_p = J_0 + Ml^2$ is the pendulum mass moment, pendulum mass M, pendulum length l, linear viscous friction b_0 and control input τ. According to [1] and [2] , $y = q$, qualifies as flat output, therefore, there is the endogenous output that parameterizes the system (4), where

$$\begin{aligned} q &= y \\ \dot{q} &= \dot{y} \\ \ddot{q} &= \frac{\tau}{J_p} - \frac{b_0}{J_p}\dot{y} + \frac{Mgl\sin y}{J_p} \end{aligned} \tag{5}$$

We have a non–linear model in (4), the physics system is simple, nevertheless, the dynamic equation of the CMk, is complex because it considers all mechanical and electrical/electronic constrained effects, data quantization, actuator modeling. Now, we need to build a mechanical design CAD like the real prototype and final assembly on Mechanical Desktop® software, see Figure 4.

3.2. *Integration of Mechatronics Constrains to CAD Model*

The next CMk stage includes the application of all mechatronics constrains to CAD model, as follows in the next items:

(1) Mechanical systems constrains are applied in VisualNastran® software, pendulum has a one DOF, then we must apply a revoluta joint with viscous linear friction, see next Figure 5.

(2) Electrical and electronic comprise: minimal resolution optical decoder ($\pi \times 10^{-3}$ rad), and signal quantization (12 bits).
(3) Initial point error is considered at each simulation by delta-differentiation time transitory [7], thus numerical error must be included (we do not have a velocity sensor).
(4) Torque control that use a current sensor with 9 % current sense linearity (up to 3 A), saturation torque (0.1 Nm).
(5) Rating data acquisition (at 3 ms is considered).

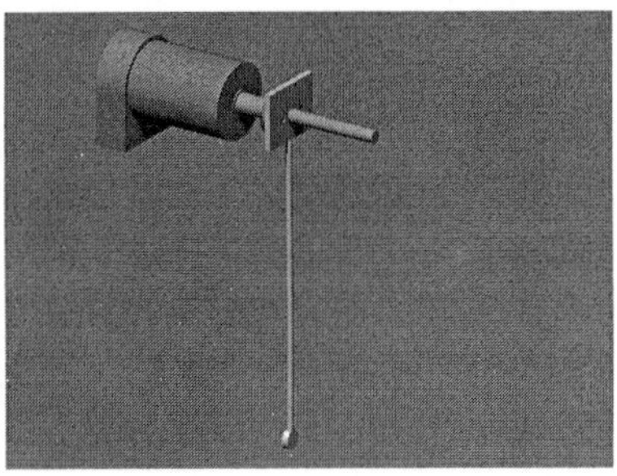

Fig. 4. Pendulum mechanical design.

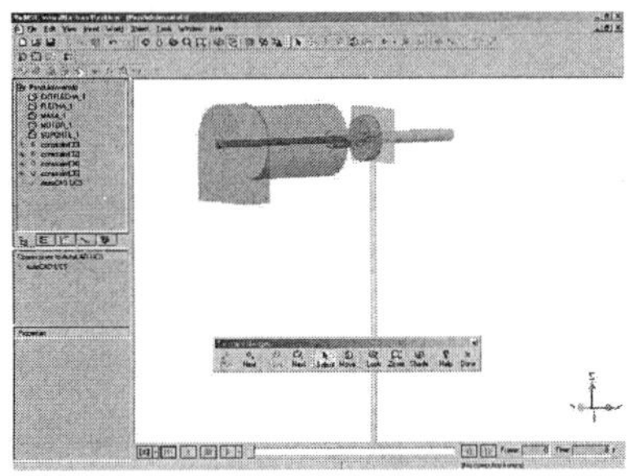

Fig. 5. Pendulum mechanical design with constraints.

3.3. *Stabilizing Control and Planning Trajectories*

Using the coordinate transformation

$$y^{(2)} = v \tag{6}$$

the following equation appears

$$\tau = J_a v + b_0 \dot{y} - Mgl \sin y \tag{7}$$

For an effective tracking control, fast convergence with robustness against endogenous disturbances is required. To this end, a flatness-based controller is proposed. This can be achieved if v is designed as follows

$$v = \ddot{y}^*(t) - k_1(\dot{y}(t) - \dot{y}^*(t)) - k_i \tanh(\lambda S) - k_d S_e \tag{8}$$

Defining the sliding surface $S_e = S + k_i \int_0^t sgn(S)dt$ and $S = (\dot{y}(t) - \dot{y}^*(t)) + k_1(y(t) - y^*(t))$, sliding modes appears at $S = 0, \forall t$ if $k_i > ||\dot{S}_e||$, for further information see [5]. Also, we need to define two points: initial rest position at $y(t_0) = -\pi/2$ and final rest position at $y(T) = \pi/2$. Trajectory path planning is given by

$$\varphi(t,t_0,T) = \Delta^5(t)[\gamma_1 + \gamma_2\Delta^1(t) + \gamma_3\Delta^2(t) + \gamma_4\Delta^3(t) + \gamma_5\Delta^4(t) + \gamma_6\Delta^5(t)] \quad (9)$$

satisfying at initial and final conditions in flat output space, we have $\Delta(t) = \frac{t-t_0}{T-t_0}$, obtaining $\gamma_1 = 130.3688, \gamma_2 = -441.8440, \gamma_3 = 583.6879, \gamma_4 = -358.6879, \gamma_5 = 91.8440$, $\gamma_6 = -4.3688$, where finally, the tracking path is given as follows

$$y^*(t) = y(t_0) + \{y(T) - y(t_0)\}\, \varphi(t,t_0,T) \quad (10)$$

Equation (10) is a smooth rest-to-rest trajectory eliminating every oscillations.

3.4. *Hardware and Software Integration: CMk Platform Results*

Now, we are going to integrate all steeps based on CMk scheme. Using control input (8) with a smooth tracking trajectory (10) at time interval $[t_0, T]$, we have the relationship of flat output conditions $y(t_0) = -\pi/2$, and, $y(T) = \pi/2$. Defining $k_d = 40$, $k_i = 5$, $k_1 = 25$ for the input controller, and physical parameters $M = 0.090$ kg, $l = 0.20$ m, $J_0 = 0.003$ kg m^2, $b_0 = 0.010$ Nm.s/rad, $T = 3$ s, $g = 9.81$ m/s^2. Experimental results delivers small average tracking error of 4×10^{-3} rad , see Figure. 6(d). The angular position and velocity perfectly follow the path planning in Figure. 6.(a), and Figure. 6.(b).

3.5. *Experimental Results vs CMk-based Results*

Finally, consider the pendulum experimental platform in Figure 7. Applying similar geometric conditions, electronics devices, and MF604 acquisition board controlled by Matlab® Real–Time Windows Target(RTWT) under PENTIUM IV personal computer at 600 MHZ , using the same control and path planning of virtual prototype, we have the experimental results of Figure 8. Good matching results of experimental real pendulum platform and CMk can be seen in Figure. 6. This confirms the high potential of CMk since the simulator delivers very close results to reality using a variety of conditions and situations for electromechanical systems.

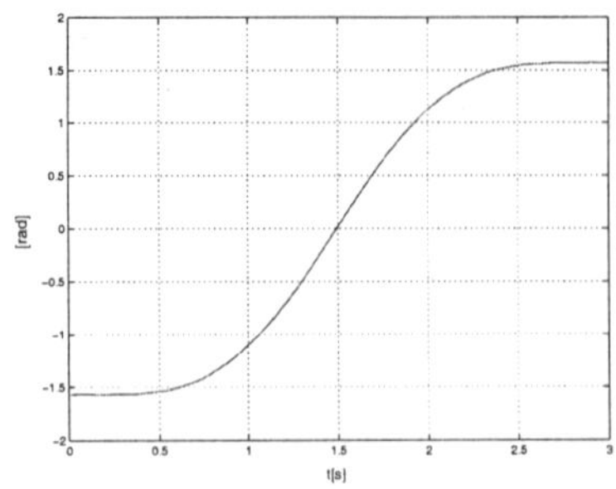

(a) Angular position (rad).

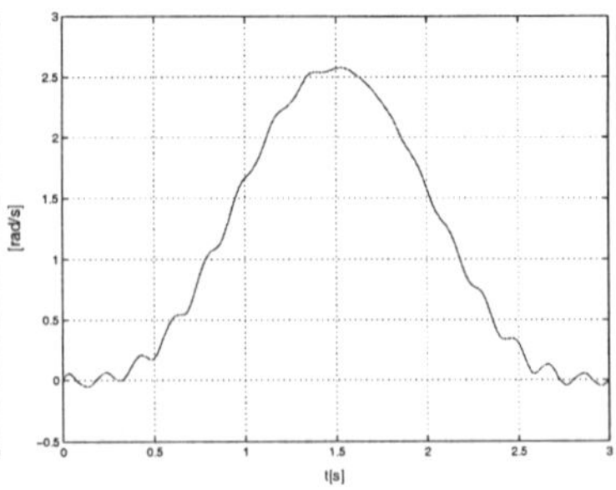

(b) Angular velocity (rad/s).

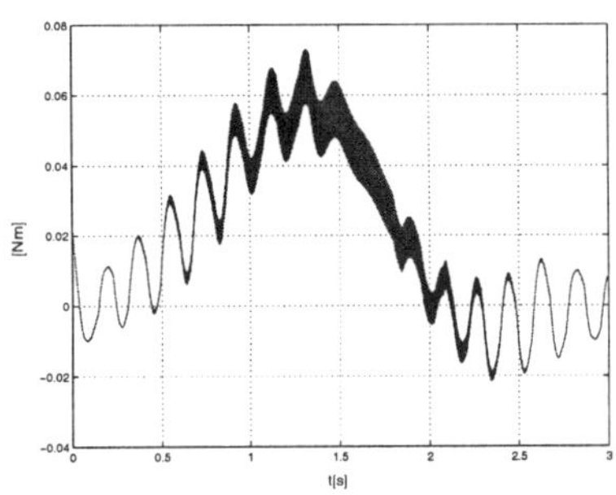

(c) Applied torque (Nm).

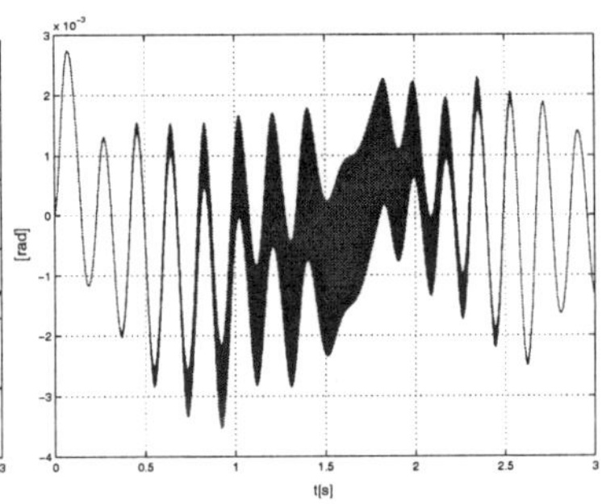

(d) Tracking error(rad).

Fig. 6. Computational mechatronics results: Pendulum closed-loop response.

Fig. 7. Photography of pendulum architecture.

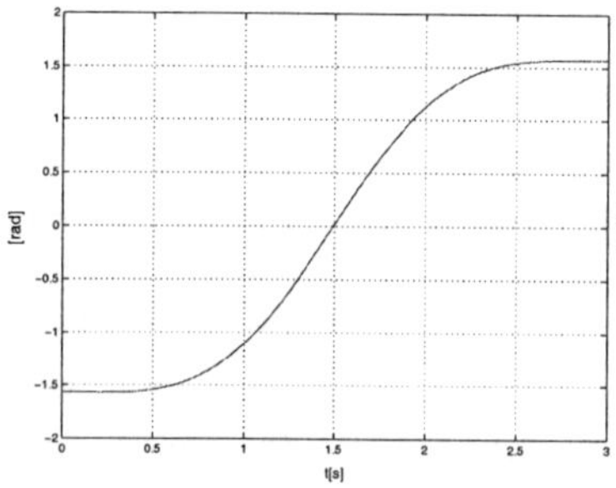

(a) Angular position (rad).

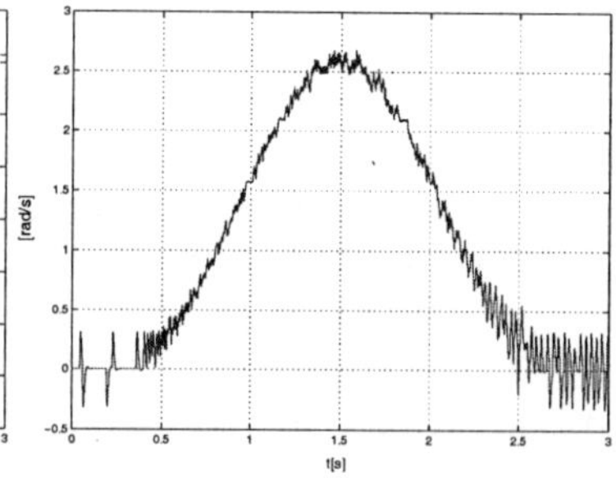

(b) Angular velocity (rad/s).

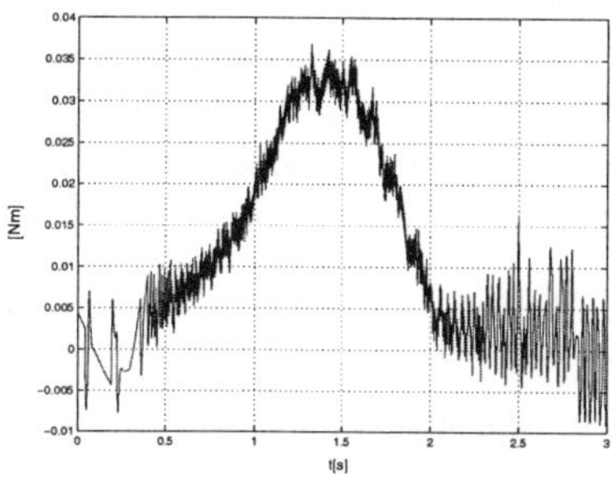

(c) Applied torque (Nm).

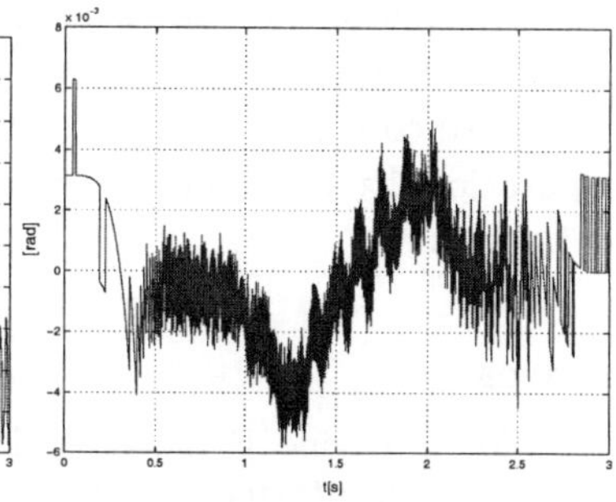

(d) Tracking error (rad).

Fig. 8. Experimental Results: Pendulum platform closed-loop response.

4. Discussions

If we carry out only simulations in Matlab® ODEs-solver, analytically we get results more idealized far away from real behavior: high torques, low noise and smooth planning trajectories, however far from real results. CMk lets us incorporate the mechatronic scheme using integration of 3D virtual prototype, which ODEs approach do not comprises totally; such as mechanical properties of CAD, gravity, friction and reaction forces, CAD kinematics constraints (all of them on three-dimensional space), different types of electronic and electric constraints. If we could include all factors in ODEs approach, modeling and solving stage will have higher level of difficulty.

5. Conclusions

A fundamental tool for mechatronics design is proposed as a CAE simulator, integrating dynamically a mechatronics approach, coined here a CMk advanced analysis, synthesis, control and CAD tools can be employed. In the case study, we use Matlab®, Simulink®, VisualNastran® and Mechanical Desktop® packages running in a standard desktop computer. The experimental validation included RTWT of Matlab®. The case study rendered the following: the design of very simple biped robot for continuous active walking stands as quite complex mechanical system and the simple model proposed in the literature is not practical to realize, in fact, the previous model is only relevant as an academic exercise, but with very poor implications for practitioners, CMk results shows a good matching with reality using standard cots hardware and software, using a very simple 'philosophy' of computational mechatronics for modeling and simulation of electromechanical systems.

References

1. Fliess, M., J. Levine, P. Martin and P. Rouchon, *Flatness and defect of non-linear systems: introductory and examples.*,Int J. Control , 61(6), 1327–1361, 1995.
2. Fliess, M., J. Levine, P. Martin and P. Rouchon, *lie–bäcklund approach to equivalence and flatness of nonlinear systems.* 11, IEEE Trans. Automatic Control , 44(5), 992–937, 1999.
3. Kieffer, J. and R. Bale. *Walking viability and gate synthesis for a novel class of dynamically simple bipeds.* Informatica, No. 17, pp. 144–155, 1993.
4. Lugo–Villeda L.I and Parra–Vega V.*Computational mechatronics approach for analysis, synthesis and design of a simple active biped robot: theory and experiments.* Journal of Applied Bionics and Biomechanics, Volume 3 No. 2. ISNN 1176 2322, Woodhead Publishing Limited 2006 .
5. Parra-Vega V., *Second order sliding mode control for robots arms with time base generators for finite tracking.* Dynamics and Control 11,pp. 175–186, 2001.
6. Rouchon, P. and H. J. Sira-Ramírez , *Control of the walking toy: A flatness approach.* IEEE American Control Conference. USA 2003.
7. Sira-Ramírez Hebertt and Lugo-Villeda L.I, *Sliding Modes, delta–modulation, and output feedback control of dynamical systems.* , Chapter 8: Variable Structure Systems: From Principles to Implementation, 157–175, ISBN 0 86341 350 1, IEE, UK 2004.
8. Spong, M. W., R. Lozano, and R. Mahony *An almost linear biped.*, IEEE Conference on Decision and Control 2000.

A COMPOSITE APPROACH TO THE ADAPTIVE NEURAL NETWORKS CONTROL OF UNKNOWN FLEXIBLE JOINT ROBOTS

HAN YAO, WENFANG XIE*

Department of Mechanical & Industrial Engineering, Concordia University

1515 St. Catherine Street, Montreal, QC, H3G 2W1, Canada

E-mail: wfxie@encs.concordia.ca

CANG YE

*Department of Applied Science, University of Arkansas at Little Rock**

2801 S. University Ave, ETAS 575, Little Rock, AR, USA

E-mail: cxye@ualr.edu

Abstract-In this paper, a composite approach to the adaptive neural networks (NN) controller is proposed for a rigid link flexible joint (RLFJ) robot manipulator with unknown nonlinearities. Based on composite control method, NNs are used to approximate the complicate nonlinear functions in both fast and slow control components so that the developed NN controllers make the link position of the robot follow the desired trajectory. By using Lyapunov theorem extension, the stability of the whole system has been proved and the output of the system is guaranteed to converge to the desired trajectory with bounded errors. The simulation results are presented to show the effectiveness of the approach.

1. Introduction

Many control strategies have been developed for the control of n-link robot manipulators, such as exact compensation of nonlinearities, robust adaptive algorithms, variable structure theory etc. [1, 11]. These control methods share the common feature that the robot dynamics are modeled by the rigid link rigid joint (RLRJ) equations of motion. Unfortunately, experimental evidence indicates that the assumption of perfect rigidity is never satisfied exactly in the real world. The joint flexibility should be taken into account in both modeling and control [13]. From the modeling point of view, a flexible-joint manipulator can be treated as rigid-links interconnected by elastic joints [2]. Normally the joints of robots are made of the harmonic drives that are gear boxes with high-

* CORRESPONDING AUTHOR

ratio and compact torque-transmission. However, the harmonic drive is plagued with friction and its unique mechanical design and assembly cannot deliver sufficient high stiffness. These characteristics pose challenges to the controller design since the joint flexibility may cause instability or resonant behavior in the system [13]. To deal with these problems, a number of control schemes based on the flexible models have been developed to control flexible-joint robots. These methods include feedback linearization [13], singular perturbation techniques [2], sliding mode [3], and robust adaptive controller approaches [4].

In the category of singular perturbation techniques, the integral manifold scheme in the context of composite control has been investigated in [5, 9]. The approaches use integration of composite control and corrective control methods to cope with flexible joint robot with unknown parameters—the so called "adaptive integral manifold" approach. The seeming drawbacks of the integral manifold method are its complexity in deriving the expression of the slow control and the computational cost of implementation. These problems are more pronounced in the adaptive integral manifold method [2]. Although the current advances in symbolic software and parallel computing technologies have facilitated the computationally intensive control algorithm, the symbolic computation remains intractable as it hinges on the robot's nonlinear model that is hard to be identified and verified. Moreover, the symbolic computation of controller has to be carried out again whenever the RJFL robot is changed.

Recently, many NN controllers with closed-loop stability [11, 12] have been proposed for various control applications. Due to its ability of universal function approximation, NN has been successfully used to design controllers for flexible-joint robots [3, 6, 7]. In [3, 6], NNs are used to approximate the inverse nonlinear function to compensate the flexible nonlinearities. In the above work, off-line training is used to obtain the preliminary weights. Kwan [11] proposed a robust NN backstepping control method for nonlinear systems and applied it to RLFJ robots without weak elasticity assumptions. The controller does not require either a linear parametrizable model or an off-line learning phase. However, it needs three NNs to approximate three very complicated nonlinear functions in order to guarantee the uniformly ultimately bounded (UUB) stability of tracking errors.

In this paper, a NN-based controller is developed for the control of a RLFJ robot manipulator using the idea of universal approximation of NN and the concept of composite control. The proposed composite controller comprises fast and slow controllers in which a rigid-based component and additional corrective terms are included. Motivated by the NN controller in [11], we use two NNs in each of fast and slow controllers to approximate two explicit nonlinear functions in rigid model and flexible joint model to alleviate the symbolic computational burden. For the flexible-joint based component of the fast controller, a fictitious variable is introduced in the design of the fast NN controller to provide sufficient damping for the fast dynamics. The NNs' weight matrix update rules are designed using the Lyapunov theorem extension [8] to ensure the unknown RJFL robot's stability. It has been proven that the proposed NN controller guarantees the boundedness of tracking errors and weight updates.

This paper is organized as follows. In section 2, the model of RLFJ robot manipulator in singular perturbation form is introduced. Some properties of the robot manipulator and the basic idea of NN model are presented. In section 3, the development of the adaptive NN based controllers for both rigid and flexible joint robots is detailed and the system stability is proved. In section 4, the numerical implementation of the controller for a two-link flexible manipulator is given. Section 5 concludes the paper.

2. System Models

A real industrial manipulator is driven by the actuators in the joints and its behavior can not be fully captured by the rigid model. Normally an n-link RLFJ robot is modeled by a chain of rigid links interconnected by elastic joints [10]:

$$M(q)\cdot\ddot{q}+V_m(q,\dot{q})\cdot\dot{q}+G(q)+F(\dot{q})+T_L+K\cdot(q-q_f)=0 \tag{2.1}$$

$$J\cdot\ddot{q}_f+B\cdot\dot{q}_f-K\cdot(q-q_f)=\tau \tag{2.2}$$

with $q,\dot{q},\ddot{q}\in R^n$ referring to the link position, velocity and acceleration, $q_f,\dot{q}_f,\ddot{q}_f\in R^n$, the motor shaft angle, angular velocity and angular acceleration, respectively, $M(q)\in R^{n\times n}$ the inertia matrix, $V_m(q,\dot{q})\in R^{n\times n}$ the coriolis and centripetal term, $G(q)\in R^n$ the gravity vector, $T_L\in R^n$ the load disturbance, $K\in R^{n\times n}$ stiffness coefficients matrix , $J\in R^{n\times n}$ the motor inertia, $B\in R^{n\times n}$ the joint damping term, τ the control torque and $F(\dot{q})\in R^n$ the friction with the form:

$$F(\dot{q})=[\alpha_0+\alpha_1\cdot e^{-\beta_1|\dot{q}|}+\alpha_2(1-e^{-\beta_2|\dot{q}|})]\cdot\mathrm{sgn}(\dot{q}) \tag{2.3}$$

where $\alpha_0+\alpha_1$ represents static friction; α_2 represents the viscous friction.

Since joint stiffness is large compared with other parameters, we assume

$$K=K_1/\gamma^2 \tag{2.4}$$

where γ is a small parameter representing the inverse of stiffness and K_1 is on the order of 1. Suppose J and B are very small and on the same order of γ. The rigid model can be derived from Eq. 2.1 and 2.2 by assuming no elasticity at the joints (i.e. $\gamma=0$) and is given by:

$$(M(q)+J)\cdot\ddot{q}+(V_m(q,\dot{q})+B)\cdot\dot{q}+G(q)+F(\dot{q})+T_L=\tau \tag{2.5}$$

3. Control Strategy

3.1 Control objective

The control objective is to develop a position tracking controller for an unknown RLFJ robot dynamics (Eq. 2.1) so that the link position follows a desired trajectory, which is continuous and its derivatives up to higher order are bounded. The filtered error of the robot is defined as

$$r = \dot{e} + \Lambda \cdot e \tag{3.1}$$

where $e(t) = q_d(t) - q(t)$, $\Lambda = \Lambda^T > 0$, $q_d(t) \in R^n$ is the desired trajectory

The elasticity at the joints is large enough so that the system can be decomposed into a "slow" subsystem and a "fast" subsystem. From [2], the control signal τ for the whole system has the form as

$$\tau = \tau_s + \tau_f \tag{3.2}$$

where τ_s is the slow part and τ_f is the fast part, which is defined as:

$$\tau_f = K_f(\dot{q} - \dot{q}_f) \tag{3.3}$$

Usually, we choose $K_f = K_2/\gamma$ with K_2 on the order of 1. Define z as the difference between the link and motor position

$$z = q_f - q\,. \tag{3.4}$$

Substituting control signal τ (Eqs. 3.2, 3.3 and 3.4) into system, and define an integral manifold as $h = K \cdot z$, one obtains:

$$J \cdot \ddot{h} + (B + K_f) \cdot \dot{h} + K \cdot h = K(\tau_s - J \cdot \ddot{q} - B \cdot \dot{q}) \tag{3.5}$$

where $h = h(t, \gamma, q, q_f)$.

An approximate reduced-order flexible model can be derived by using a power series expansion of the integral manifold h and control τ_s around $\gamma = 0$. It is found that the slow control component τ_s is independent of fast control component τ_f. Let us denote:

$$\begin{aligned} h &= h_0 + \gamma\, h_1 + O(\gamma^2) \\ \tau_s &= \tau_0 + \gamma\, \tau_1 + O(\gamma^2) \end{aligned} \tag{3.6}$$

where τ_0 is the control input to the rigid model, τ_1 is the corrective torque term for compensating the effects of γ, the vector h_0 represents a zero-order approximation of h and the h_1 represents the first order correction to h_0.

As a result, we get

$$\begin{aligned} M(q) \cdot \ddot{q} + V_m(q, \dot{q}) \cdot \dot{q} + G(q) + F(\dot{q}) + T_L = \tau_s - J \cdot \ddot{q} - B \cdot \dot{q} \\ - \gamma \cdot K_1^{-1}[\gamma \cdot J \cdot \ddot{h}_0 + (\gamma \cdot B + K_2) \cdot \dot{h}_0] + O(\gamma^2) \end{aligned} \tag{3.7}$$

The variables $\ddot{h}_0$ and $\dot{h}_0$ are "fast" variables; the link variables q and $\dot{q}$ are "slow" variables. Moreover, the rigid model (Eq. 2.5) is obtained by setting $\gamma = 0$.

The control task is to design τ_0 and τ_1 so that the link position of robot follows the desired trajectory. In [9], both τ_0 and τ_1 are derived with complicate expression especially in the adaptive integral manifold method. In control application, NN is usually used as a tool for modeling nonlinear function due to their universal function approximation capability. In order to alleviate the symbolic computational burden in calculating τ_0 and τ_1, two three-layer MLP neural networks are utilized to approximate two complicate nonlinear functions to form the control signals τ_0 and τ_1.

3.2 Controller design

Introducing tracking error (Eq. 3.2) into system (Eq. 3.7), we get

$$M \cdot \dot{r} = -V_m \cdot r + F_0 - \tau_s + T_L + J \cdot \ddot{q} + B \cdot \dot{q} + \gamma \cdot K_1^{-1}[\gamma \cdot J \cdot \ddot{h}_0 + (\gamma \cdot B + K_2) \cdot \dot{h}_0] \tag{3.8}$$

where F_0 is a complicated nonlinear function defined as

$$F_0 = M(q) \cdot (\ddot{q}_d + \Lambda \dot{e}) + V_m(q, \dot{q}) \cdot (\dot{q}_d + \Lambda e) + G(q) + F(\dot{q}) \tag{3.9}$$

Then design a nonlinear function

$$F_1 = K_1^{-1}[\gamma \cdot J \cdot \ddot{h}_0 + (\gamma \cdot B + K_2) \cdot \dot{h}_0] + K_2^{-1} \cdot K_f [J \cdot \ddot{q} + B \cdot \dot{q}] \tag{3.10}$$

We derive the error dynamics as:

$$M \cdot \dot{r} = -V_m \cdot r + F_0 - \tau_0 - \gamma \cdot \tau_1 - O(\gamma^2) + \gamma \cdot F_1 + T_L. \tag{3.11}$$

To implement F_1 (Eq. 3.10), we need to compute $\dot{h}_0$ and $\ddot{h}_0$. As a result F_1 is highly complex nonlinear function of τ_0, q, $\dot{q}$ and $\ddot{q}$.

We utilize two first-layer-fixed MLP neural networks to approximate the nonlinear function F_0 (Eq. 3.9), F_1 (Eq. 3.11).

$$\hat{F}_0 = \hat{f}_0(x) = \hat{W}_0^{\ T} \cdot \sigma(V^T \cdot x) \qquad \hat{F}_1 = \hat{f}_1(y) = \hat{W}_1^{\ T} \cdot \sigma(V^T \cdot y) \tag{3.12}$$

$$x = \begin{bmatrix} \ddot{q}^T & \dot{q}^T & q^T & \ddot{q}_d^{\ T} & \dot{q}_d^{\ T} & \operatorname{sgn}(\dot{q})^T & 1 \end{bmatrix}^T \qquad y = \begin{bmatrix} \tau_0^{\ T} & q^T & \dot{q}^T & \ddot{q}^T & 1 \end{bmatrix}^T$$

input-layer weight matrix V^T is pre-fixed and $\hat{W}_0^{\ T}$, $\hat{W}_1^{\ T}$ is the estimated output-layer weight matrix $W_0^{\ T}$, $W_1^{\ T}$.

The corrective term is designed as:

$$\tau_0 = \hat{F}_0 + K_v \cdot r \qquad \tau_1 = \hat{F}_1 + K_u \cdot \varphi \tag{3.13}$$

where r is the filtered error , φ is a fictitious variable, which will be designed later; and K_v K_u is a gain matrix

Substituting the control strategy (Eq. 3.13) into the error dynamics (Eq. 3.11), one obtains:

$$M \cdot \dot{r} = -(V_m + K_v) \cdot r - \gamma \cdot K_u \cdot \varphi + \widetilde{F}_0 + \gamma \cdot \widetilde{F}_1 + T_L \tag{3.14}$$

Design fictitious variable as

$$\dot{\varphi} = \tau_0 - K \cdot z \tag{3.15}$$

Using Eq. 3.10 and Eq. 3.15,

$$\dot{\varphi} = \gamma \cdot \widetilde{F}_1 - \gamma \cdot K_u \cdot \varphi \tag{3.16}$$

Now, we have two different NNs based controllers—one is the first slow part τ_0 based on a rigid NN F_0 function and the other is the second slow part τ_1 based on the corrective NN F_1 function. The composite control scheme is shown as:

$$\tau = \tau_s + \tau_f = (\hat{F}_0 + K_v \cdot r) + \gamma \cdot (\hat{F}_1 + K_u \cdot \varphi) - K_f \dot{z} \ . \tag{3.17}$$

Choose the update rule for those weight matrix respectively as

$$\dot{\widetilde{W}}_0 = -\dot{\hat{W}}_0 = -\Gamma_0 \cdot \sigma_0(V^T \cdot x) \cdot r^T + k \cdot \Gamma_0 \cdot \|\xi\| \cdot \hat{W}_0 \tag{3.18}$$

$$\dot{\widetilde{W}}_1 = -\dot{\hat{W}}_1 = -\Gamma_1 \cdot \sigma_1(V^T \cdot y) \cdot (\varphi^T + r^T) + k \cdot \Gamma_1 \cdot \|\xi\| \cdot \hat{W}_1 \tag{3.19}$$

where $\xi = \begin{bmatrix} r^T & \varphi^T \end{bmatrix}^T, \Gamma = \Gamma^T > 0$ and $k > 0$,

Theorem 3.1

For a RJFL robot (Eq. 2.1), the NN controller (Eq. 3.17) and update rule (Eqs. 3.18, 3.19) are applied. For a desired trajectory $q_d(t)$, it is assumed that its time derivatives up to third order are continuous and bounded. The controlled system's filtered error $r(t)$ and fictitious variable $\varphi(t)$ are bounded and the tracking error $e(t)$ will converge to a small neighborhood around zero by appropriately choosing suitable gain matrix K_f, K_v, and K_u.

Proof

Define the Lyapunov function as

$$L = \frac{1}{2} r^T \cdot M \cdot r + \frac{\gamma}{2} \varphi^T \cdot \varphi + \frac{1}{2} tr(\widetilde{W}_0^T \cdot \Gamma_0^{-1} \cdot \widetilde{W}_0) + \frac{\gamma}{2} tr(\widetilde{W}_1^T \cdot \Gamma_1^{-1} \cdot \widetilde{W}_1) \geq 0 \tag{3.20}$$

Differentiating the above function then Introducing Eq. 3.16, 3.18 and 3.19 yields

$$\begin{aligned} \dot{L} = & -r^T \cdot K_v \cdot r - r^T \cdot K_u \cdot \varphi + r^T \cdot (\varepsilon_0 + \varepsilon_1) - \gamma \cdot \varphi^T \cdot K_u \cdot \varphi + \gamma \cdot \varphi^T \cdot \varepsilon_1 \\ & + tr(\widetilde{W}_0^T \cdot k \cdot \|\xi\| \cdot \hat{W}_0) + \gamma \cdot tr(\widetilde{W}_1^T \cdot k \cdot \|\xi\| \cdot \hat{W}_1) \end{aligned} \tag{3.21}$$

Defining $Q=\begin{bmatrix} K_v & K_u \\ 0 & \gamma \cdot K_u \end{bmatrix}$ and $W = diag\{W_0 \quad \gamma \cdot W_1\}$, $\widetilde{W} = W - \hat{W}$, we acquire:

$$\dot{L} \le -\|\xi\| \cdot [\lambda_{Q\min} \cdot \|\xi\| - \varepsilon_N - \frac{k \cdot W_N{}^2}{4} + k \cdot (\|\widetilde{W}\|_F - \frac{W_N}{2})] \tag{3.22}$$

where $\lambda_{Q\min}$ is the minimum eigenvalue of Q and $\varepsilon_N = \max(|\varepsilon_0 + \varepsilon_1| \quad \gamma \cdot |\varepsilon_1|)$,

Define W_N the bound of the ideal weight matrix W . If we have

$$\|\xi\| > \frac{k \cdot W_N{}^2/4 + \varepsilon_N}{\lambda_{Q\min}} \text{ or } \|\widetilde{W}\|_F > W_N/2 + \sqrt{W_N{}^2/4 + \varepsilon_N/k} \tag{3.23}$$

we can prove $\dot{L}$ negative. Inequality shows that if the control gains K_f , K_v, and K_u are chosen large enough so that

$$\frac{k \cdot W_N{}^2/4 + \varepsilon_N}{\lambda_{Q\min}} < b_\xi$$

where $b_\xi > 0$ represents the radius of a ball inside the compact set C_ξ of filtered error $\xi(t)$.

Thus, any trajectory $\xi(t)$ starting in compact set $C_\xi = \{\xi | \|\xi\| \le b_\xi\}$ converges within C_ξ and is bounded. According to the standard Lyapunov theorem extension [8], it demonstrates the UUB (uniformly ultimately bounded) of both $\xi(t)$ and $\widetilde{W}$.

4. Simulation

The effectiveness of the proposed control scheme is demonstrated on a two-link manipulator, which can be described in the form of Eq. 2.1 and 2.2 by [10]

$$M(q) = \begin{bmatrix} a + b \cdot \cos(q_2) & c + \frac{b}{2} \cdot \cos(q_2) \\ c + \frac{b}{2} \cdot \cos(q_2) & c \end{bmatrix}$$

$$V_m(q, \dot{q}) = \begin{bmatrix} -\frac{b}{2} \cdot \dot{q}_2 \cdot \sin(q_2) & -\frac{b}{2} \cdot (\dot{q}_1 + \dot{q}_2) \cdot \sin(q_2) \\ \frac{b}{2} \cdot \dot{q}_1 \cdot \sin(q_2) & 0 \end{bmatrix}$$

$$G(q) = \begin{bmatrix} d \cdot \cos(q_1) + e \cdot \cos(q_2) \\ e \cdot \cos(q_1 + q_2) \end{bmatrix}$$

$$F(\dot{q}) = \begin{bmatrix} \{35 + 1.1 \cdot e^{-50|\dot{q}_1|} + 0.9(1 - e^{-65|\dot{q}_1|})\} \cdot \mathrm{sgn}(\dot{q}_1) \\ \{38 + 1.0 \cdot e^{-55|\dot{q}_2|} + 0.95(1 - e^{-60|\dot{q}_2|})\} \cdot \mathrm{sgn}(\dot{q}_2) \end{bmatrix}$$

$$a = l_2{}^2 \cdot m_2 + l_1{}^2 \cdot (m_1 + m_2), \quad b = 2 \cdot l_1 \cdot l_2 \cdot m_2,$$

$$c = l_2^{\,2} \cdot m_2, \quad d = (m_1 + m_2) \cdot l_1 \cdot g_0, \quad e = m_2 \cdot l_2 \cdot g_0$$

The parameter values are chosen as $l_1 = l_2 = 1$, $m_1 = 0.8$, $m_2 = 2.3$. And the flexible joint parameters are $J = diag\{0.3 \quad 0.3\}$, $B = diag\{0.02 \quad 0.02\}$, $K = diag\{100 \quad 100\}$. The inputs to the NNs are given by

$$x = \begin{bmatrix} \ddot{q}^T & \dot{q}^T & q^T & \ddot{q}_d^{\,T} & \dot{q}_d^{\,T} & \text{sgn}(\dot{q})^T & 1 \end{bmatrix}^T, \quad y = \begin{bmatrix} \tau_0^{\,T} & q^T & \dot{q}^T & \ddot{q}^T & 1 \end{bmatrix}^T$$

Two input reference signal are chosen as desired two joints positions: $q_{1d} = 2 \cdot \sin(0.1\pi \cdot t)$ and $q_{2d} = 3 \cdot \sin(0.1\pi \cdot t)$. The control objective is defined as to make the flexible-joint robot joint angle $q = [q_1 \quad q_2]^T$ follow the given desired joint trajectory $q_d = [q_{1d} \quad q_{2d}]^T$. The gains are selected as: $\Lambda = [20 \quad 1]^T$, $K_u = diag\{5 \quad 5\}$, $K_u = diag\{5 \quad 5\}$, $K_f = diag\{3 \quad 3\}$, $\Gamma_0 = diag\{10 \quad 10\}$, $\Gamma_1 = diag\{20 \quad 20\}$, and $k = 0.1$. The system responses under the control of the proposed NN-based controller are shown in Fig. 1.

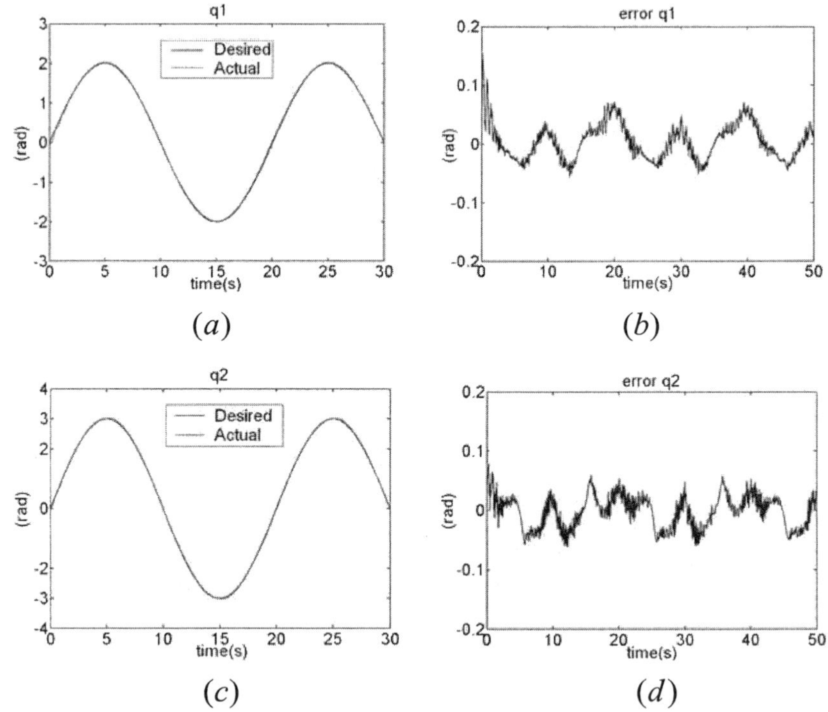

Figure 1 Performance of NN controller with $K = diag\{100 \quad 100\}$

(a) Actual (dashed line) and desired (solid) joint q_1 angles (b) Error $q_1 - q_{1d}$

(c) Actual (dashed line) and desired (solid) joint q_2 angles. (d) Error $q_2 - q_{2d}$,

5. Conclusion

In this paper, an adaptive NN controller is designed for a rigid link flexible joint (RLFJ) robot manipulator with unknown nonlinearities by using composite control approach. Two NNs are used to approximate two complicated unknown nonlinear functions in both fast and slow control components. No off-line training is required for NNs. The control algorithm and the weight matrix update rule are derived from Lyapunov theorem extension. The stability and the boundedness of tracking error of this unknown RLFJ robot manipulator have been proved. Simulation results show that the proposed NN controller outperforms the adaptive composite control method and can be applicable to unknown flexible robots with a larger range of stiffness.

References

[1] S. Nicosia and P. Tomei, "Robot Control by using only Joint Position Measurement," *IEEE Transactions on Automatic Control*, vol. 35, no. 9, pp.1058-1061, 1990.

[2] F. Ghorbel, J. Y. Hung, and M. W. Spong, "Adaptive Control of Flexible-Joint Manipulators," IEEE Control System Magazine, vol. 9, no. 7, pp. 9-13, 1989.

[3] O. Aboulshamat and P. Sicard "Position Control of A Flexible Joint with Friction Using Neural Network Feedforward Inverse Models," *Proceeding of the Canadian Conference on Electrical and Computer Engineering*, Toronto, Canada, 2001, pp. 283-388.

[4] F. Abdollahi, H. A. Talebi, and R.V. Patel, "State Estimation for Flexible-Joint Manipulators using Stable Neural Networks," *Proceeding of the IEEE International Symposium on Computation Intelligence in Robotics and Automation*, Kobe, Japan, 2003, pp. 25-29.

[5] M. Moallem, K. Khorasani, and R. V. Patel, "An Integral Manifold Approach for Tip-Position Tracking of Flexible Multi-Link Manipulators," *IEEE Transactions on Robotics and Automation*, vol. 13, no. 6, pp. 823-837, 1997.

[6] V. Zeman, R.V. Patel, and K. Khorasani, "Control of a Flexible-Joint Robot Using Neural Networks," *IEEE Transactions on Control Systems Technology*, Vol. 5, No. 4, pp.453-462, 1997

[7] W. Chatlatanagulchai and P. H. Mechl, "Motion Control of Two-Link Flexible-Joint Robot, Using Backstepping, Neural Networks and Indirect Method," *Proceeding of the IEEE International Conference on Control Applications*, Toronto, Canada, 2005, pp. 601-605.

[8] K. S. Narendra and A. M. Annaswamy, *Stable Adaptive Systems,* Englewood Cliffs, New Jersey: Prentice Hall, 1989

[9] F. Ghorbel and M. W. Spong, "Adaptive Integral Manifold Control of Flexible Joint Manipulators," *Proceedings of the IEEE International Conference on Robotics and Automation*, Nice, France, pp.707-714, 1992

[10] F. L. Lewis, C. T. Abdallah, and D. M. Dawson, *Control of Robot Manipulator,* New York, New York: Macmillan Publishing Company, 1993

[11] C. Kwan, F. L. Lewis, and D. M. Dawson, "Robust Neural-Network Control of Rigid-Link Electrically Driven Robots," *IEEE Transactions on Neural Networks*, vol. 9, no. 4, pp.581-588, 1998.

[12] S. S. Ge and C. Wang, "Direct Adaptive NN Control of a Class of Nonlinear System," *IEEE Transactions on Neural Networks*, Vol. 13, No. 1, pp.214-221, 2002.

[13]. M.W. Spong, "Modeling and Control of Elastic Joint Manipulators", Journal of Dynamic System, Measurement and Control, Vol. 109, pp.310-319, 1987.

CHARACTERIZATION OF RATE DEPENDENT HYSTERESIS

MOHAMMAD AL JANAIDEH, CHUN-YI SU, SUBHASH RAKHEJA
aljanaideh@yahoo.com, cysu@alcor.concordia.ca, rakheja@vax2.concordia.ca

Department Mechanical and Industrial Engineering, Concordia University
1455 de Maisonneuve Blvd. W. Montreal, Quebec, H3G 1M8, Canada

A modified Prandtl-Ishlinskii model is proposed for characterizing the rate dependent variations in the hysteresis behavior. A rate dependent operator and density functions are proposed and integrated into the Prandtl-Ishlinskii model to predict the rate dependent hysteresis phenomenon. The fundamental properties of the proposed rate dependent operator are presented in details and discussed. The resulting modified model could also account for the rate of change of the output with respect to the input as a function of the amplitude and time rate of the input. Simulations results show that the proposed operator and density functions allow for accurate prediction of the rate dependent hysteresis under dynamically varying inputs.

1. Introduction

Hysteresis has been widely observed in physical systems, materials and devices, such as electro-magnetic fields, smart materials, mechanical or smart actuators, and electronic relay circuits. Hysteresis acts as a phase lag, which has the potential to cause inaccuracy and oscillations in the systems' responses that may lead to instability of the closed-loop systems. The phenomena of hysteresis in various fields have been extensively studied through experimental and analytical means. The most widely cited models include: Preisach model [1], Krasnosel'skii-Pokrovskii operator [2,8], Prandtl-Ishlinskii model [3,4], Duhem model [3], Bouc-Wen model [3] and backlash hysteresis model [3]. These models have been formulated to characterize the hysteresis behavior of different systems in order to facilitate the design of controllers to compensate for the hysteresis effects [11,16].

The classical models generally provide reasonably accurate characterizations of the hysteresis behaviors in different materials and devices [1,5,16], while the rate dependence of the hysteresis is entirely ignored. A number of studies involving experimental characterization of hysteresis behavior of smart actuators and magnetic materials have shown that the hysteresis is a rate dependent property. Hysteresis in magnetic materials is strongly affected by the frequency of the input magnetic field. The width and the area of the hysteresis loop increase when the frequency of the input magnetic field is increased [1,10, 15]. Diebolt [15] showed that the coercive field of a soft magnetic material increased nearly 50 folds when the input frequency was increased from 0 to 100 kHz. Mrad and Hu [17] performed experiments to characterize the hysteresis of a Piezoceramic actuator under inputs at different discrete frequencies and showed that the width of the hysteresis loop increased from 15% of the peak-peak output amplitude under the 0.1 Hz input to 38.6% under the input at 800 Hz.

Owing to the strong rate dependence of the hysteresis effects, the classical Preisach model, Krasnosel'skii-Pokrovskii operator, and Prandtl-Ishlinskii models could not be applied particularly under dynamically varying inputs. A few studies have proposed modifications to these models in order to enhance their prediction abilities of the hysteresis effects under dynamically varying inputs. The validity of such modified models, however, has been mostly demonstrated under relatively low frequency inputs [1,5,16].

Mayerqoyz [9] proposed a dynamic Preisach model by introducing the speed of the output in the Preisach function. Hodgdon [18] proposed a modified formulation of the Hodgdon magnetics model, comprising a rate dependent function of the current input and its time derivative, for characterizing the rate dependent hysteresis in soft magnetic materials. A dynamic generalization of the Preisach model has been used by Bertotti [10] to model dynamic hysteresis in magnetics. The proposed model replaced the Preisach operator by an operator comprising variations in the output at different finite rates of the input. The study presented the dynamic hysteresis loops at three different frequencies (1, 10 and 100 Hz) and concluded that both the area and the width of the hysteresis loop increases with the input frequency.

Song and Li [7] applied a Neural Network model in conjugant with the classical Preisach model for describing the rate dependent hysteresis behavior of a piezoceramic actuator subject to a harmonic excitation at two distinct frequencies (2 and 32 Hz). Yu et al. [13] proposed and integrated a modified density function in time rate of the input to the Preisach model to predict dynamic hysteresis in a piezoceramic actuator. The study presented the results under a harmonic excitation at two distinct frequencies (0.05 and 5 Hz), while its validity under other frequencies was not attempted. Mrad and Hu [17] presented a dynamic hysteresis model based on the Preisach model and demonstrated its validity using the measured responses at 6 selected points in the hysteresis loop of a piezoceramic actuator.

Ang et al. [14] proposed a function in the time rate of change of the input and integrated it to the Prandtl-Ishlinskii model for characterizing the rate dependent hysteresis in a Piezoceramic actuator. The study presented the results under a pure harmonic input at 10 Hz and a complex harmonic signal comprising 5, 20, and 35 Hz components, and concluded that the proposed modification reduced the peak error from 10.7% to 4.4% at 10Hz. The model validity at higher frequencies, however, was not attempted. Tan and Bras [12], in a similar manner, proposed and integrated a dynamical system model to the Preisach operator to characterize and to compensate for the rate dependent hysteresis in a magneto-restrictive actuator. The study showed model validity in predicting the major loops under inputs up to 200 Hz, and minor loops up to 50 Hz. However, this model could not be applied under inputs at low frequencies where the hysteresis is rate independent.

The reported modified models have shown their validity under inputs at particular or limited ranges of frequencies. The vast majority of the models have been applied under inputs at relatively lower frequencies, where the rate dependence of the hysteresis is generally lower. This paper proposes a rate dependent hysteresis operator that may be applied in conjunction with the density function to characterize the rate dependent hysteresis under dynamic inputs over a wide range of frequencies.

2. Hysteresis Model

Classical models describe the hysteresis effect through different operators, such as the Preisach operator, KP operator, and play and stop operators. The rate dependent effects could also be effectively described through these hysteresis operators. The fundamental properties of the hysteresis operators are briefly examined for exploring their potential to include the rate dependence.

2.1 Play and Stop Operators

Play and stop operators are continuous, invertible and rate independent hysteresis operators [4], as illustrated in Fig. 1. These operators are used to characterize the elastic-plastic behavior in continuum mechanics. In the elastic domain, stress w being less than the yield stress r, the strain v is related to w through the Hook's law [4]. Let $C_m[0,t_E]$ represent the space of piecewise monotone continuous functions. For any input $v(t) \in C_m[0,t_E]$, the stop operator, E_r and the play operator, F_r for any threshold $r \geq 0$, can be defined by :

$$E_r[v](0) = e_r(v(0))$$

$$E_r[v](t) = e_r(v(t) - v(t_i) + E_r[v](t_i));$$

for $t_i < t \leq t_{i+1}$ and $0 \leq i \leq N-1$ with $e_r(v) = \min(r, \max(-r, v))$ (1)

$$F_r[v](0) = f_r(v(0),0)$$

$$F_r[v](t) = f_r(v(t), F_r[v](t_i));$$

for $t_i < t \leq t_{i+1}$ and $0 \leq i \leq N-1$ with $f_r(v,w) = \max(v-r, \min(v+r, w))$ (2)

Where $0 = t_0 < t_1 < ..... < t_N = t_E$ are the partitions in $[0, t_E]$ such that the function v is monotone on each of the sub-intervals $[t_i, t_{i+1}]$. The argument of the operator is written in square brackets to indicate the functional dependence, since it maps a function to a function. The operator $F_r[v]$ is the complement of the operator $E_r[v]$, and the two are related in the following manner [3, 4]:

$$F_r[v](t) + E_r[v](t) = v(t) \tag{3}$$

For any piecewise monotone input function v and $r \geq 0$

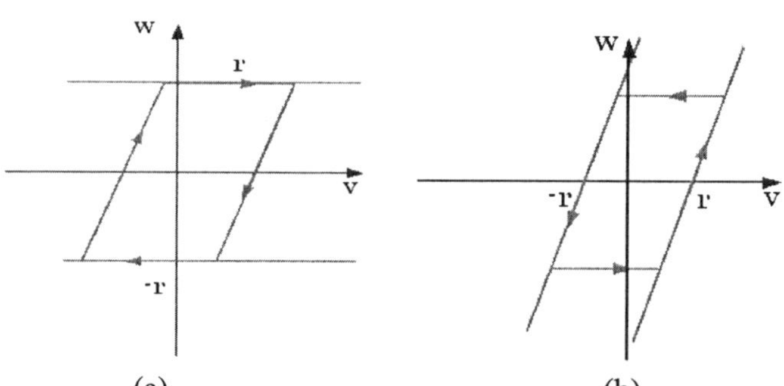

Fig. 1: Stop and play hysteresis operators (a) stop operator (b) play operator

2.2 Prandtl-Ishlinskii Model

Prandtl-Ishlinskii model is based on the above play and stop hysteresis operators [6], such that:

$$w_p(t) = qv(t) + \int_0^{\infty} p(r)F_r[v](t)dr \tag{4}$$

$$w_s(t) = \int_0^{\infty} p(r)E_r[v](t)dr \tag{5}$$

Where p is a nonnegative continuous weighting function and q is a positive function ($q > 0$) expressed by:

$$q = \int_0^{\infty} p(r)dr \tag{6}$$

The fundamental properties of the Prandtl-Ishlinskii model have been described in details in [4]. The parameters of the weighting functions, and stop and play operators, are identified from the measured data.

3. Rate Dependent Hysteresis Operators

The width of the hysteresis loop generally increases with increasing time rate of the input, while the hysteresis magnitude decreases. The measured properties of smart actuators and magnetic circuits invariably suggest that the width of the hysteresis loop generally increases with increasing time rate of the input, while the hysteresis magnitude decreases [10,13,15,17]. The typical properties of a piezoceramic actuator are shown in Fig. 2, which clearly show variations in the width and the amplitude of hysteresis loops with frequency of the input.

The outputs of the play and the stop operators of the Prandtl-Ishlinskii model are not limited to a certain value. A relationship between the output of the operator, the threshold r and the time rate of the input $\dot{v}(t)$, could thus be defined. A rate dependent operator P is thus proposed as:

$$P(t) = f(\dot{v}, v, r, t) \tag{7}$$

Where $v \in R$ is the input, $\dot{v} \in R$ is the time rate of the input, $r \in R^+$ is the dynamic threshold and $P(t) \in R^3$ is the rate dependent operator.

On the other hand, based on the properties of the threshold r in the play and stop operator:

- Measure the width of the hysteresis loop is directly related to the threshold r. An increase in the threshold r yields of the hysteresis loop.
- The amplitude of the play hysteresis operator depends on the quantity $(v - r)$ and $r > 0$ is always positive, and then increasing the

threshold r leads to decrease the gain and the amplitude of the hysteresis operator.

A relation between the time rate of the input $\dot{v}(t)$ and the threshold r could be expressed as:

$$r(t) = f(\dot{v}(t)) \tag{8}$$

The dynamic threshold $r : R \to R^{+}$ is always positive. Then Eq. (7) is defined by:

$$P(t) = f(v, r(\dot{v}(t)), t) \tag{9}$$

Where $P \in R^{2}$ is the modified rate dependent operator. For the rate dependent stop and play operators based on dynamic threshold r:

$$e_{r(\dot{v}(t))}(v) = \min(r(\dot{v}(t)), \max(-r(\dot{v}(t)), v)) \tag{10}$$

$$f_{r(\dot{v}(t))}(v, w) = \max(v(t) - r(\dot{v}(t)), \min(v(t) + r(\dot{v}(t)), w)) \tag{11}$$

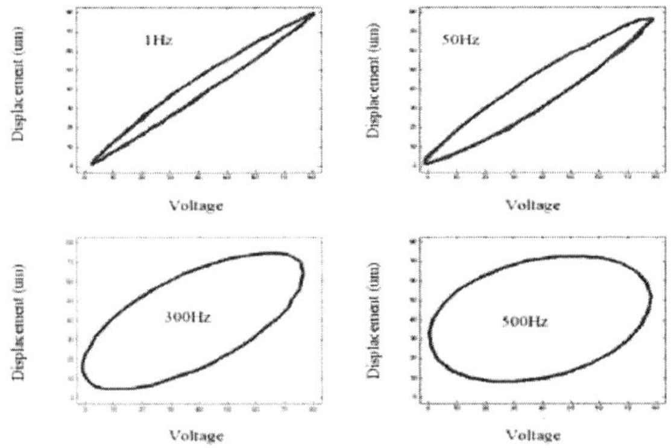

Fig. 2: Measured responses of a piezoceramic actuator at different frequencies.

3.1 Properties of the Rate Dependent hysteresis Operators

The modified play and stop operators are nonlinear operator (linearity here means the shape of the boundary of the hysteresis loop [4]). The time rate of the input $\dot{v}(t)$ is parameter in the modified operator and it varies between $\dot{v}_{\max}(t)$ and $\dot{v}_{\min}(t)$.

The properties of the rate dependent play and stop hysteresis operators include:

(1) The rate dependent play and stop operators are related through the following Eq.:

$$F_{r(\dot{v}(t))}(v(t)) + E_{r(\dot{v}(t))}(v(t)) = v(t) \tag{12}$$

(2) The width of the rate dependent operator is monotonic with respect to the time rate of the input $\dot{v}(t)$

(3) The width of the rate dependent is monotonic with respect to the amplitude of the input, as the amplitude increases the width of the rate dependent operator increases.

(4) The rate independent hysteresis operator is a special case of the rate dependent hysteresis operator.

(5) The Range of the rate dependent play operator is presented by the following expressions:

- $$\max_{t\in[0,t_E]} F_{r(\dot{v}(t))}(v(t); w_{-1}) = f_r(\max_{t\in[0,t_E]} v(t), \min_{t\in[0,t_E]} r(\dot{v}(t), w_0)) \tag{13}$$
- $$\min_{t\in[0,t_E]} F_{r(\dot{v}(t))}(v(t); w_{-1}) = f_r(\min_{t\in[0,t_E]} v(t), \min_{t\in[0,t_E]} r(\dot{v}(t), w_0)) \tag{14}$$

(6) The rate dependent play and stop operators are Lipschitz-continuous.

4. Rate Dependent Prandtl-Ishlinskii Model

Rate dependent Prandtl-Ishlinskii model using rate dependent stop operator and density functions is presented in the following Eq.:

$$y_s(t) = \int_0^\infty K(v,\dot{v})p(r)E_{r(\dot{v}(t))}(v(t))dr \tag{15}$$

Based on the rate dependant play operator, the rate dependent Prandtl-Ishlinskii model is:

$$y_p(t) = w(v,\dot{v})qv(t) + \int_0^\infty k(v,\dot{v})p(r)F_{r(\dot{v}(t)}(v(t))dr \tag{16}$$

The proposed dynamic threshold r and density functions are presented in the following Eqs. :

$$r(\dot{v}) = a\ln(b + c|\dot{v}(t)|^{c_1})\ln(d + e|\dot{v}(t)|^{e_1}) \tag{17}$$

$$w(\dot{v}, v) = ge^{-m\dot{v}}\, e^{m_1 v} \tag{18}$$

$$k(\dot{v}, v) = ne^{-s\dot{v}}\, e^{s_1 v} \tag{19}$$

Where $w(v,\dot{v})$ and $k(v,\dot{v})$ are positive continuous functions and g, b, a, c, c_1, d, e, e_1, n, m, m_1, s_1 and s are positive constants.

5. Simulation Results

Consider v(t)=200sinc(6.2832ft+4.7124) is the input signal and the nominal values of the rate dependent Prandtl-Ishlinskii model are: $g = 0.0196$, $b = 1.4$, $a = 8$, $c = 3/120000$, $d = 8$, $e = 9/123$, $m = 0$, $m_1 = 0$, $s_1 = 0$, $s = 0$, $c_1 = 1$, $e_1 = 1$, $n = 0.0191$, $p(r) = 1$ and $q = 1$. Simulations results of rate dependent Prandtl-Ishlinskii model ($\dot{v}(t) \neq 0$) and rate independent Prandtl-Ishlinskii model ($\dot{v}(t) = 0$) using rate dependent play operator at six distinct frequencies: 0.1, 1, 10, 100, 200, and 300 Hz are presented in the following figures. These excitation frequencies include rate dependent and rate independent hysteresis.

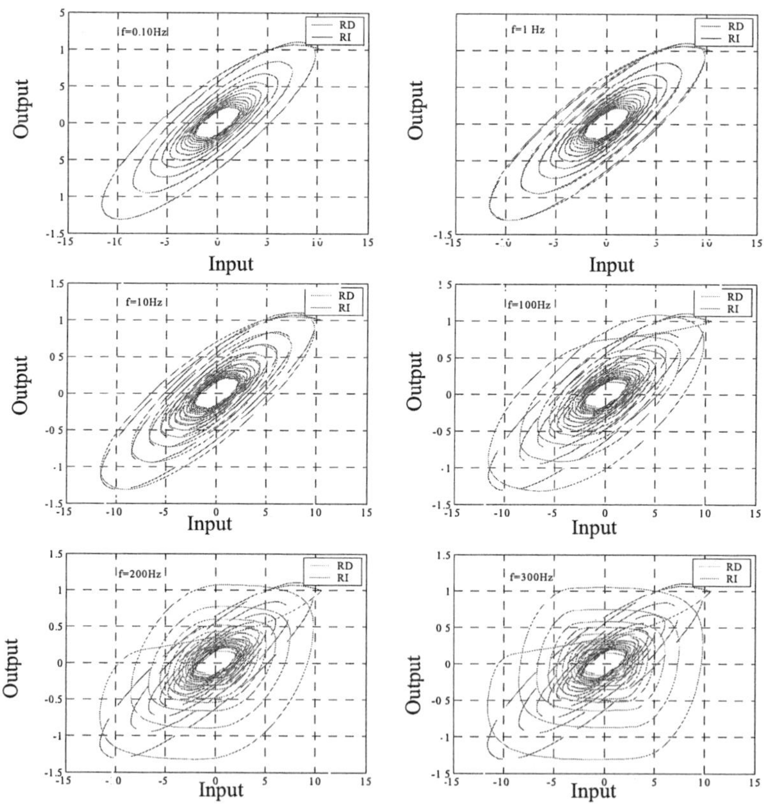

Fig. 3: Simulation results for the rate dependent and rate independent Prandtl-Ishlinskii models

Simulation results show that the rate independent hysteresis model is a special case of the rate dependent hysteresis model and this model satisfies the behavior of the rate independent hysteresis model when the excitation frequency is low(less than 10Hz). In other words, the rate dependent hysteresis model can be used for characterizing rate independent hysteresis. As the excitation frequency increases beyond 10Hz the width of the hysteresis loop increases and the amplitude of the simulated output decreases. On the other hand, Experimental results of hysteresis in smart actuators and magnetic materials show that the hysteresis is rate independent at low frequencies and rate dependent at high frequencies. The width increases and the amplitude output deceases as the frequency increases beyond certain frequency.

6. Conclusion

In this paper, Prandtl-Ishlinskii model is proposed to model rate dependent hysteresis. A rate dependent operator and density functions are used for the Prandtl-Ishlinskii model. A dynamic threshold, which is function of the rate of the input is established for the play and stop operators. Properties of the rate dependent play and stop operators are presented. Simulations results show the ability of the modified model to model rate dependent hysteresis under dynamically varying input. This model will be very useful for the control applications with dynamically varying input, because the rate dependent Prandtl-Ishlinskii model is invertible.

References

1. Mayergoyz, Mathematical Models of Hysteresis and Their Applications, Elsevier, 2003.
2. H.T. Banks, A.J. Kurdila, G.Webb, *Math. Prob. Eng.* **3**, 287(1997).
3. A.Visitian, Differential Models of Hysteresis, Springer, Berlin, 1994.
4. M. Brokate, J. Sprekels, Hysteresis and Phase Transitions, Springer, New York, 1996.
5. P. Ge, M. Jouaneh, *Prec. Eng.* **17**, 211(1995).
6. H. Janocha, K. Kuhnen, *Sens. Actua* . **2**, 83(2000).
7. D. Song, C. J. Li, Mechatronics **9**, 391.
8. M. Krasnoselskii and A. Pokrovskii, Systems with Hysteresis, Nauka, Moscow, 1983; Springer, Heidelberg , 1989.
9. Mayergoyz, *IEEE Tran. Magn.* **24**, 2925(1988).
10. G, Bertotti, *IEEE Tran. Magn.* **28** , 2599(1992).
11. R. B. Gorbet, PhD dissertation, Control of hysteresis systems with Preisach representations, University of Waterloo, Canada. 1997.

12. X. Tan, J. S. Baras, *Automatica* **40**, 1469(2004).
13. Y. Yu, Z. Xiao, N. G. Naganathan, R.V. Dukkipati, *Mech. Mach. Theo.* **37**,75(2002).
14. W.T. Ang, F.A. Garmon, P.K. Khosla, C.N. Riviere, *Proceedings of the 2003 IEEE/RSJ International Conference on Intelligent Robots and systems,* Las Vegas, Nevada, October 2003.
15. D.M. Diebolt, *Proceedings of the IEEE Applied Power Electronics Conference and Exposition*, Boston, Massachusetts, February 1992.
16. W. Galinaities, Two methods for modeling scalar hysteresis and their using in controlling actuators with hysteresis, PhD dissertation, Blacksburg, Virginia, USA. 1999.
17. R. Mrad, H. Hu, *IEEE Trans. Mech.***7**, 479(2002).
18. M.L. Hodgdon, *IEEE Tran. Magn.***24***,* 218(1988).

MODELING THE TORQUE-SPEED HYSTERESIS BEHAVIOR OF AN ULTRASONIC MOTOR

Carlos C. Cuevas Gutiérrez, S. Rakheja, C.Y. Su

Mechanical Engineering Department, Concordia University,
1455 de Maisonneuve Blvd. W., EV11-215, Montreal, Quebec, Canada, H3G 1M8
E-mail: c_cuevas@alcor.concordia.ca
http://concave.concordia.ca/

Several laboratory measurements are carried out to characterize the torque-speed hysteresis behavior of a traveling wave ultrasonic motor under several excitation conditions. A generalized model is proposed to describe the rotor subsystem in order to characterize the friction torques involved in the model. The model's validation is determined by comparison of simulation results with experimental data. The simulation results exhibit a good correlation with the experimental data. Therefore, the results suggest that the proposed model could be effectively applied to characterize the torque-speed hysteresis.

Keywords: Ultrasonic motor, hysteresis, friction torque.

1. Introduction

Ultrasonic motors are a new family of actuators, which have shown outstanding characteristics and very promising applications in micro-position fields. The ultrasonic motor (USM), however, exhibits a torque-speed hysteresis behavior that affects its performance for positioning applications. The magnitude of the hysteresis depends upon the friction properties at the contact interface and the nature of the excitation magnitude and frequency. From the wide variety of ultrasonic motors, the traveling wave type is selected for its controllability characteristics. The traveling wave ultrasonic motor (TWUSM) uses high frequency vibrations to produce a traveling wave on the surface of an annular metallic ring. The wave is produced by a superposition of two sinusoidal waves where one of them has a phase shift of 90 degrees. Although simple in concept, serious challenges are associated with the practicality of the concept mainly due to the high non-linear friction phenomena. Since the first traveling-wave ultrasonic motor made by Sashida in 1982 [1], the research on this type of motor has been increasing and many companies have directed their efforts in using this actuator for positioning and for its silencing operation. Several authors have studied the motor's components from different techniques. For example, through analytical formulation, through finite element modeling, through hybrid models as well as experimental approaches [2–4]. Nevertheless, none of them have studied in detail the hysteresis phenomena. Moreover, the hysteresis behavior has not been addressed in control schemes because most of the ultrasonic motor's controllers are based on black-box models with neural networks or fuzzy logic control schemes. Furthermore, the development of an effective model capable of describing the torque-speed hysteresis in the USM is highly desirable to realize the real-time control and improvement. In the present study, a series of laboratory experiments are conducted in order to characterize the ultrasonic motor's hysteretic behavior. The experimental data is used to synthesize a generalized rotor model on the basis of a friction scheme. The effectiveness of the model is verified with hardware-in-the-loop (HiL) simulations.

2. Rotor Model

The traveling wave ultrasonic motor consists of an annular ring with piezoceramic elements bonded at its lower side and a circular disk as a rotor. The rotor is pressed against the stator through a spring disk, K, which creates a pre-load force between 250 to 300N. A contact layer, which greatly affects the motor performance, is bonded at the lower side of the rotor and it makes contact with the stator. The contact layer has a friction coefficient μ. An schematic of the TWUSM is illustrated in Fig.1.

The torque τ, produced at the contact interface through frictional forces (F_N, F_B, and F_T), drives the rotor in both directions clockwise (CW) and counterclockwise (CCW). The transmission of friction torque to the load involves consideration of the dynamics responses of the rotor and its supports. Furthermore, it has been reported that a low speed operation yields relative higher friction force and thus the torque [5,6]. This behavior has not been in depth addressed in the reported models, while several authors have described noticeable differences in the torque-speed characteristics in the low speed range. The proposed model thus includes the Stribeck effect in order to account for the variations at low speeds. The rotor model also describes the torque-speed hysteresis in the USM, although it involves identification of some of the parameters from the measured data. The rotor subsystem is considered as a rigid disk mounted on a solid circular shaft of negligible mass and inertia, as illustrated in Figure 2, where J_0 is the polar

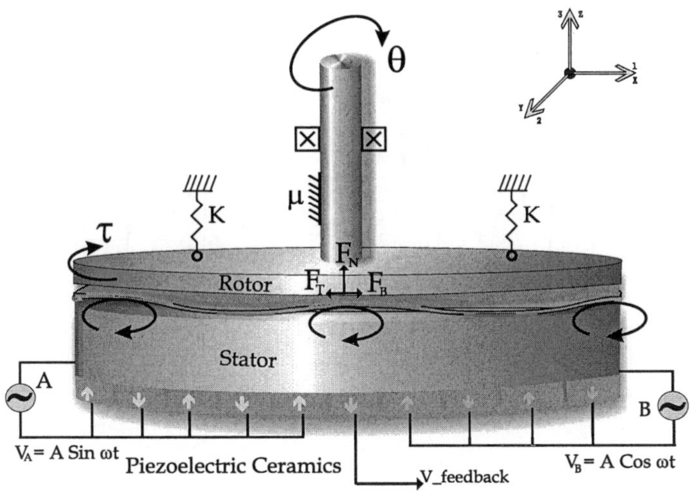

Fig. 1. Graphical representation of the stator-rotor interactions in a TWUSM.

mass moment of inertia of the disk.

The simple rotor model is refined to reproduce the experimental setup described in Sec 4.1. An inertia load J is thus added to represent the load due to the torquemeter and a magnetic brake used in the experiment. Assuming the shaft as a rigid element, the equivalent inertia J_{eq}, is subjected to the rotor torque $\tau_{\text{rotor}}(t)$ developed along the contact points between the stator and the rotor as derived in [6].

The load torque T_f applied to the motor is mainly produced by the magnetic brake. As a result, the loading torque is proposed as:

$$T_f = \alpha_0 \text{sign}(\omega) + \alpha_1 \, \omega \tag{1}$$

where α_0 represents the load produced by the magnetic brake as a function of the electric current applied to the reticulated pole structure (coil), α_1 represents the friction torque associated with the viscous damping effect of the brake and the supporting bearing, and ω is the angular velocity.

The Stribeck effect is incorporated to account for variations in the drive torque at low speeds. Using the Stribeck effect formulation, described in [7], the equation of motion for the rotor can be expressed as:

$$J_{\text{eq}} \, \dot{\omega} = \tau_{\text{rotor}} - \alpha_2 e^{-(\omega/\omega_s)^2} \text{sign}(\omega) - T_f \tag{2}$$

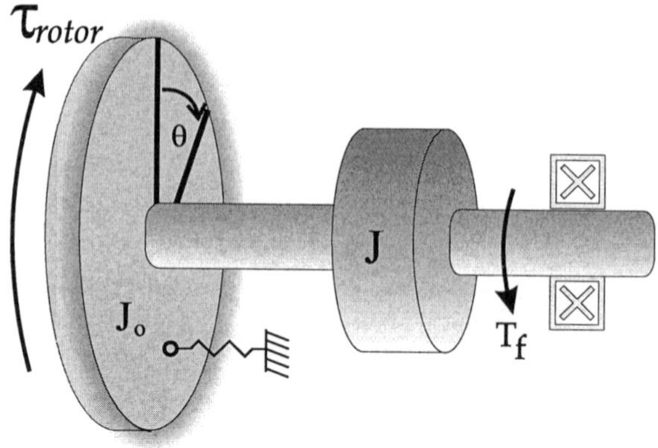

Fig. 2. The rotor model.

where α_2 is the friction torque associated with the Stribeck effect and ω_s is the Stribeck critical velocity.

Upon substituting for the load torque from Eqn.(1), the equation of motion for the rigid rotor can be written as:

$$J_{\text{eq}}\dot{\omega} = \tau_{\text{rotor}} - [(\alpha_0 + \alpha_2 e^{-(\omega/\omega_s)^2})\text{sign}(\omega) + \alpha_1\, \omega] \tag{3}$$

3. Method of Analysis

The analysis of the proposed model is conducted through the numerical solution of the equation of motion. The analysis, however, requires that the model parameters must be identified through a parameter identification process which involves experimental data and a HiL simulation. Consequently, a series of experiments must be conducted on the ultrasonic motor under certain operation conditions. In those experiments, two variables are considered: the speed command signal $u(t)$, and the load signal V_{load}. The experiment strategy consist of holding one of the two variables ($u(t)$ or V_{load}) while studying the dependence of the output variable (ω) on the other.

3.1. *Parameter Range*

The effect of two variables on the motor's behavior are considered. The speed command signal $u(t)$, and the the load signal V_{load}. Their operation ranges are shown in the next table.

Parameter	Operation Range
$u(t)$	0 to 3.2V (0 to 15 rad/s)
V_{load}	0 to 5V (0 to 1N-m)

In order to identify the parameter contributions to the motor model, it is necessary to hold one of the input variables constant ($u(t)$ or V_{load})while studying the dependence of the output variable (ω) on the other.

3.2. *Parameter Identification*

The parameters of the proposed model in Eqn. (3) are identified through a minimization of the following error function:

$$\min_{\omega} \frac{1}{2} |\omega_{\text{data}} - \omega_{\text{model}}|_2^2 \tag{4}$$

where ω_{data} is the experimental and ω_{model} is the simulation result. The error minimization is performed using MATLAB Optimization Toolbox using the multi-objective function method under different operation conditions, including excitation frequencies (0.9, 1.2 and 1.56 Hz), speed amplitudes (±6.25%, ±12.5%, ±22%, and ±28%), and speed bias (Low speed: 37.5%, Medium speed: 56%, and High speed: 75%).

4. Model Validation

In order to validate the proposed model in Eqn.(3), several model simulations are carried out with the same excitation conditions (amplitude and frequency) that were used for the experimental stage.

The use of a real torque signal provides a good test input to validate the rotor's model since the small variations on the motor's operation are reflected on the torque signal. As a result, the comparison between simulation results and test data reveals a sound agreement over the entire range of test conditions considered.

4.1. *Experiment Setup*

The ultrasonic motor used for modeling has a maximum speed of 150 R.P.M. with a rated torque of 0.5Nm and 5W of mechanical power. The USM is assembled in a test bench together with a non-contact torquemeter, a magnetic brake, and a high resolution encoder as it is illustrated in Fig.3. The motor is installed on a rigid plate where the non-contact torquemeter is connected to the motor' shaft whereas the high resolution encoder (50,000 pulses/rev) is connected to the USM' secondary shaft. In addition, the magnetic brake is coupled to the torquemeter. A very important issue that must be considered is the phase shift between the encoder and the torquemeter signals. A 180 degrees phase shift is caused by the connection assembly of both sensors with the USM. As a result, it is necessary a phase shift compensation for one of the two signals. Herein, the encoder signal is chosen as the compensating signal. The equivalent inertia J_{eq} is estimated as 0.0008 Kg-m^2.

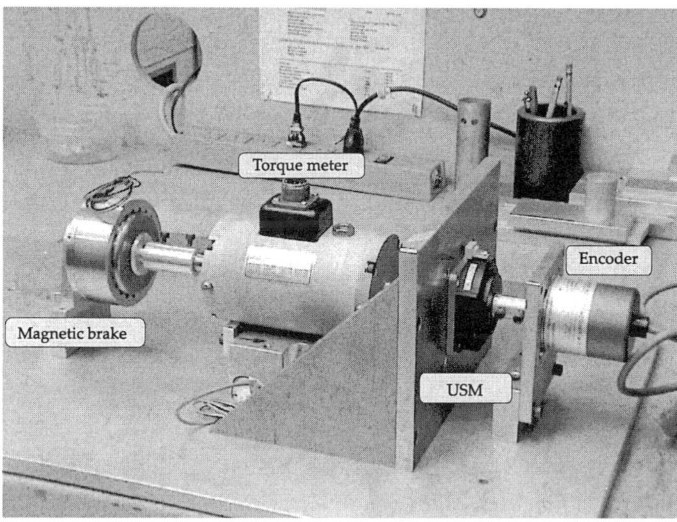

Fig. 3. Experiment setup developed at CONCAVE Research Center.

4.2. *Results and Discussion*

Figure 4 shows the torque-speed characteristics of both simulation results and measured data. The graphic shows the sinusoidal variation of the command speed signal $u(t)$ with a fixed load signal. The speed offset is 8.5 rad/s (1.8V), the speed amplitude is 4.5 rad/s (0.9V), and the excitation frequency is 1.46Hz.

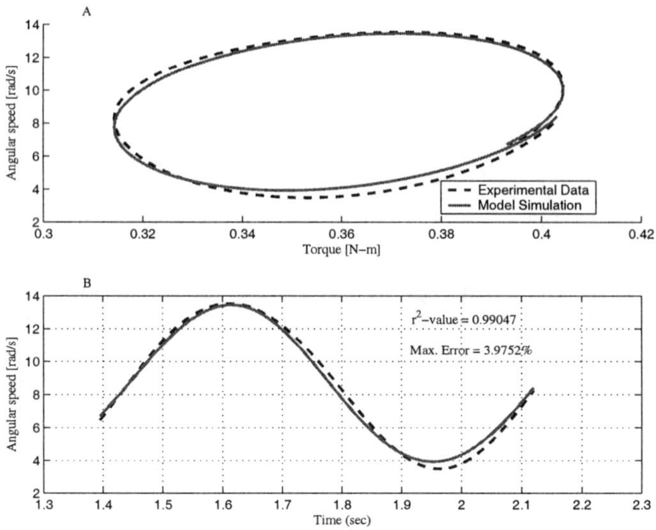

Fig. 4. Hysteresis validation. A) Torque-speed hysteresis loop. B) Time history.

Excellent correlations between the model response and experimental data are evident from Fig. 4 in the proposed range of speed and torque. The quality of the curve fit is estimated with the *r-squared value* [8], which is 0.9904 while the maximum error $e_{\max}$ is 3.97%. It is also observed that there is a small speed shifting caused by the increase of temperature in the contact interface. The increasing of temperature causes a speed drift that in consequence, moves down the hysteresis loop.

Figure 5A shows several hysteresis loops at medium speed offset with different speed amplitude levels. The excitation frequency is 1.56Hz for all the samples and the speed offset is 8.5 rad/s. The fixed level of the magnetic brake (V_{load}) is 0.3606N-m (3.0V), however, a small deviation of $\pm$0.07N-m is observed on the excitation torque. This variation is caused by the sinusoidal excitation.

The parameters of the proposed model are estimated through the measured data and the identification process already discussed in Sec. 3.2. The comparison shows a sound agreement between the model and measured data, although some deviations are evident during the change between the deceleration and acceleration of the motor, which is attributed to the inertia effect. The magnitude of error increases when the motor changes from acceleration to deceleration and viceversa.

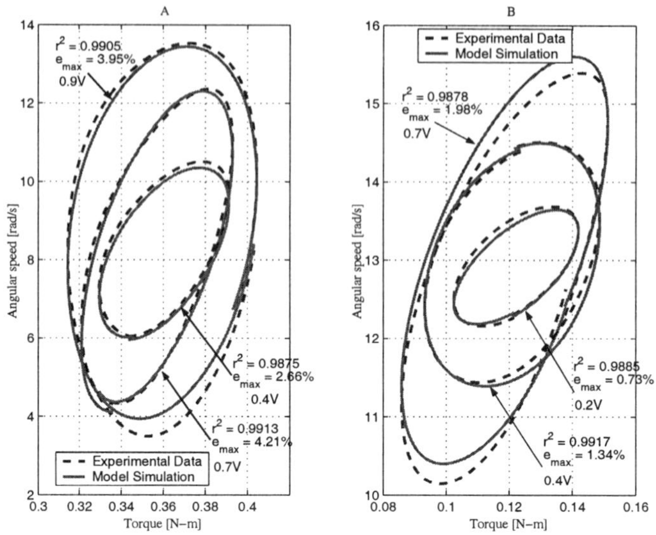

Fig. 5. Speed amplitude variations. A) medium speed, and B) high speed. The loop direction is counterclockwise.

Figure 5B shows hysteresis loops with different speed amplitude levels at a high speed offset. The excitation frequency is 2.22Hz for all samples and the speed offset is 12.9 rad/s. The fixed level for the magnetic brake (V_{load}) is 0.120N-m (2.4V), however, a small deviation of ±0.057N-m is observed. The results attained from the model with high speed data and the corresponding experimental data show a sound agreement.

Fig.6A shows the torque-speed hysteresis in the low-speed range. The graphic also illustrates a clear distortion at the lower speed range as a consequence of the increasing friction torque. The excitation frequency is 0.9Hz and the speed offset is 5.5rad/s (1.2V). Moreover, the load level is set at 2.5V which represents a mean load torque of 0.22N-m.

Speed variations are shown in Fig.6B. The speed offset is set at three different operation points and the speed amplitude is fixed at $\pm 0.4V$ for the three experiments. Additionally, the load level is fixed at 2.0V (0.116N-m). The results show a width increase of the hysteresis loop as a direct result of the speed increasing. This phenomena is mainly attributed to the viscous friction properties of the system.

Comparing the hysteresis results of Figs. 5A, 5B, and 6A it is revealed that the hysteresis loops at medium and high speed do not show the asymmetric distortion that it is exhibited at the low speed range. Therefore, it

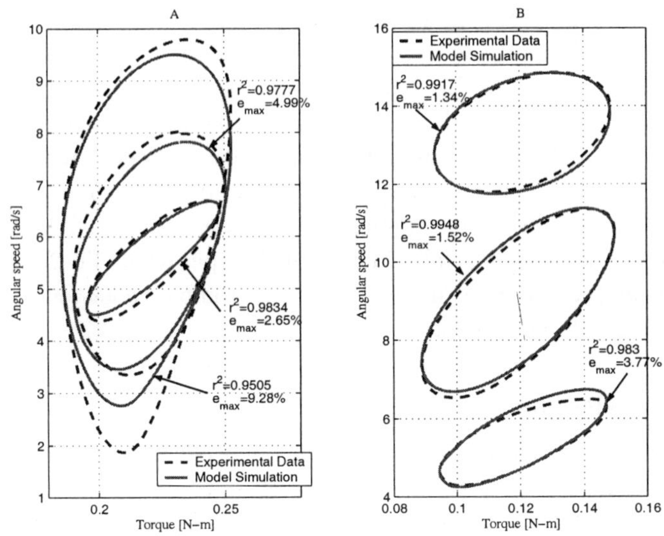

Fig. 6. Speed amplitude variations. A) Low speed, and B) Different speed offsets.

is plausible that the Stribeck effect is accounted for this friction torque increment when the motor is operated at low speed ranges since the Stribeck effect operates at low speed ranges. The parameter identification of the proposed model is shown on the next table where the maximum and minimum values indicate the parameter range.

Model Parameters	Minimum value	Maximum value	Observations
α_0	0.029	0.451	Load torque caused by the brake
α_1	0.0024	0.0174	Torque associated with speed
α_2	0.050	0.050	The Stribeck effect parameters are
ω_s	0.150	0.150	not affected by medium or high speed

5. Conclusions

A generalized model is proposed to characterize the torque-speed hysteresis behavior of a traveling wave ultrasonic motor under a considerable range of excitation conditions for speed and torque. The model integrates a friction torque mechanism for low speed operation based based on the Stribeck effect. Moreover, several measurements were performed at the laboratory under different excitations conditions in order to characterize the hysteresis behavior in the USM. As a result, a set of hysteresis loops are obtained, and the experimental the data is compared with simulations results of the proposed rotor model. The results show a sound agreement between simulation results and experimental data. *The r-squared value* and the plots confirm such a claim. Although the temperature effect is not considered in the present work, it is necessary to integrate its effect when a model-based controller is designed. The rising in temperature strongly influences the motor's performance specially in continuous operation. Therefore, it is concluded that the proposed model shows remarkable consistency throughout all the range of experimentation.

6. Acknowledgments

The present research is supported by the México National Council of Science and Technology (CONACYT), the Querétaro State Council of Science and Technology (CONCYTEQ), and the CONCAVE research center of Concordia University. The author also would like to thank José Esteves support from Concordia's machine shop.

References

1. Sashida, T., *Introduction to Ultrasonic Motors*, 1st edition (Oxford:Claredon Press., 1993).
2. Maeno, T.; Tsukimoto, T. and Miyake, A. *The Contact Mechanism of an Ultrasonic Motor Int. Conf on Applications of Ferroelectrics*, pp. 535-538, 1990.
3. Kenji Uchino, *Piezoelectric Actuators and Ultrasonic Motors*, (Kluwer Academic Press, 1997).
4. Hagood IV, Nesbitt and McFarland, Andrew. *Modeling of a Piezoelectric Rotary Ultrasonic Motor IEEE Transactions on Ultrasonics, Ferroelectrics, and Frequency Control.* **42**, n.2 (March 1995), pp. 210 -224
5. Brian Armstrong-Hélouvry, Pierre Dupont,Carlos Canudas de Wit *A Survey of Models, Analysis Tools and Compensation Methods for the Control of Machines with Friction, Automatica*, **30**, n.7 pp.1083-1138, 1994.
6. Jurgen Wallaschek and X. Cao, *Estimation of the Tangential Stresses in the Stator/Rotor Contact Travelling Wave Ultrasonic Motors Using Viscoelastic Foundation Model, Proc. 2nd Int. Conf. on Contact Mechanics* (Ferrara, 1995), pp. 53 - 61.
7. Canudas de Wit, Carlos, Olsson, H., Astrom, K, J., and Lischinsky,P. *A New Model for Control of Systems with Friction, IEEE Transactions on Automatic Control* **40 n.3** pp. 419-425, March 1995.
8. William J. Palm III, *Modeling, Analysis and Control of Dynamic Systems*, (John Wiley & Sons, Inc., 2000).

MODELLING AND EXPERIMENTATION THE ACCRETING MEDIUM IN THE 1D SEMI-INFINITE MOVING SOLID FOR HEAT TRANSFER WITH A NOVEL CONTROL VOLUME CONDUCTANCE METHOD *

LUIS DEL LLANO VIZCAYA†

ALEJANDRO CASTAÑEDA-MIRANDA
Engineering & Industrial Development Center (CIDESI)
Av. playa pie de la cuesta no. 702. Desarrollo San pablo
Santiago de Querétaro, Querétaro, Zip 76130, México.

A control volume conductance method is discussed in this paper. The method is designed for materials that exhibit heat transfer. Particular attention is put to problems where convection overpowers the mechanism of conduction. The semi-infinite solid which moves with arbitrary imposed velocities along the X-axis and has various surface conditions at x=0 is a classical problem where convection instantaneously overpowers conduction. The analytical solution for this problem becomes physically unrealistic when the strength of convection is high which is defined by the Peclet number. For small Peclet numbers, the diffusion behavior is reasonably described by linear diffusion coefficients, but at large Peclet numbers lineal behaviors become incorrect and hence bad. The true is that diffusion in the analytical solution has indeed an exponential behavior. The exponential behavior in the convection-diffusion exact solution has an exponential behavior. Here false diffusion which is related to the Peclet number corresponds to the energy being supplemented by continuous falls such as a snowfield. As standard numerical schemes do not have this exponential feature, they eventually cross the zero dividing line. The result is unrealistic solution in the form of numerical oscillations. In this paper, this problem is circumvented with a new augmented conductivity term, where false diffusion is added to the true diffusion via exponential relationships with no need of curve fitting procedures. The novelty of the approach is that convection effects are embedded into the conductivity term. This originates new equivalent governing equations for heat transfer. The control volume numerical solution of the method is similar to that of standard parabolic heat conduction. The method is shown to yield exact solutions, to be accurate and computationally competitive.

Key words: capacitance; convection; heat transfer; control volume; coefficients

---

† Corresponding author: e-mail: lvizcaya@cidesi.mx
Ph. 52 -442-2119800 x 283, Fax.52-442-2119839

*Work supported by Engineering & Industrial Development Center (CIDESI) www.cidesi.com

1. Introduction

The Photo thermal techniques are widely used for the investigation of material properties. Their non contact and non-invasive character make these techniques very useful for measuring thermal diffusivity (α), carrier diffusion coefficient (Dn), minority carrier lifetime (τ) and front surface recombination velocity (S1)[1,2]. The importance of composite materials for electronic applications has grown in recent years. Composite materials are used in the electronics industry for resistors, sensors, transducers and packaging materials as well as substrates.[3,4,5] Composite materials have been developed in many forms with a variety of particle sizes, volume fractions, shapes, and topologies that depend on the particular processing route used to fabricate the materials. In this work, we study the thermal diffusivity and electrical resistivity of silicon-epoxy composite materials as a function of the silicon volume fraction. We also perform a qualitative analysis of the PTR signal amplitude and phase. At low carrier densities, the thermal conductivity of (isotropic) crystalline silicon can readily be described in terms of phonon conductivity where the phonon mean-free-path is limited mainly by scattering with other phonons, various impurities and imperfections and by crystal boundaries[6].

Conversely, thermal conduction in epoxy involves the lattice. In composite materials the electrical and thermal transport properties have a spatial dependence and thus it is necessary to refer to these properties as "effective". Phenomenological percolation theory has been used to explain the electrical behavior of composites as a function of the included phase in a material formed by two phases (insulator-conductor).[7] It is clear that in the case of electric carriers the existence of a continuous path is necessary in order to have current continuity. At the percolation threshold value the existence of this path allows carrier movement through the bulk, resulting in a dramatic change in the electrical resistivity. However, in the case of thermal transport in composite materials, it is necessary to take into account the character of each type of carrier (electrons, holes, ambipolar, phonons) associated with thermal transport. It is known that the interfacial thermal contact resistance (ITCR) between different constituent phases in a composite can arise from the combination of poor mechanical or chemical adhesion at the interface, as well as from thermal expansion mismatch.5 The existence of such thermal barriers leads to a lowering of the effective thermal diffusivity (conductivity) of the composite.

2. Theoretical Model

For the case of optically opaque and thermally thick solids, it has been shown[8] that photo acoustic detection can yield the thermal diffusivity (α), of a sample through a frequency scan and fitting of the PA amplitude data to a simple expression:

$$PA = \frac{A}{f} e^{-G\sqrt{f}} \tag{1}$$

Where G $=\sqrt{\pi l^2/\alpha}$, l is the thickness of the sample, f is the modulation frequency and A includes all other factors such as the thermal properties of the gas, laser power and so on. Electrical measurements in these composites were also made using a conventional four point probe system. The physical foundations of signal generation in laser photothermal radiometry of semiconductors have been described elsewhere.[2,9] Thus, it is possible to obtain information about the thermal and electronic transport properties by means of PTR detection. The thermal contribution for this type of samples is strongest at the low frequency end of the spectrum of the experimental data. The PTR signal is influenced by the optical properties of the sample and thus it is important to study how these properties change as the volume percentage changes. A theoretical model[10] that takes into consideration the optical and thermal properties of the sample is used to fit the low frequency end of the PTR data.

3. Experimental Results And Discussion

The Photo acoustic Characterization resulting value of thermal diffusivity from the experimental data fit to Eq. 1 for all the samples is shown in table 1. For samples with low silicon volume fraction (x < 10 vol. % fraction) it was necessary to add a thin aluminum foil on the surface, due to the optical semi transparency of these samples. The thermal diffusivity of the epoxy-silicon composites as a function of silicon volume fraction is shown in Figure 1. In the thermal analysis of these composites three regions were observed: a region with x <10 % volume fraction (Maxwell dilute limit)3, followed by a region with 10 % < x < 20 % vol. and then by region III for x > 20 vol. %. This division is determined by the average number of contacts per particle (M) [3,4]. For x < 10 vol. % it is expected that there isn't any contact among the particles of silicon embedded inside the epoxy matrix, M<1. However, the thermal diffusivity values in this region are not governed by the epoxy matrix. The inclusion of Si particles at these concentrations increases the effective thermal diffusion value by an order of magnitude 0.0021 cm^2/s for pure epoxy to 0.0178 at 9.35 vol. % of silicon. For 10 % < x < 20 Vol. %, two situations are possible: a) the system remains in dilute phase, and silicon particles are randomly well mixed on the epoxy matrix. In this case we have M < 1; or b) the system displays the formation of chains or clusters. If the volume fraction of silicon particles increases, a process of random chain formation arises, and/or the inter-particle conductivity also increases, giving rise to the formation of silicon chains. Theoretically at the percolation limit there have an infinite interconnectivity among the silicon chains and clusters, although there is no electrical conductivity. This means that the silicon particles are percolating through the epoxy matrix. When the silicon volume fraction increases from 10 to 15 vol. %, it is possible that M may increase thus allowing the formation of iron chains and clusters. In this way heat can diffuse through the matrix. This results in the observed jump in the diffusivity value at x > 10 vol. % in Fig. 1. Finally for a high volume fraction (region III), x > 20 vol. %, the addition of iron particles increases the average number of contacts per particle above 2^3. In this case, the thermal behavior is controlled by iron particle distribution. According a high volume fraction of diamond, in Iron, the interconnectivity among the diamond particles appears to be very anisotropic (x >30 vol. %) as a group and theoretically the calculation of the thermal conductivity (diffusivity) for this system requires extensive information about the correlation between the relative positions of particles. This implies that different morphologies can take place as a function of the volume fraction and the effective thermal diffusivity values exhibit a non-linear shape for high volume fraction values.

Table 1. Values of physical parameters for iron, FeO_2, epoxy and composite materials

Sample	Volume fraction (%)	Thickness (μm)	α (cm^2/sec.)	β (cm^{-1}) λ=632 nm.
Iron		300	0.900	
FeO_2			0.010	
Epoxy		420	0.0163	
Iron-Epoxy -1	6.0	398	0.0178	30
Iron-Epoxy -2	9.0	400	0.0169	60
Iron-Epoxy -3	12.0	427	0.0275	60
Iron-Epoxy -4	15.0	427	0.0340	130
Iron-Epoxy -5	19.0	433	0.0375	200
Iron-Epoxy -6	22.0	467	0.0390	260
Iron-Epoxy -7	32.0	490	0.0523	700

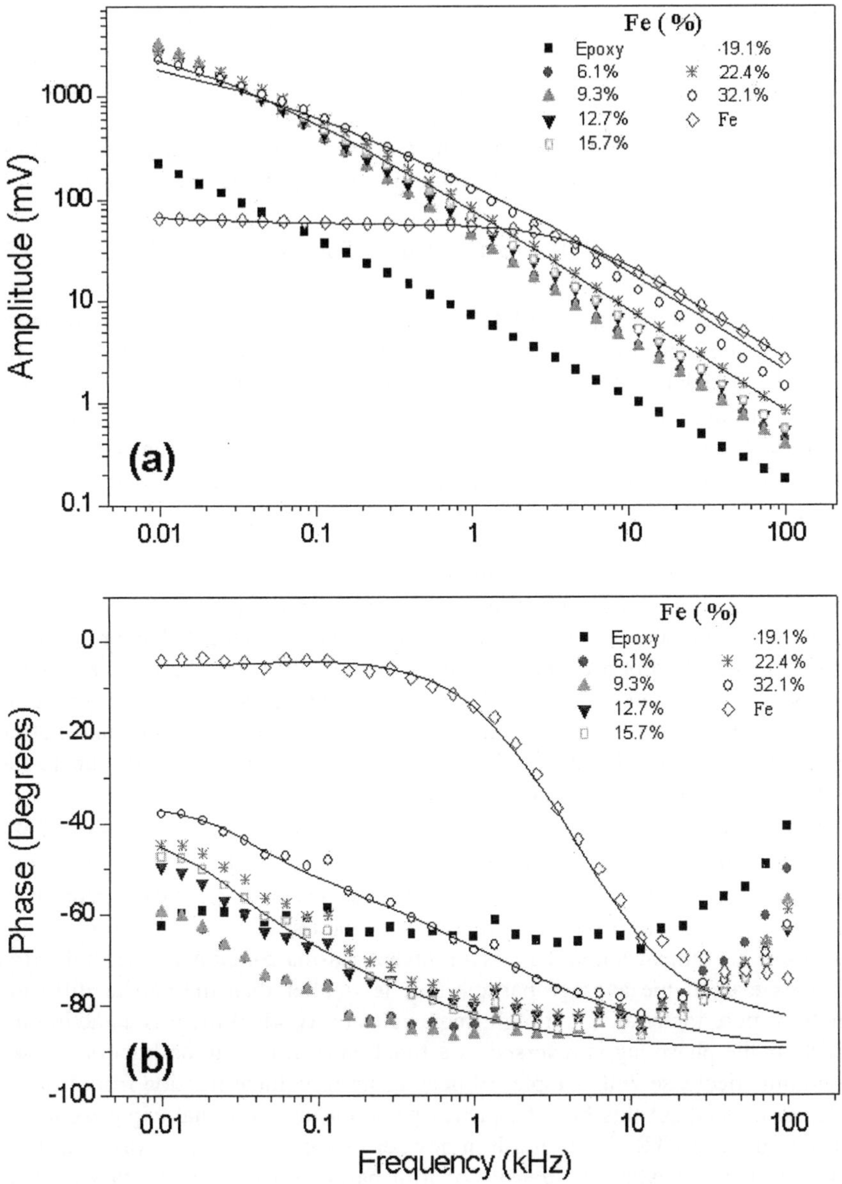

Fig. 1. Thermal diffusivity values of various iron as extracted from best fits of data such as those in Region I, using eq. 1.

The PTR characterization in 3D photo thermal radiometric[2,12] experiments for iron and 1D PTR for iron-epoxy composites and iron powder were carried out between 10 Hz. and 100 kHz as described above. It was found that the total PTR signal consists of two contributions: thermal-wave (SThermal) due to the lattice absorption contribution of iron, and carrier plasma-wave contribution (SPlasma) due to the electronic contribution, resulting in the vectorial sum[12]:

$$\text{SPTR} = (\text{S Thermal}) + (\text{S Plasma}) \qquad (2)$$

In figure 2 shows the 3-D PTR normalized signal amplitude (a) and phase (b) for the above mentioned oxidized iron obtained between 10 Hz and 100 kHz. Using the 3-D PTR model[12], the multiparameter best fit values found for the thermolectronic parameters were: τ = 95 μs, Dn = 4.2 cm2/s, Iron = 300 cm/s and α = 0.90 cm2/s. The carrier diffusion length for the iron sample was estimated to be around 199 μm. Figure 2 shows the PTR normalized signal amplitude (a) and phase (b) of the entire set of iron composite samples, for pure epoxy (square), and iron. In the case of pure epoxy (solid squares) only the thermal wave contribution is present as expected[1, 12]. The normalized PTR signal amplitude exhibits a pure thermal frequency dependence of f -0.5 with a flat phase for frequencies up to 10 kHz. This is the expected behavior for one-dimensional PTR. For frequencies higher than 10 kHz the phase, which is the more sensitive channel, increases. This behavior could be due to micro particle dispersion. In the case of the crystalline sample (diamond) the main contribution due to electronic carrier diffusion and recombination is seen to dominate the PTR signal for the entire frequency range.[10, 12]

The PTR signal amplitude for the iron-epoxy composites approaches the pure iron signal as the iron volume fraction is increased. In the PTR phase two kinds of behavior are present: 1) At low frequencies (<10 kHz) the signal phase tends toward the pure iron signal phase as a function of increasing iron volume fraction. This indicates that there is a remnant plasma component due to the iron particles even if the average particle size is smaller than the carrier diffusion length of pure material. 2) At high frequencies (> 10 kHz) there is an additional effect, as the phase lag is reversed as a function of iron volume fraction, with a monotonic decrease with samples of decreasing in volume fraction from iron. In order to understand this high frequency behavior an additional experiment was carried out. The PTR signal of a iron powder on top of pure iron was measured. It was found that when the density of iron particles increases the phase signal decreases at high frequencies and approaches the iron behavior. When there are any particles on top of the iron, at low frequencies it is possible to see the remaining plasma component. For the low frequency (<10 kHz) PTR signal the PTR thermal model with optical absorption was used to fit the data. The values obtained from photoacoustic detection for thermal diffusivity were used in the photothermal model to obtain the optical properties of the sample assuming a constant infrared coefficient for epoxy10. In fitting the data it is found that for frequencies lower than 100 Hz three-dimensional effects are introduced in the sample. The full three-dimensional theoretical fitting[12] (solid lines) is shown in figure 2.

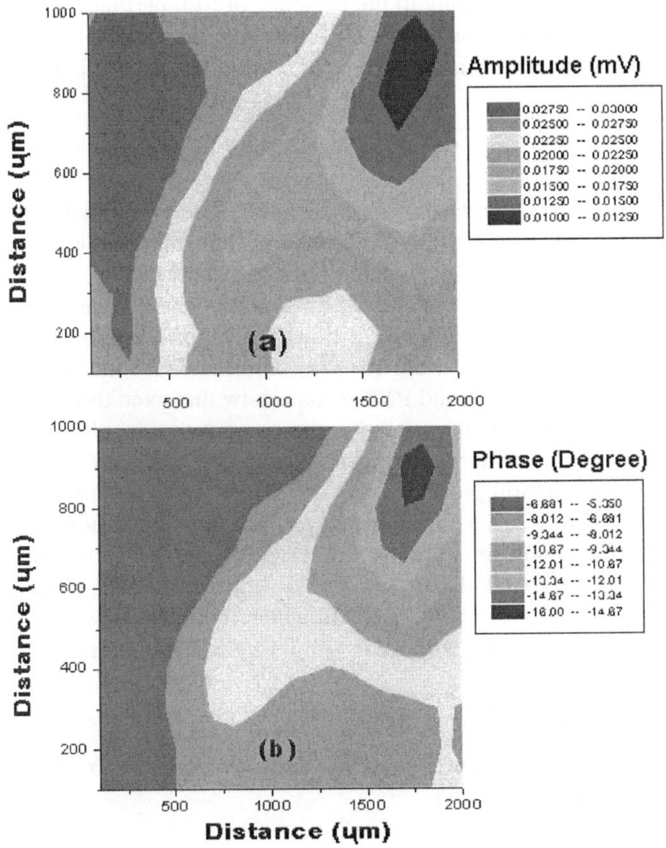

Fig. 2. PTR signal amplitude and phase for iron material.

The optical absorption coefficient values for the iron composites are shown in table 1 where the optical absorption of iron is also documented[12]. It is found that as the iron volume percentage is increased the optical absorption increases. At a 15 vol. % the optical absorption is lower that the 12 vol. %. This can be attributed to the fact that at this vol. % different morphology of iron could have been formed, thus lowering the local optical absorption In summary, using PTR measurements of iron particles, it is possible to establish their contribution to the plasma component of the signal even if their radii are smaller than the dc carrier diffusion length, $L = (D\tau)^{1/2}$. Furthermore, the volume electronic excitation process can contribute to the effective thermal diffusivity even if the main contribution arises from direct lattice iron-epoxy absorption and non radioactive conversion. The measured increase in the effective thermal diffusivity of the iron composites could be due mainly to the presence of iron particles acting localized volume heat sources. It is interesting to note that using PTR it is possible to establish the existence of an electronic contribution as well as the thermal contribution.

4. Conclusion

In conclusion, the iron composite system exhibits strong thermal and electronic barrier formation, presumably via the insulating thin films grown on the surface of the oxygen particles, as is evident from our electrical measurements. The effective thermal diffusivity values of this composite exhibit an anomalous behavior around 16 % iron volume fraction, which corresponds to the existence of a particle percolation threshold for three dimensional random close packed materials. The combined PA and PTR results show that even though there has an electrical barrier of very high resistivity, the composite nevertheless allows heat conduction by means of random lattice vibrations (conduction) and heat release by electronic deexcitation processes that are converted in to translational energy as heat at the particle surface. According to the foregoing results, the existence of a percolation phenomenon similar to electrical percolation can be monitored both thermally and optically in iron-epoxy composites via combined photo thermal radiometric and photo acoustic detection. The problem of thermal barriers could be further studied by changing the particle size and distributions.

Acknowledgements

This work was supported by CONACYT, the financial support of CONCYTEQ is also gratefully acknowledged.

References

1. Tuchin, V.V., Tissue Optics: Light Scattering Methods and Instruments, SPIE Tutorial Texts in Optical Engineering, T138, SPIE Press, Bellingham, 2000.
2. Bicout, D., Brosseau, c., Martinez, A.S., and Schmitt, J.M., Depolarization of multiply scattering waves by spherical diffusers: influence of the size parameter, Phys. Rev. E., 49, 1767, 1994.

3. S. E. Bialkowski, en: Photothermal spectroscopy methods for chemical analysis, Vol 13, John Willey & sons, New York, cap 2 (1996).
4. Fishkin, J.B. and Gratton, E., Propagation of photon-density waves in strongly scattering media containing an absorbing semi-infinite plane bounded by a straight edge, J. Opto. Soc. Am. A, 10, 127, 1993.
5. B. Fraden, Handbook of Modern Sensors Physics, Designs, and Application, American Institute of Physics. ISBN 1-56396-538-0
6. thyam, U.S. and Prahl, S.A., Limitations in measurement of subsurface temperatures using pulsed photothermal radiometry, J. Biomed. Opt., 2, 251, 1997.
7. H. Nakumara, K. Tsubouchi, N. Mokoshiba y T. Fukuda, Jpn. J. Appl. Phys. 24, L876 (1985).
8. Sathyam, U.S. and Prahl, S.A., Limitations in measurement of subsurface temperatures using pulsed photothermal radiometry, J. Biomed. Opt., 2, 251, 1997.
9. D. F. Edwards y P. D. Maker, J. Appl. Phys., 33, 2466 (1962). H. Kachare, W. G. Spizer, F. K. Euler y A. Kahan, J. Appl. Phys., 45, 2938 (1974)
10. E. A. Ulmer and D. R. Franck. Proc. Ixth Int. Conf. Physics, Semiconductors, Nauka, p 170 (1968).
11. Masters, B., Ed., Selected Papers on Optical Low-Coherence Reflectometry and Tomography, MS 165, SPIE Press, Bellingham, 2002.
12. Alejandro Castañeda , Luis del Llano Vizcaya, A novel system for measuring optical properties in arterial blood of man, journal of Applied Bionics and Biomechanics, 3, 1-6, 2006.

A NOVEL HYBRID REPRESENTATION AND CONTROL OF CONVECTIVE SPATIALLY DISTRIBUTED SYSTEMS

J. PAULO GARCIA-SANDOVAL AND BERNARDINO CASTILLO-TOLEDO*
CINVESTAV-IPN Unidad Guadalajara,
Av. Científica No. 1145,
44019 Zapopan, Jal., México
E-mail: pgarcia@gdl.cinvestav.mx, toledo@gdl.cinvestav.mx

and VICTOR GONZALEZ-ALVAREZ
Chemical Engineering Department of the University of Guadalajara,
Blvd. Marcelino García Barragán and Calzada Olímpica,
44460 Guadalajara, Jal., México
E-mail: victor.gonzalez@cucei.udg.mx

In this work we study spatially distributed convective systems described by first order partial differential equations and we show how this systems in a fixed spatial point can be exactly represented by a hybrid system with two commutable ordinary differential equations subsystems, where the first one is dominated by the initial condition, whereas the second one is governed by the boundary condition. Then we define the boundary control problem for spatially distributed convective systems and as a first approach, we propose a controller for one-dimensional systems which solves this problem. Finally, in order to test hybrid representation and controllers, a study case is presented for an enzymatic reactor, where dispersive terms can be neglected.

1. Introduction

First orden partial differential equations (FOPDE) arise naturally in the model of any situation that is dominated by convective rather that dissipative or dispersive phenomena. For many systems the convective model is the simplest nontrivial one, the solutions are much easier to obtain, at the same time, elucidating the basic structure of the solution space. For example, FOPDE's arise most naturally from the simplest models of exchange processes and other physical conservation laws. We can also obtain FOPDE's from variational problems and by the introduction of genera-

*Work partially supported by grants 36687-a and 163390 of the CONACyT.

ting functions or limiting assumptions in kinetic equations, representative chemical processes modeled by such systems include exchangers[1], plug-flow reactors, fixed-bed reactors[2], pressure swing adsorption processes[3] and so forth. If a quantity $X(z,t)$ is a function of position and time in a flow in which velocity is V, then the convective derivative $V(\partial X/\partial z) + (\partial X/\partial t)$ will appear. Because the properties of FOPDE's systems, it is impossible to use modal (spectral) decomposition techniques to derive low-order ordinary differential equations models that approximately describe the dynamics of the partial differential equations system[4]. However, some times solutions are easy to obtain using the method of characteristics[5] and control problems can be solved as has been reported in some works[6,7].

The conventional approach to the control of FOPDE's is based on the direct spatial discretization (typically using finite-difference method) of the partial differential equation model followed by the controller design on the basis of the resulting (linear or nonlinear) ordinary differential equation (ODE) model. However, there are certain well-known disadvantages associated with this approach. For example, fundamental control theoretical properties, like controllability and observability, which should depend only on the location of control actuators, specification of control objectives, and measurements sensors, may also depend on the discretization method used and the number and location of the discretization points. Moreover, by neglecting the infinite dimensional nature of the original system may lead to erroneous conclusions concerning the stability properties of the open-loop and/or the closed-loop system[4].

In this work we define the "boundary control problem for spatially distributed convective systems (BCPSDCS)" and we assume that the dynamic behavior of a convective system in a fixed specific axial point can be exactly represented by means of an ODE's system, which may supply an easier way to design controllers capable of solving the boundary control problem for spatially distributed systems. Using the method of characteristics, we show that the convective system can be described by an hybrid system with two commutable subsystems of ODE's. In particular, we analyze the BCPSDCS for constant distributed initial conditions and constant semi-continuous boundary conditions through time for n-dimensional systems, and as a first approach we propose a controller for one-dimensional systems. Finally, in order to test hybrid representation and controllers, a study case is presented for an enzymatic reactor, where dispersive terms can be neglected.

2. Problem statement

Consider a convective spatially distributed system described by

$$\frac{\partial X(z,t)}{\partial t} = -\alpha u(t)\frac{\partial X(z,t)}{\partial z} + f(X(z,t)) \tag{1}$$

with the following boundary and initial conditions

$$X(0,t) = g(t) \tag{2}$$

$$X(z,0) = h(z) \tag{3}$$

In where $X(z,t)$, denotes the state vector. $X(z,t) \in \mathcal{H}^n[(0,L),\mathbb{R}^n]$, with $\mathcal{H}^n[(0,L),\mathbb{R}^n]$ being the infinite dimensional Hilbert space of n-dimensional-like vector functions defined on the interval $[0,L]$ whose spatial derivatives up to n-th order are square integrable, $z \in [0,L] \subset \mathbb{R}$ and $t \in (0,\infty]$ denote position and time, respectively, while $\alpha > 0$ is a scalar and the manipulated variable is the flow function $u(t) \in [a,b] \subset \mathbb{R}^+$. $f(X)$ is a sufficiently smooth vector function, $g(t)$ is a column vector which is assumed to be a sufficiently smooth function of time, and $h(z) \in \mathcal{H}^n[(0,L),\mathbb{R}^n]$. Equation (2) is a boundary condition, which may be considered as a disturbance, while the initial condition is described in (3). The formal mathematical classification of equation (1) is a "semilinear first order partial differential equation".

We state the control problem, called *Boundary Control Problem for Spatially Distributed Convective Systems (BCPSDCS)*, as the problem of control the value of an output $y(t) = \phi(X(L,t))$, which is a function of the states in the upper boundary $z = L$, manipulating the flow function $u(t)$, while the boundary condition (2), $g(t)$ and initial condition (3), $h(z)$, are considered disturbances.

3. Hybrid representation for convective spatially distributed system

3.1. *Solution for convective spatially distributed system*

By using the method of characteristics[5], the solution of system (1) together with conditions (2) and (3) can be split in two parts; the first one is handled by initial conditions, while the second one is dominated by the boundary condition. For both of them, the solution is obtained by solving the ODE's

$$\frac{dz}{dt} = \alpha u(t), \qquad z(t_0) = z_0 \tag{4}$$

$$\frac{dX}{dt} = f(X), \qquad X(t_0) = X_0 \tag{5}$$

whose solution is given by

$$z = z_0 + v(t) - v(t_0),$$
$$X(t, t_0) = \Phi^f_{t-t_0}(X_0).$$

where $v(t) = \alpha \int_0^t u(\lambda)\, d\lambda$, follows the dynamic behavior

$$\frac{dv(t)}{dt} = \alpha u(t), \qquad v(0) = 0. \tag{6}$$

and $\Phi^f_{t-t_0}(x)$ is the flow vector field of the function f satisfying

$$\frac{\partial}{\partial t}\left(\Phi^f_{t-t_0}(x)\right) = f\left(\Phi^f_{t-t_0}(x)\right)$$

for a given fixed x.

In this case the solution handled by initial condition occurs when $z > v(t)$, and it must hold that $t_0 = 0$ and $X_0 = h(z_0)$. On the other hand, when $z < v(t)$, the boundary condition handle the behavior and it must hold that $z_0 = 0$, and $X_0 = g(t_0)$. Hence the global analytic solution for (1) and conditions (2) and (3) is

$$X(t, z) = \Phi^f_t(h(z - v)), \qquad z > v(t) \tag{7a}$$
$$X(t, z) = \Phi^f_\tau(g(t - \tau)), \qquad z < v(t), \tag{7b}$$

where $\tau(t) = t - t_0$, represents the resident time, whose dynamic behavior may be obtained by solving

$$\frac{d\tau}{dt} = 1 - \frac{u(t)}{u(t-\tau)}, \qquad \tau(t_1) = t_1,$$

where t_1 is the time wherein $z = v(t_1)$. In solution (7a) it is evident the z dependence, however, in solution (7b) the z dependence is related to τ, which depend on t and z through the relationship

$$v(t - \tau) = v(t) - z.$$

3.2. *Hybrid representation for convective spatially distributed system*

In the next theorem we present the main result of this work which states that a spatially distributed convective system described by first order partial differential equations in a fixed spatial point can be exactly represented by a hybrid system with two commutable ordinary differential equations subsystems.

Theorem 3.1. *Consider the system described by PDE (1) together with conditions (2) and (3). The dynamic behavior in a fixed axial point,* $z = z^*$*, can be described by a hybrid system with two commutable EDO's subsystems of the form*

$$\begin{array}{c}\textit{Subsystem 1:}\\ \textit{when } v(t) \leq z^*,\end{array} \begin{cases} \dot{x}_1 &= f(x_1) + \omega_1 u(t) \ , x_1(0) = h(z^*) \\ \dot{x}_2 &= f(x_2) \qquad\qquad\quad , x_2(0) = g(0) \\ \dot{v} &= \alpha u \qquad\qquad\qquad\;\; , v(0) \;\; = 0 \\ \dot{\tau} &= 1 \qquad\qquad\qquad\quad\; , \tau(0) \;\; = 0 \\ X(t, z^*) &= x_1(t) \end{cases} \tag{8a}$$

$$\begin{array}{c}\textit{Subsystem 2:}\\ \textit{when } v(t) > z^*,\end{array} \begin{cases} \dot{x}_1 &= 0 \ , \\ \dot{x}_2 &= [f(x_2) + \omega_2]\left(1 - \frac{u(t)}{u(t-\tau)}\right) , \\ \dot{v} &= 0 \ , \\ \dot{\tau} &= 1 - \frac{u(t)}{u(t-\tau)} \ , \\ X(t, z^*) &= x_2(t) \end{cases} \tag{8b}$$

where

$$\omega_1(t) = \left[\left(\Phi_t^f\right)_\star \circ h(z^* - v)\right]\left[\frac{dh}{dz} \circ (z^* - v)\right],$$

$$\omega_2(t) = \left[\left(\Phi_t^f\right)_\star \circ g(t - \tau)\right]\left[\frac{dg}{dt} \circ (t - \tau)\right],$$

and $(M)_\star = \frac{\partial M}{\partial x}$.

Proof. It is straightforward to prove Theorem 3.1 by taking the derivative of (7) with respect to time in a fixed axial point, z^*. □

Remark 3.1. Commutation between subsystem 1 and 2 only happens once, at $t = t_1$ when $v(t_1) = z^*$. On the one hand, state $x_1 \in \mathbb{R}^n$ in subsystem 1 represents $X(t, z^*)$, while in subsystem 2 is meaningless; on the other hand, $x_2 \in \mathbb{R}^n$ in subsystem 1 represents the evolution for an element which is initially in the boundary $z = 0$, flows through the system and in a given instant t it is in $z = v(t)$, while in subsystem 2 represents $X(t, z^*)$. Variable $v \in \mathbb{R}^+$ is only necessary when subsystem 1 is activated and defines the instant wherein commutation occurs. The equation $\dot{\tau} = 1$ in subsystem 1 guarantees that condition $\tau(t_1) = t_1$ at commutation instant holds, which is a requirement for $\tau \in \mathbb{R}^+$ in subsystem 2.

In particular, when we have unidimensional systems, $X \in \mathcal{H}^n[(0, L), \mathbb{R}]$, because $\left[\left(\Phi_{t-t_0}^f\right)_\star f\right] \circ x_0 = f\left(\Phi_{t-t_0}^f(x_0)\right)$, it holds that $\left(\Phi_{t-t_0}^f\right)_\star \circ x_0 =$

$f(x)/f(x_0)$, and vectors ω_1 and ω_2 can be simplified to

$$\omega_1(t) = \frac{f(x)}{f(h(L-v))}\left[\frac{dh}{dz}\circ(L-v)\right],$$

$$\omega_2(t) = \frac{f(x)}{f(g(t-\tau))}\left[\frac{dg}{dt}\circ(t-\tau)\right].$$

It is important to remark that when initial condition $h(z)$ is constant, $\omega_1 = 0$, and subsystem 1 becomes not controllable; however we can only regulate the instant t_1 when commutation occurs by changing the flow velocity. By the same token if boundary condition $g(t)$ is constant, $\omega_2 = 0$, but in this case when $v(t) > z^*$ system keeps controllable. Then, for constant boundary and initial conditions system (8) in $z^* = L$ becomes

$$\text{Subsystem 1: when } v(t) \le L, \quad \begin{cases} \dot{x}_1 & = f(x_1),\ x_1(0) = h(L) \\ \dot{x}_2 & = f(x_2),\ x_2(0) = g(0) \\ \dot{v} & = \alpha u \quad,\ v(0) \ = 0 \\ \dot{\tau} & = 1 \quad\ ,\ \tau(0) \ = 0 \\ X(t,L) & = x_1(t) \end{cases} \tag{9a}$$

$$\text{Subsystem 2: when } v(t) > L, \quad \begin{cases} \dot{x}_1 & = 0 \quad , \\ \dot{x}_2 & = f(x_2)\left(1 - \frac{u(t)}{u(t-\tau)}\right), \\ \dot{v} & = 0 \quad , \\ \dot{\tau} & = 1 - \frac{u(t)}{u(t-\tau)} \quad , \\ X(t,L) & = x_2(t) \end{cases} \tag{9b}$$

In the next section we consider the control problem for the specific case for $X \in \mathcal{H}^n[(0,L),\mathbb{R}]$ and constant boundary and initial conditions.

4. Control for a convective one-dimensional system with constant conditions

In this section, we analyze the case of controlling the output of system of the form

$$\frac{\partial X(z,t)}{\partial t} = -\alpha u(t)\frac{\partial X(z,t)}{\partial z} + f(X(z,t)) \tag{10}$$

$$y(t) = X(L,t) \tag{11}$$

with boundary and initial conditions

$$X(0,t) = g \tag{12}$$

$$X(z,0) = h \tag{13}$$

where $X(z,t)$ denotes the state variable. $X(z,t) \in \mathcal{H}^n[(0,L),\mathbb{R}]$, while $\alpha > 0$ is a unknown scalar and the manipulated variable is the flow function $u(t) \in [a,b] \subset \mathbb{R}^+$. The output of the system is $y(t)$, the boundary condition, g, is a piece-wise constant function of time, and $h \in \mathbb{R}$ is the constant initial condition through z. We consider that α, g and h, are unknown while $f(X)$ is a uncertain sufficiently smooth function with well defined sign, $\text{sign}(f(X)) = \pi$. As presented in the last Sec., a system described by (10)-(13) can be exactly represented by (9) hence the BCPSDCS can be solved controlling (9).

As discussed before, when subsystem 1 is activated, the system is not controllable and we can only modify the commutation instant by increasing or reducing the flow rate, $u(t)$. Hence, the control problem will be only studied for the case when subsystem 2 is activated, then the control problem is reduced to control a system of the form

$$\dot{x} = f(x)\left(1 - \frac{u(t)}{u(t-\tau(t))}\right)$$
$$\dot{\tau}(t) = 1 - \frac{u(t)}{u(t-\tau(t))}$$

since α is unknown, so is the transition time t_1 as well as $\tau(t)$, then the control system can be described for

$$\dot{x} = f(x)\left(\frac{\vartheta(t) - u(t)}{\vartheta(t)}\right) \tag{14}$$

where $\vartheta(t) = u(t-\tau(t))$, is unknown but bounded (*i.e.* $a \leq \vartheta(t) \leq b$). The next theorem presents a controller which solves the control problem for system (14).

Theorem 4.1. *Consider a system of the form (14), where $x \in \mathbb{R}$, is the state, $u \in [a,b] \in \mathbb{R}^+$ is the input variable and $\vartheta(t) \in [a,b] \in \mathbb{R}^+$ is a bounded unknown function satisfying* $\lim_{t\to\infty}\vartheta(t) = \vartheta^*$ *and $f(X)$ is an uncertain sufficiently smooth function with well defined sign,* $\text{sign}(f(X)) = \pi$. *We can regulate system (14) around the point $x = x^*$, where not necessarily $f(x^*) = 0$ with a controller of the form*

$$\dot{\xi} = -\Gamma\left|\xi(u-\xi)\right|\text{sign}(e)$$
$$u = \xi - \psi(e) \tag{15}$$

where $e = x - x^*$, ξ is an estimation of $\vartheta(t)$, function $-\pi\gamma\psi(e)$ with $\gamma > 0$ must be a monotonically increacing function and $\psi(0) = 0$, while $\Gamma > 0$ is an adjustable parameter.

Proof. Considering that $f(X)$ has a well defined sign, $\operatorname{sign}(f(X)) = \pi$ and $-\pi\gamma\psi(e)$ with $\gamma > 0$ is an increacing function, then it holds

$$\pi(u - u^*)\operatorname{sign}(e) < 0$$
$$\pi\psi(e)\operatorname{sign}(e) < 0$$

where $u^* = \vartheta^*$ is the steady state input when $x = x^*$. Then by defining the Lyapunov function

$$V = \frac{1}{2}[\gamma\psi(e)]^2 + \gamma^2\psi(e)(u - u^*) + \frac{\rho + \gamma^2}{2}(u - u^*)^2 > 0$$

with $\rho > 0$ and choosing a large enought γ and Γ and a small enought ρ, its derivative with respect to time

$$\begin{aligned}\dot{V} &= \pi\gamma^2\Gamma\,|\xi(u-\xi)|\,\psi(x)\operatorname{sign}(x) \\ &\quad + (u-u^*)\left[\pi\left(\rho+\gamma^2\right)\Gamma\,|\xi(u-\xi)|\operatorname{sign}(x) - \rho\frac{\partial\psi}{\partial e}\frac{f(x)}{z}(z-u)\right]\end{aligned}$$

is negative defined. □

Remark 4.1. In particular, when $\psi(e) = Ke$ and there is no saturation in the input variable, the controller becomes

$$\dot{\xi} = -\Gamma\,|K\xi|\,e$$
$$u = \xi - Ke$$

wich can be seen as a modified PI controller with variable error integration time.

5. Study case

In order to test the hybrid representation (8) of a system described by (1)-(3) we consider an enzymatic reactor whose model is

$$\begin{aligned}\frac{\partial s}{\partial t} &= -\alpha u(t)\frac{\partial s}{\partial z} - \frac{\mu_m x_0 s}{K_m + s} \qquad \begin{array}{c}0 < s < L\\ t > 0\end{array} \\ s(z,0) &= s_0(z) \\ s(0,t) &= s_{in}(t)\end{aligned} \tag{16}$$

The analytic solution for (16) is

$$s(z,t)\exp\left[\frac{s(z,t)}{K_m}\right] = s_0(z - v(t))\exp\left[\frac{s_0(z-v) - \mu_m x_0 t}{K_m}\right] \qquad v(t) \le z$$
$$s(z,t)\exp\left[\frac{s(z,t)}{K_m}\right] = s_{in}(t-\tau)\exp\left[\frac{s_{in}(t-\tau) - \mu_m x_0 \tau}{K_m}\right] \qquad v(t) > z$$

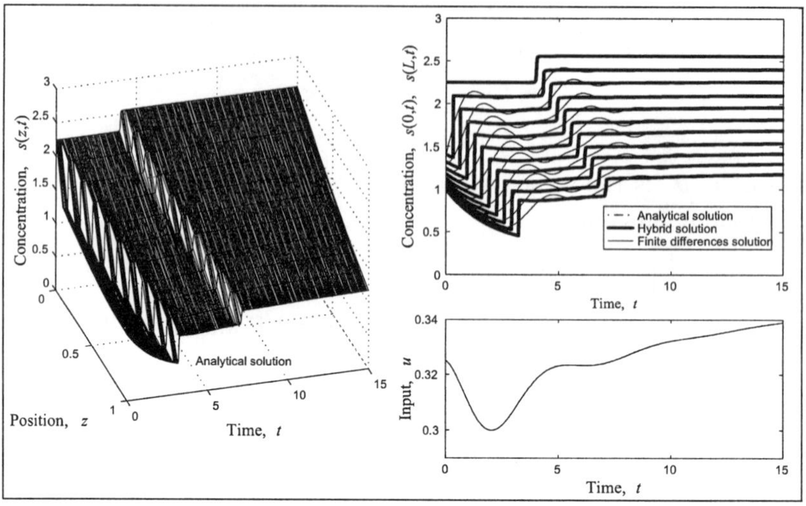

Figure 1. Comparison between analytic, hybrid and finite differences solutions (Parameters: $\mu_m x_0 = 0.8$, $K_m = 1.3$ and $\alpha = 1$).

where $v(t) = \alpha \int_0^t u(\lambda)\, d\lambda$, and τ can be obtained solving the equation $z = v(t) - v(t - \tau)$. The hybrid commutable representation can be obtained using (8) with $f(X) = -\mu_m x_0 X / (K_m + X)$, which has a well defined sign.

Figure 1 shows the comparison between analytic solution, hybrid commutable solution and the solution obtained using direct spatial discretization (using finite differences method with 10 discretization points). As can be seen there are no differences between the analytical and hybrid commutable solutions, in contrast, the finite difference solution exhibits a dynamic behavior quite different to those obtained by both the analytical and hybrid commutable solutions. Hence we can conclude that hybrid commutable representation produces an exact solution and does not increase the dimension of the model as the direct spatial discretization does.

A controller of the form (15) is tested, and Fig. 2 shows system response. It can be seen that controller tracks the reference despite input concentration disturbances and reference changes.

6. Conclusions

In this work, we present a hybrid commutable system with two ODE's subsystems which describes exactly, in a fixed spatial point, a spatially distributed convective system where the input is the flow rate. This hybrid re-

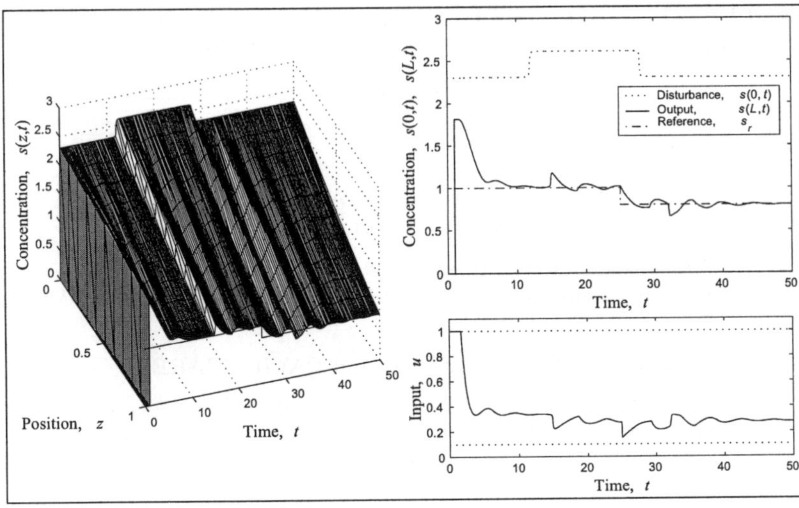

Figure 2. Controled system (Controller parameters: $\Gamma = 1.5$, $\psi(e) = Ke$ and $K = 0.6$).

presentation shows some advantages over the approximated model obtained by spatial discretization, such as the smaller dimension and the accuracy. Because the hybrid representation is composed by ODE's, the regulation problem of the output in a specific axial point for the spatially distributed convective system is easier to achieve using this hybrid representation and as a fist approach, we have developed a controller for unidimensional systems. As a future work, we pretend to extend the controller to n-dimensional systems and to use this methodology in order to obtain easier representations for others classes of systems described by partial differential equations, as well as to develop its suitable controllers.

References

1. W. H. Ray, *Advanced Process Control.* McGraw-Hill, New York, (1981).
2. B. E. Stangeland and A. S. Foss, *I & EC Res.* **9**, 38, (1970).
3. D. M. Ruthven and S. Sircar, *Proc. of 4th Int. Conf. on Found. of Comp.-Aided Procss. Design.* 29, (1996).
4. P. D. Christofides, *Nonlinear and Robust Control of PDE systems: Methods and Applications to Transport-Reaction Processes*; Birkhäuser, Boston, (2001).
5. H.-K. Rhee, R. Aris and N. R. Amundson, *First-Order Partial Differential Equations: vol. 1*; Dover, New York , (2000).
6. H. Shang, J. F. Frobes and M. Guay, *Ind. Eng. Chem. Res.* **43**, 2140 (2004).
7. J. Choi and K. S. Lee, *Ind. Eng. Chem. Res.* **44**, 1812 (2005).

A TEST-BASED METHODOLOGY FOR PARAMETER ESTIMATION FOR A PILOT PLANT DISTILLATION COLUMN

CARLOS ASTORGA, ADRIAN SANTIAGO, FRANCISCO LÓPEZ,
and VICTOR M. ALVARADO

Centro Nacional de Investigación y Desarrollo Tecnológico
Interior Internado Palmira S/N, Palmira 62050, Cuernavaca, Mor., Mexico

ALFREDO HERNÁNDEZ, and DAVID JUÁREZ

Centro de Investigación en Ingeniería y Ciencias Aplicadas, UAEM,
Av Universidad 1001 Col. Chamilpa, Cuernavaca 62209,
e-mail: djuarezr7@gmail.com, Tel 777+3297084

A methodology for parameter estimation of unmeasurable parameters in a distillation process is presented. In this process the basic mass and energy balances are known, thus most of the unknowns are related to hydrodynamic equations, and with heat transfer parameters (condenser heat transfer coefficient, Weir coefficient, vapor pressure drop coefficient, Murphree efficiency, ratio of liquid-vapor in plate). As a result, instead of a black-box or a white-box, the model can be perceived as a ***spotted-box***. This methodology consist of selective tests which are devised to isolate the effect of the unknown parameters. This is exemplified in a pilot distillation column of a two-input, two-output implementation for model predictive control.

1. Introduction

The attraction of models obtained from fundamental principles is that they are globally valid, therefore adequate for optimization and control which some times require extrapolation out of the range where the data were obtained to fit the model . Lee[2] has mentioned the requirements of an adequate model (1998): In nonlinear model predictive control a model should enable to predict accurately effects of both known and unknown changes on the system (output) behavior possibly in a closed-loop setting.

The distillation column is instrumented with temperature sensors, feed input measurement, and actuators for reboiler heat, preheating and condenser recycle. Table 1.1 shows the operating conditions for a methanol-

Table 1.1	Operating conditions
Property	Value
X_B	0.3339 mol
Q_B	100-2500 W
P_B	96.15 kPa
T_B	67.18 ^oC
V_B	1.6862 mol/min
B	0.8970 mol/min
Z_F	0.5002 mol
F	1.25 mol/min
Q_F	100-500 W
T_F	32.8 ^oC
Y_D	0.9233 mol
ΔP_{Tot}	3.63 kPa
T_D	63.02 ^oC
L_D	1.4106 mol/min
D	0.3527mol/min

ethanol distillation. A mathematical model was built to apply model predictive control of a two-input, two-output process, with the following assumptions:

- Liquid Vapor Equilibrium can be represented by Peng-Robinson-Stryjek-Vera equation of state (PRSV) [6] with 2 mixing parameters. This EOS was also used to approximate the liquid-vapor equilibrium, and the thermodynamic properties which are required in the mass, composition and energy balances.
- Each plate contains two phases which affect the hydraulic behavior [1]
- The heat supplied is completely absorbed by the fluid in the reboiler.
- The distillate leaves the condenser as saturated liquid.

The conservation equations are:

Mass balance

$$\frac{dM}{dt} = L_{p+1} + F_p - L_p - V_p + V_{p-1} \tag{1}$$

Composition balance in tray p

$$\frac{d(Mx)}{dt} = L_{p+1}x_{p+1} + F_p z_p - L_p x_p - V_p y_p + V_{p-1}y_{p-1} \tag{2}$$

Energy balance in tray p

$$\frac{d(M_p E_p)}{dt} = L_{p+1}E_{L,p+1} + F_p E_{F,p} - L_p E_{L,p} - V_p E_{V,p} + V_{p-1}E_{V,p-1} + Q_p \tag{3}$$

This assumptions and the associated balances produce a set of sparse algebraic-differential equations whose mathematical representation can be grouped as

$$\frac{d\mathbf{x}}{dt} = f\left(t, \mathbf{p}, \tau, \mathbf{x}, \mathbf{u}, \mathbf{v}\right)$$

$$y = g\left(\mathbf{p}, \tau, \mathbf{x}, \mathbf{u}, \mathbf{v}\right)$$

The equations describing this model are presented in appendix A. This model contains parameters which cannot evaluate directly, thus a methodology for parameter estimation is proposed.

2. Methodology

This methodology is similar to the scheme proposed for the control analysis [4]time domain analysis helps to evaluate steady state parameters and time constants, frequency domain analysis applied with set of different frequencies evaluate time delays in input-output. Some methods are complementary, for instance, while state analysis eliminates time-dependent parameters, frequency analysis [3]is more sensitive to time dependent parameters.

Also, through the design of experiments we can produce a behavior whose model has a reduced number of unknown parameters. Our strategy is to isolate as much as possible the effect of every parameter by designing selective tests obtaining the parameters via model inversion; thus instead of the orthogonality of the tests, the goal of experiments is to observe the isolated effect of uncorrelated parameters. Sensitivity analysis helps us to obtain information about the value or range of the parameters and the propagation of uncertainty to the output variables.

Finally, a global parameter estimation is used for the remaining parameters. The benefit of using these selective tests previous a global parameter estimation is that the problem dimension is reduced, or promising starting values for global parameter estimation are suggested. Some tests are redundant, but they help to verify the value of parameters.

2.1. *Steady State Analysis*

Test 1. Heat transfer in condenser. The condenser acts as an inherent regulator for the column. Since the actual column is made out of glass, vibrations should be avoided during operation (see Fig. 1). As a result this column operates as a double pass condenser with an excess of cooling water. Thus both temperatures (cooling water and condensing mixture) have only small changes.

Since input and output cooling water temperature are known, we can evaluate V_C from

$$Q_C = L_W\, Cp\,(T_I - T_O) = V_C\, \Delta H_{Cnd,C}$$

also,

$$Q_C = U\, A_h LM\Delta T$$

where at different condenser flows, U can be correlated as:

$$U \approx K_h\, L_W^{\gamma}$$

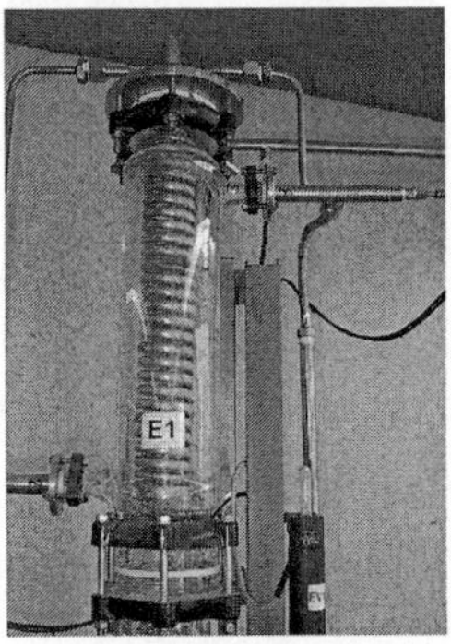

Figure 1. Condenser

Test 2. Heat losses along the column.

The column is not insulated, thus the glass walls are open to the environment (see Fig.2).

Selective Conditions $L_F = 0$; $B = L_n = 0$; $D = 0$

The main heat losses happen along the column body, since the condenser operates at atmospheric conditions, and the boiler transfer area is small (see next section),thus a global balance for all the trays produces:

$$L_C E_{L,Cond} + V_B E_{V,B} = V_C E_{V,Cond} + L_{B+1} E_{L,B} + Q_A$$

With the Q_A approximated by Newton's cooling law for the column body.

$$\frac{dQ_A}{dt} = \sum_{2}^{Np-1} K_{p,A} * (T_p - T_A)$$

Evaluating $K_{p,A}$ heat loses can be evaluated at different environmental conditions.

Figure 2. Condenser

Test 3. Reflow valve
Reflow valve is operated by a set on/off pulses with frequency variation
Selective Conditions $L_{B+1} \approx V_B$; $L_B = 0$
Assumption

$$V_B \approx V_C$$

Then in all the trays flows steam to maintain the operating pressure drop.

$$L_{B+1} = L_F + L_C$$
$$L_C = V_B - D$$
$$L_{B+1} = L_F + V_B - D$$

then $L_F \approx D$ With this equation the behavior of reflow valve can be correlated as: $L_C = f(\omega_R, M_C, T, P)$

2.2. *Time Domain analysis*

Time constant are static, or have a small dependence on geometry of the piece of equipment at given operating conditions. Some of this constants can be tuned by the following set of tests

Test 4. Time constants

Selective Conditions

$$D = 0$$

Assumptions The main elements for heat for heat transfer are condenser and boiler, as a result we have

Then we have

$$\frac{dM_B}{dt} = F_f + L_{B+1} - V_B - B$$

also

$$\begin{aligned}
L + D &= \frac{Q_C}{\Delta H_{Cnd,C}} \\
V_B &= \frac{Q_B}{\Delta H_{Vap,B}} \\
B &= \sqrt{\frac{\rho_B\, g M_B}{Kv_L\, A}} \\
\frac{dM_B}{dt} &= F_f + \frac{Q_C}{\Delta H_{Cnd,C}} - \frac{Q_B}{H_{Vap,B}} - \sqrt{\frac{\rho_B\, g M_B}{Kv\, A}} \\
\frac{dM_C}{dt} &= \frac{Q_B}{H_{Vap,B}} - \frac{Q_C}{\Delta H_{Cnd,C}}
\end{aligned}$$

Here we can analyze the effect of mass holdup and the liquid discharge valve.

As noticed earlier [7], distillation process behaves like a first order dynamics. From a linearized model it is possible to obtain the time constants $\tau \frac{dX}{dt} = -x$

Test 5. Reboiler heat holdup.

To speed the boiling response, the electric devise is inserted in a separate vessel with mass holdup M_b where smaller mixture is heated (see Fig.3).

Selective Conditions

$B = 0$ $F_f = 0$

Figure 3. Boiler

Given the small value of V_B compared with the mass of the container, M_B, and neglecting diffusion effects, the boiler dynamics is represented as:

$$M_b \frac{dH_b}{dt} = -V_B \, \Delta H_{Vap,B} + Q_B$$

As a result of this configuration, mass holdup and concentration have little change, but energy response is fast. The estimated values of M_b can be corroborated with the time constants of the linearized model.

Table 2.2 shows the fitted parameters

Table 2.2	**Estimated parameters**
Parameter	meaning
Q_{AB}	Heat loses
U	heat transfer coefficient
$K_{v,L}$	Liquid Discharge coefficient
$K_{v,V}$	Vapor Discharge coefficient
M_b	Mass holdup in the reboiler vessel

Other parameters like ϕ need to be correlated globally.

3. Conclusions

In building a model useful for predictive control attention was placed to the actual geometry of the distillation column The purpose of this work is to "find ways to make identification experiment efficient to generate data that are rich in the information" [2] relevant to the model used for predictive control. Some of these test can be inferred from the incidence matrix of $[\frac{dx}{dt}, \mathbf{y}] vs. \mathbf{p}$; but when an experimental test is devised, the feasibility of operation is guaranteed. With this selective test we aim to tune the parameters, then we could aim to obtain a "systematic certainty" [Balmes, *El Criterio*]

References

1. Benett D.L, R Agrawal, P.J. Cook (1983), *New Pressure Drop Correlation for Sieve Tray Distillation Columns*, AICHE J, V 29, No 3 pp434-442
2. Lee J. H. (1998) *Modeling and identificatin for nonlinear model predicitive control:Requirements, current status and future research needs*, in Nonlinear Model Predictive Control, Volume 26 of Progress in Systems and Control Theory Series, Birkhauser Verlag, Basel, Switzerland.
3. Lee H, Rivera D. E. (2005) *An Integrated Methodology for Plant-Friendly Input Signal Design and Control-Relevant Estimation of Highly Interactive Processes*, AIChE Meeting, (Cincinnati, OH)
4. Luyben W.L.(1990), *Modeling, Simulation and Control for Chemical Engineers*,McGraw Hill, USA.
5. Skogestad S. and Morari M. (1998), *Understanding the Dynamic Behaviour of Distillations Columns*Ind. Eng. Chem. Res. 27 pp 1848-1862.
6. Stryjek, R. Vera, J.H. (1986), *An Improved Peng-Robinson Equation of State ofr Pure Compounds and Mixtures*Can. J. Chem. Engn. 64, pp 334-340.
7. Wittgens and S. Skogestad (1995), *Evaluation of Dynamic Models of Distillation Columns with Emphasis on the Initial Response* DYCORD'95, Denmark.

Nomenclature

B = Boiler flow
C = Condenser
Cp = Heat capacity
E = Energy
F = Feed flow
H = height
K = equilibrium constant
L = Liquid flow
M = mass holdup
Mw = Molecular Weight
R = Reflow
t = time
u = vector of manipulated variable
V = Vapor flow
v = vessel
v = vector of measured disturbances
w = vector of unmeasured disturbances.
x = vector of states
y = vector of measured variables
Q = heat
x = liquid composition
y = vapor composition
z = feed composition

Greeks

Dep =departure
Δ = increment
Tot = Total
ρ = density
γ = fitting exponent
φ = holdup effectiveness
τ = time related parameters
η =Murphree plate efficiency
ω = frequency

Subscripts

A = Atmospheric conditions
a = algebraic
b = bubble point
B = bottom
d = differential
D = Distillate
Eff = effective
f = feed
h = Heat transfer
g = Gravity
i = component
I = input
L = Liquid
O = output
p = plate number
v = valve
V = vapor
W = Water
Wer = Weir

Superscripts

* = equilibrium
Ig = ideal gas

Appendix A. Model Equations

Equilibrium compositions are obtained by equating the fugacities. Thus for bubble point

$$f_{L,i}(T, P_b, x^*) = f_{V,i}(T, P_b, y^*)$$

$$\Phi(T, P, x) = P_b(\sum_{1}^{n} K_i x_i^* - 1)$$

$$E = E^{ig}(T, x*) - E^{dep}(T, P, x*)$$

Murphree plate efficiency

$$y_{p,i} = y *_{p,i} \eta_V + (1 - \eta_V) y*_{p-1,i}$$

Hydraulic equations for plates

$$L = \rho_L \frac{\sqrt{h_{eff}^3}}{K_{vL}}$$
$$h_{eff} = \max(0, h - h_{wer})$$
$$h = M/A\rho_L$$
$$V = \frac{\sqrt{\Delta P \rho_V M w_V}}{K_{vV}}$$
$$\Delta P = P_{p+1} + \varphi g h \rho_L - P_p$$

Condenser

$$Q_c = U\, A\; LM\Delta T$$
$$\Delta T_i = (T_D - T_i)$$
$$\Delta T_o = (T_D - T_o)$$
$$LM\Delta T = \frac{(\Delta T_i - \Delta T_o)}{\log(\Delta T_i / \Delta T_o)}$$
$$T_o = T_i - Q_c / W_{H2O} Cp_{H2O}$$

PERFORMANCE MONITORING OF HEAT EXCHANGERS USING ADAPTIVE OBSERVERS

C.-M. ASTORGA, A. SANTIAGO AND R.-M. MÉNDEZ

Centro Nacional de Investigación y Desarrollo Tecnológico
Interior Internado Palmira S/N, Palmira 62050, A. P. 5-164, Cuernavaca, Mor., Mexico
E-mail: astorga@cenidet.edu.mx

A. ZAVALA

Instituto Potosino de Investigación Científica y Tecnológica, Apartado Postal 2-66, Lomas 4a. Sección 78216, San Luis Potosí, S. L. P., Mexico
E-mail: azavala@ipicyt.edu.mx

In this paper, a method for monitoring the performance degradation in a heat exchanger is presented. This method is based on the use of an adaptive observer which estimates the overall heat transfer coefficient U. The monitoring of this parameter can be useful to decide when the heat exchanger needs preventive or corrective maintenance. A simplified mathematical model of the heat exchanger is used to synthesize the adaptive observer. The effectiveness of the proposed method is demonstrated via numerical simulations and through experimental results.

1. Introduction

One of the main problems of heat exchangers is the deterioration of the heat transfer surface due to the accumulation of a fouling film. This most often leads to increased energy consumption. In general, fouling is accepted as an unavoidable problem but many efforts are made to try to detect, mitigate and/or correct its occurrence[1,2]. This work is devoted to propose a way to detect performance degradation in a heat exchanger by means of an adaptive observer. Observers are used to estimate unknown parameters or unmeasured state variables from on-line and/or off-line measurements (see e.g. [3,4]). Much of the work done in the area of observer design has been based in the application of Kalman filters, extended Kalman filters (EKF)[5] or Luenberger observers. These observers are used only for state estimation. Nevertheless, it is often the case that some parameter values of the processes are physically unavailable for measurement. When such is the case, it is possible to use adaptive observers for their estimation. An adaptive observer is one in which both the parameters and state variables of the system are estimated simultaneously[7]. This work is devoted to propose a method based on an adaptive observer that can be used to track the overall heat transfer coefficient U of a countercurrent heat exchanger. The knowledge of this coefficient can be useful to determine when the equipment needs a preventive or corrective maintenance.

This paper is organized as follows. Sec. 2 presents the nomenclature and the mathematical model of a countercurrent heat exchanger. In Sec. 3, the problem of estimating the state for a class of nonlinear systems is considered. The observer synthesis for the heat exchanger is based on the mathematical model described in Sec. 2. Finally, concluding remarks are given in Sec. 4.

2. Nomenclature and mathematical model

The following nomenclature is defined for its use throughout this work:

T_{ci}, T_{hi}	inlet temperatures in the cold and hot side, ^oK
T_{co}, T_{ho}	outlet temperatures in the cold and the hot side, ^oK
U	heat transfer coefficient, $J/(m^2 \cdot {}^oK \cdot s)$
A	heat transfer surface area, m^2
c_{pc}, c_{ph}	specific heat in the cold and the hot side, $J/(kg \cdot {}^oK)$
ρ_c, ρ_h	density of the cold and the hot fluid, kg/m^3
V_c, V_h	volume in the cold and the hot side, m^3
v_c, v_h	flow rate in the cold and the hot side, m^3/s
$(\cdot)^T$	transpose of a matrix
$\|\cdot\|$	norm of a matrix
$(\hat{\cdot})$	estimated value of the variable in question

The mathematical model takes into account the following assumptions:

- A1 equal inflows and outflows, implying constant volume in both tubes
- A2 U is related to the temperatures of the fluids
- A3 there is no heat transfer with the surroundings
- A4 the thermophysical properties of the fluids are constant
- A5 there is no energy storage in the walls
- A6 the inlet temperatures are constant.

The dynamic system is obtained through an energy balance rule applied to every element of a lumped model [8]. Over a time interval Δt, the application of such an energy balance rule considering a single element per fluid (covering the whole tube length), gives rise to [9]:

$$\begin{aligned} \rho_c c_{pc} V_c \left[T_c|_{t+\Delta t} - T_c|_t \right] &= (\rho_c c_{pc} \Delta t)\, v_c T_{ci} - (\rho_c c_{pc} \Delta t)\, v_c T_{co} \\ &\quad + (UA\Delta t)\Delta T \\ \rho_h c_{ph} V_h \left[T_h|_{t+\Delta t} - T_h|_t \right] &= (\rho_h c_{ph} \Delta t)\, v_h T_{hi} - (\rho_h c_{ph} \Delta t)\, v_h T_{ho} \\ &\quad - (UA\Delta t)\Delta T \end{aligned} \tag{1}$$

T_c and T_h respectively represent the cold and hot fluid bulk (average) temperature. ΔT is the (mean) temperature difference among the fluids. Since the lumping procedure assumes that every element behaves like a perfectly stirred tank [10], the fluid temperature at each of such elements is generally considered to be uniformly distributed. As a consequence, T_{co} and T_{ho}, and the outlet temperature difference, $T_{ho} - T_{co}$, may be taken to respectively estimate T_c, T_h, and ΔT in (1), *i.e.* $T_c = T_{co}$, $T_h = T_{ho}$, and $\Delta T = T_{ho} - T_{co}$ [11]. However, Steiner[12] shows that the less inaccurate model is the one obtained by approaching ΔT in (1) through the *logarithmic mean temperature difference* (LMTD), typically expressed as

$$\Delta T = \Delta T_\ell \triangleq \frac{(T_{ho} - T_{ci}) - (T_{hi} - T_{co})}{\ln\left(\frac{T_{ho} - T_{ci}}{T_{hi} - T_{co}}\right)} \tag{2}$$

The model was further refined by Alsop and Edgar[13] through an additional consideration: the bulk temperatures in the accumulation terms (left-hand side of (1)) were taken to be the (arithmetic) average among the inlet and outlet temperatures, *i.e.* $T_c = (T_{co} + T_{ci})/2$ and $T_h = (T_{ho} + T_{hi})/2$, through which the transient response time is improved (under Assumption A6). Thus, taking into account all the above mentioned considerations, multiplying both sides of Eqs. (1) by $2/(\rho_c c_{pc} V_c \Delta t)$, respectively $2/(\rho_h c_{ph} V_h \Delta t)$, and letting $\Delta t \to 0$, the system takes the form:

$$\begin{cases} V_c \dot{T}_{co} = 2\left[v_c\,(T_{ci} - T_{co}) + UA\Delta T / c_{pc}\rho_c \right] \\ V_h \dot{T}_{ho} = 2\left[v_h\,(T_{hi} - T_{ho}) - UA\Delta T / c_{ph}\rho_h \right] \end{cases} \tag{3}$$

Let us further note that the LMTD expression in (2) reduces to an indeterminate form when $T_{ho} - T_{ci} = T_{hi} - T_{co}$, which poses a serious problem to model (3). Such an inconsistency is overcome by taking the LMTD as[14]

$$\Delta T = \Delta T_L \triangleq \begin{cases} \Delta T_\ell & \text{if } T_{ho} - T_{ci} \neq T_{hi} - T_{co} \\ \Delta T_0 & \text{if } T_{ho} - T_{ci} = T_{hi} - T_{co} = \Delta T_0 \end{cases} \tag{4}$$

3. Adaptive observer design

An adaptive observer is a recursive algorithm that is used to estimate the state of a system with unknown parameters. Besançon[15] proposed the unifying adaptive observer form (5) which emphasizes properties allowing some asymptotic state estimation in spite of unknown parameters:

$$\begin{cases} \dot{\mathbf{y}}(t) = \boldsymbol{\alpha}(\mathbf{y}(t), \mathbf{z}(t), \mathbf{u}(t)) + \boldsymbol{\beta}(\mathbf{y}(t), \mathbf{z}(t), \mathbf{u}(t))\boldsymbol{\theta}(t) \\ \dot{\mathbf{z}}(t) = \boldsymbol{\gamma}(\mathbf{y}(t), \mathbf{z}(t), \mathbf{u}(t)) \end{cases} \tag{5}$$

where $\mathbf{y}(t) \in \mathbb{R}^p$ is the output vector of the system (the measurable states), $\mathbf{z}(t) \in \mathbb{R}^q$ is the vector of the unmeasurable states, $\mathbf{u}(t) \in \mathbb{R}^m$ is the measurable bounded input vector and $\boldsymbol{\theta}(t) \in \mathbb{R}^r$ is a vector of unknown parameters. $\boldsymbol{\alpha}(\mathbf{y}(t), \mathbf{z}(t), \mathbf{u}(t))$ and $\boldsymbol{\beta}(\mathbf{y}(t), \mathbf{z}(t), \mathbf{u}(t))$ are two globally Lipschitz functions with respect to $\mathbf{z}$.

3.1. *The proposed adaptive observer*

An adaptive observer for system (5) is proposed as follows:

$$\begin{cases} \dot{\hat{\mathbf{y}}}(t) = \boldsymbol{\alpha}(\hat{\mathbf{y}}(t), \hat{\mathbf{z}}(t), \mathbf{u}(t)) + \boldsymbol{\beta}(\hat{\mathbf{y}}(t), \hat{\mathbf{z}}(t), \mathbf{u}(t))\hat{\boldsymbol{\theta}}(t) - k_y\left(\hat{\mathbf{y}}(t) - \mathbf{y}(t)\right) \\ \dot{\hat{\mathbf{z}}}(t) = \boldsymbol{\gamma}(\hat{\mathbf{y}}(t), \hat{\mathbf{z}}(t), \mathbf{u}(t)) \\ \dot{\hat{\boldsymbol{\theta}}}(t) = -k_\theta \boldsymbol{\beta}^T\left(\hat{\mathbf{y}}(t), \hat{\mathbf{z}}(t), \mathbf{u}(t)\right)\left(\hat{\mathbf{y}}(t) - \mathbf{y}(t)\right)^T \end{cases} \tag{6}$$

such that for any $\hat{\mathbf{y}}(0)$, $\hat{\mathbf{z}}(0)$, any $\mathbf{y}(0)$, $\mathbf{z}(0)$ and any measurable bounded $\mathbf{u}$, the estimation errors $\|\hat{\mathbf{y}}(t) - \mathbf{y}(t)\|$ and $\|\hat{\mathbf{z}}(t) - \mathbf{z}(t)\|$ asymptotically go to zero when $t \to \infty$, while $\|\hat{\boldsymbol{\theta}}(t) - \boldsymbol{\theta}(t)\|$ remains bounded. Also, if $\boldsymbol{\beta}^T(\mathbf{y}, \mathbf{z}, \mathbf{u}, t)$ is persistently exciting, and its time derivative is bounded, then $\|\hat{\boldsymbol{\theta}}(t) - \boldsymbol{\theta}(t)\| \underset{t\to\infty}{\longrightarrow} 0$. Constants $k_y > 0$, $k_\theta > 0$ are the observer gains. Observer (6) is a modified version of the observer proposed by Besançon[15]. In this case, the observer uses $\hat{\mathbf{y}}(t)$ instead of the measured vector $\mathbf{y}(t)$ in order to compute $\boldsymbol{\alpha}(\hat{\mathbf{y}}(t), \hat{\mathbf{z}}(t), \mathbf{u}(t))$, $\boldsymbol{\beta}(\hat{\mathbf{y}}(t), \hat{\mathbf{z}}(t), \mathbf{u}(t))$ and $\boldsymbol{\gamma}(\hat{\mathbf{y}}(t), \hat{\mathbf{z}}(t), \mathbf{u}(t))$. This is a more realistic case, because the initial-condition vector of the observer $\hat{\mathbf{y}}^0(t)$ can be different than the initially measured vector $\mathbf{y}^0(t)$.

3.2. *Application to a heat exchanger*

Consider the heat exchanger model (3). The system parameters A, c_{pc}, c_{ph}, ρ_c, ρ_h, V_h, and V_c are known and constant, according to Assumption A4. The heat transfer coefficient is time-varying, according to Assumption A2. Let $k_c = A/(c_{pc}\rho_c V_c)$ and $k_h = A/(c_{ph}\rho_h V_h)$. Other assumptions are:

A7 $v_c(t)$, $v_h(t)$ are the measured inputs: $u_1(t)$, $u_2(t)$ respectively
A8 $T_{co}(t)$, $T_{ho}(t)$ are the measured outputs: $y_1(t)$, $y_2(t)$ respectively

These measurements are the only ones needed to make use of the model (3), and they coincide with the measurements available in the ideal case for a single-cell model (this is the usual case in an industrial environment). Assumptions *A7* and *A8* lead to the following representation of the model:

$$\begin{pmatrix} \dot{y}_1(t) \\ \dot{y}_2(t) \end{pmatrix} = \begin{pmatrix} \frac{2u_1(t)}{V_c}(T_{ci} - y_1(t)) \\ \frac{2u_2(t)}{V_h}(T_{hi} - y_2(t)) \end{pmatrix} + \begin{pmatrix} 2k_c \Delta T_L(y_1(t), y_2(t)) \\ -2k_h \Delta T_L(y_1(t), y_2(t)) \end{pmatrix} U(t) \tag{7}$$

which has the same form of system (5) without nonmeasurable states $\mathbf{z}(t)$. Then, an adaptive observer of the form (6) for system (7) is given by:

$$\begin{aligned} \begin{pmatrix} \dot{\hat{y}}_1(t) \\ \dot{\hat{y}}_2(t) \end{pmatrix} = &\begin{pmatrix} \frac{2u_1(t)}{V_c}(T_{ci} - \hat{y}_1(t)) \\ \frac{2u_2(t)}{V_h}(T_{hi} - \hat{y}_2(t)) \end{pmatrix} + \begin{pmatrix} 2k_c \Delta T_L(\hat{y}_1(t), \hat{y}_2(t)) \\ -2k_h \Delta T_L(\hat{y}_1(t), \hat{y}_2(t)) \end{pmatrix} \hat{U}(t) \\ &- k_y \begin{pmatrix} \hat{y}_1(t) - y_1(t) \\ \hat{y}_2(t) - y_2(t) \end{pmatrix} \end{aligned} \tag{8}$$

$$\begin{aligned} \dot{\hat{U}}(t) = &-2k_\theta \left[k_c \Delta T_L(\hat{y}_1(t), \hat{y}_2(t)) \quad - k_h \Delta T_L(\hat{y}_1(t), \hat{y}_2(t))\right] \\ &\times \begin{pmatrix} \hat{y}_1(t) - y_1(t) \\ \hat{y}_2(t) - y_2(t) \end{pmatrix} \end{aligned} \tag{9}$$

3.3. *Numerical simulations*

Suspended solids present a major problem in most heat exchangers in applications such as pellet water coolers as well as catalyst slurry heaters and coolers. If the solids begin to settle on the heat transfer surface, an insulating layer is formed that reduces the heat transfer rate. In this simulation, $U(t)$ was supposed time-varying with the purpose of illustrating this kind of degradation. The simulation was carried out using $T_{ci} = 298\ ^oK$ and $T_{hi} = 338\ ^oK$. The constants in the simulated model were $A = 0.633\ m^2$, $\rho_h = \rho_c = 1000\ kg/m^3$, $V_c = 6.05 \times 10^{-3}\ m^3$, $V_h = 3.2 \times 10^{-3}\ m^3$, $c_{pc} = 1910\ J/(^oK \cdot kg)$, $c_{ph} = 1590\ J/(^oK \cdot kg)$.
For the sake of simplicity, the variation of $U(t)$ occurs in a period of time of about 2.33 hours, from 160 $J/(m^2 \cdot {}^oK \cdot s)$ (for all $t < 40\ min$) to 120 $J/(m^2 \cdot {}^oK \cdot s)$ (for all $t > 100\ min$). Two kinds of variations were taken into account: (*i*) a ramp representing a slow degradation of $U(t)$

and (ii) a step representing an abrupt degradation of $U(t)$. These variations are shown in Fig. 1. The initial conditions of the process were: $T_{co}^0 = 306.82\ ^oK$, $T_{ho}^0 = 325.38\ ^oK$. The initial conditions of the observer were quite different from the initial conditions of the process: $\hat{T}_{co}^0 = 273\ ^oK$, $\hat{T}_{ho}^0 = 350\ ^oK$ and $\hat{U}^0 = 140\ J/(m^2 \cdot {}^oK \cdot s)$. The observer gains were tuned at $k_\theta = 50$ and $k_y = 0.5$.
Fig. 1 shows the simulation result of the estimation of $U(t)$. The estimation time is of about 10 min. It can be seen that once the observer converges, it tracks well $U(t)$ in spite of the time-varying nature of this parameter.

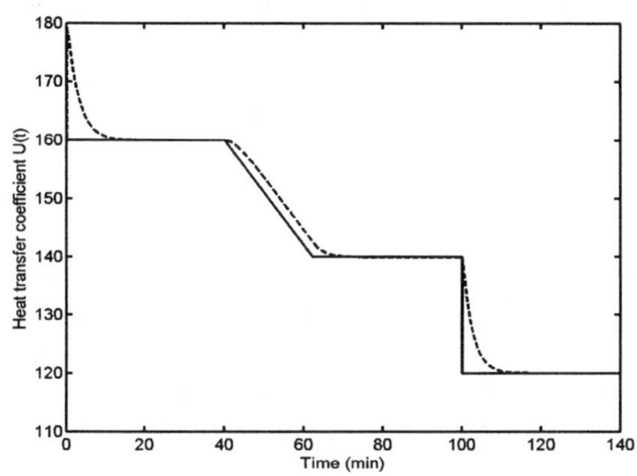

Figure 1. The simulated $U(t)$ (solid line) and its estimated value (dashed line).

3.4. *Experimental results*

In what follows, the case of a water-cooling process is presented. In this case, the hot water flows through the tube and the cooling water flows in the shell. In this experiment the inlet temperatures were kept constant at $T_{ci} = 301.5\ ^oK$ and $T_{hi} = 343.1\ ^oK$. The volumetric flowrates were $v_h = 1.333 \times 10^{-5}\ m^3/s$, and v_c was time varying between 0.6667×10^{-6} and 0.75×10^{-6} as shown in Fig. 2. T_{co} and T_{ho} were used off-line by an adaptive observer in order to estimate $U(t)$. The adaptive observer was implemented in MATLAB/Simulink®. The integration method to solve the observer equations was the Euler method. The sampling period was $t_s = 1\ min$. The constants and physical data used for the internal model of the observer were $A = 14 \times 10^{-3}\ m^2$, $\rho_h = 983.3\ kg/m^3$, $\rho_c = 991.8\ kg/m^3$,

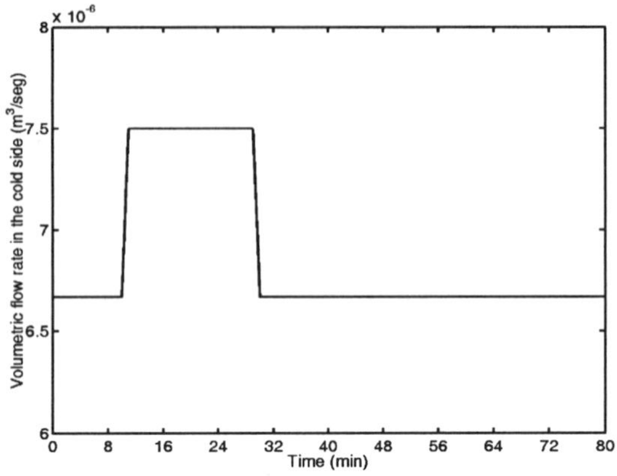

Figure 2. Variation of v_C.

$V_c = 135 \times 10^{-6}\ m^3$, $V_h = 15.51 \times 10^{-6}\ m^3$, $c_{pc} = 4174\ J/({}^oK \cdot kg)$, $c_{ph} = 4179\ J/({}^oK \cdot kg)$. The gains of the observer were tuned at $k_y = 15$ and $k_\theta = 1 \times 10^4$. The estimation of $U(t)$ is illustrated in Fig. 3. It can be seen that this parameter does not vary as a function of time, but is rather reflective of changes in temperatures. The final estimated value of U was $1049.4\ J/(m^2 \cdot {}^oK \cdot s)$. This is an acceptable value and corresponds well to the heat exchanger used for this experiment.

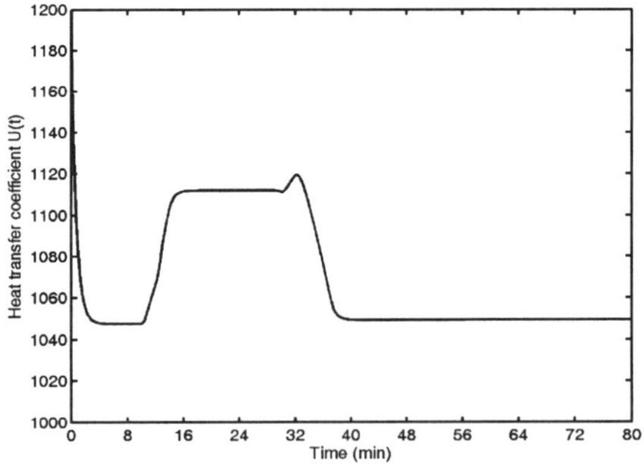

Figure 3. The estimated $\hat{U}(t)$

4. Conclusions

One of the main features of the adaptive observer developed in this work lies in the fact that its implementation and calibration is simple. On the other hand, it must be remarked that the proposed observer requires only limited knowledge on how the system behaves, and does not assume anything about the unknown dynamics of $U(t)$.

The convergence of this observer was demonstrated by means of a numerical simulation and then by means of a real-time experiment, the good agreement between the estimated states and the measured values allows us to conclude that this observer can be used to detect performance degradation on heat exchangers and then to schedule preventive or corrective maintenance.

References

1. S. Isogai and M. Nakamura, *Proc. of the 2003 ECI Conf. on Heat Exchanger Fouling and Cleaning: Fundamentals and Applications*, 221 (2003).
2. H. M. Joshi and G. Brons, *Proc. of the 2003 ECI Conf. on Heat Exchanger Fouling and Cleaning: Fundamentals and Applications*, 219 (2003).
3. C. M. Astorga, N. Othman, S. Othman, H. Hammouri and T.F. McKenna, *Control Engineering Practice*, **10**, 3 (2002).
4. M. Nadri, H. Hammouri and C. Astorga, *European Journal of Control*, **10**, 252 (2004).
5. Ph. Bogaerts, *Bioprocess Engineering*, **20**, 249 (1999).
6. G. Bastin and M. Gevers, *IEEE Transactions on Automatic Control*, **33**, 650 (1988).
7. R. Marino and P. Tomei, *Nonlinear Control Design. Geometric, Adaptive and Robust*, Prentice Hall, (1995).
8. M. H. R. Fazlur-Rahman and R. Devanathan, *Proceedings of the Third IEEE Conference on Control Applications*, **3**, 1801 (1994).
9. G. Stephanopoulos, *Chemical Process Control: An Introduction to Theory and Practice*, Prentice Hall, (1984).
10. E. I. Varga, K. M. Hangos and F. Szigeti, *Control Engineering Practice*, **3**, 1409 (1995).
11. E. Weyer, G. Szederkényi and K. Hangos, *Control Engineering Practice*, **8**, 121 (2000).
12. M. Steiner, *Proceedings of the International Symposium on District Heat Simulations*, (1989).
13. A. W. Alsop and T. F. Edgar, *Chemical Engineering Communications*, **75**, 155 (1989).
14. A. Zavala-Río, R. Femat and R. Santiesteban-Cos, *Rev. Mex. Ing. Quím.*, **4**, 201, (2005). Available at http://www.iq.itc.mx/rmiq/rmiq_contents.htm
15. G. Besançon, *Systems & Control Letters*, **41**, 271 (2000).

VERIFICATION OF NEUROFUZZY SPEED CONTROL TUNING FOR A COMBUSTION TURBOGENERATOR

LUIS CASTELO-CUEVAS RAUL GARDUNO-RAMIREZ

Electrical Research Institute, Division of Control Systems GCI 29-1.
113 Reforma Ave., Cuernavaca, Morelos, 62490, Mexico

It has been shown that a 2 degrees-of-freedom PI controller can improve the response of a turbogenerator to speed reference changes and operation disturbances at any single point of operation during startup. The 2 DOF neurofuzzy PI controller (PI-NF2DF) may spread the benefits throughout the operating space, from ignition speed to synchronization speed. To that aim, the PI-NF2DF is first designed as a linear controller equivalent to the existing PI controller so that substitution can be carried out without altering the current turbogenerator startup response. Once in the loop, the PI-NF2DF may be manually tuned to produce a higher performance non-linear controller. The tuning procedure directly modifies the input-output mappings of the neurofuzzy systems based on the operator experience and observations, improving the controller performance step by step. The suitability of the knowledge-based tuning is assessed by comparison to automatic tuning based on a numerical optimization program. The advantage of knowledge-based tuning is that it can be applied on-site and on-line for an actual turbogenerator, whereas the numerical optimization tuning procedure can only be used off-line in computer simulations.

1. Introduction

Due to the need to generate high quality electric power, reliability requirements to put combustion turbogenerators (CTGs) in service, whenever needed without any faults, have notably increased. Successful startup, synchronization, loading and stopping, strongly depends on the control system features. At startup, the main duty of the control system is that of accelerating the CTG from turning gear speed up to synchronization speed according to a predefined acceleration pattern. The speed controller calculates control actions to accelerate safely, avoiding the occurrence of stall, surge, high vibration, resonance, high temperature and combustion instabilities, in the shortest time, saving fuel and preserving the CTG duty life.

Current CTG control systems have single-input-single-output control loops based on conventional PI or PID control algorithms [1]. The ability of these controllers for regulation at any single point of operation has been widely demonstrated. Nevertheless, their use for speed control during startup or power control throughout the generation range is debatable, since the nonlinear CTG dynamics change with the point of operation. In addition, conventional controls cannot provide an optimal response to more than one control objective.

The 2 degrees of freedom PI controller (PI-2DF) was proposed to optimize both, the response to changes in the speed reference as well as to disturbances produced by normal operation events during turbogenerator startup [2]. Although the PI-2DF has more resources than the conventional PI controller to provide higher quality control signals, it still has the disadvantage of being valid in the neighborhood of the operating point at which it was tuned.

To spread the PI-2DF benefits to all the operating points at startup, the controller gains were exchanged by wide-range mappings based on neurofuzzy systems, which yield the PI-NF2DF controller. The PI-NF2DF consists of a neurofuzzy P feedforward control circuit and a neurofuzzy PI feedback control circuit, which may be independently tuned to improve both, the speed reference tracking response and the disturbance rejection response throughout the startup operating range. The PI-NF2DF is first designed as a linear controller equivalent to the existing PI controller so that substitution can be carried out without altering the current CTG startup response. Once in the loop, the PI-NF2DF neurofuzzy input-output mappings may be arbitrarily modified to improve the controller performance.

This paper reports progress about tuning the PI-NF2DF controller based on knowledge acquired by the operator to modify the neurofuzzy input-output mappings to get a non-linear controller with superior performance. The suitability of knowledge-based tuning is assessed by comparison to automatic tuning based on a numerical optimization program. Section 2 describes the conventional PI speed control strategy of a CTG and the discrete-time PI-2DF speed controller. Section 3 introduces the PI-NF2DF controller. Then, Section 4 introduces the knowledge-based procedure for on-line and on-site modification of controller mappings, as well as the automatic tuning procedure. Section 5 presents the results of simulation experiments that assess the suitability of manual tuning. Finally, Section 6 draws paper conclusions.

2. Conventional and 2DF Turbogenerator Speed Control

2.1. *Conventional Speed Control Strategy*

A typical CTG consists of five major components that operate continuously and simultaneously to produce electric power (Figure 1). The starting device can be an electric motor that provides energy to initially move the CTG, the compressor takes in atmospheric air, compresses it and sends it to the combustion chamber. In the chamber, pressurized air is mixed with fuel and burned to produce the hot flue gas that is delivered to the turbine moving blades through expansion nozzels. The exhausted flue gas is released to the atmosphere and the rotatory mechanical energy is transmitted to the electric generator, which transforms it into electric energy that is delivered to the power grid.

Essentially, the control system scheme of a typical CTG contains two control circuits: the inlet guide vane position control circuit to regulate air flow and a dual speed and power control circuit to regulate fuel flow. In the former circuit, the flue gas temperature, compressor discharge pressure and turbine speed are permanently monitored to set security levels to the fuel valve demand signal to ensure CTG physical integrity (Figure 2). At startup, the control system activates the closed loop speed control from ignition through synchronization, then the closed loop power control is activated. These control loops are normally based on conventional PI or PID algorithms.

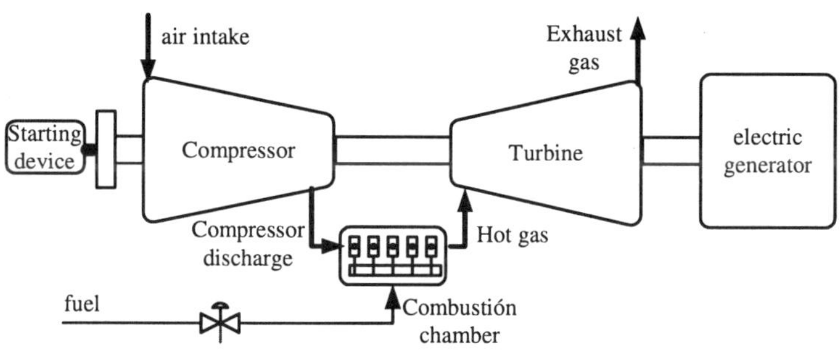

Figure 1. Major components of a typical combustion turbogenerator.

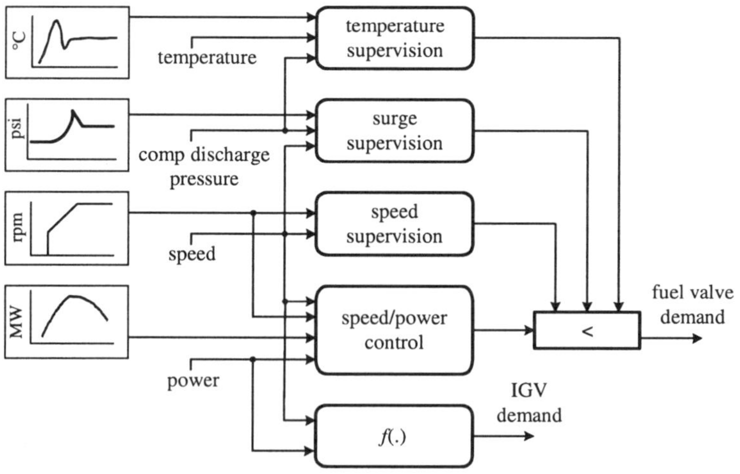

Figure 2. Speed/power control scheme.

2.2. *Discrete-time 2DF Speed Control*

The PI-2DF controller has a structure with 2 degrees-of-freedom (Figure 3), where $R(s)$ is the speed reference, $Y(s)$ is the speed measurement, $E(s)$ is the speed error, $C_{fb}(s)$ is the feedback controller, $U_{fb}(s)$ is the feedback control signal, $C_{ff}(s)$ is the feedforward controller, $U_{ff}(s)$ is the feedforward control signal, and $U(s)$ is the total control signal.

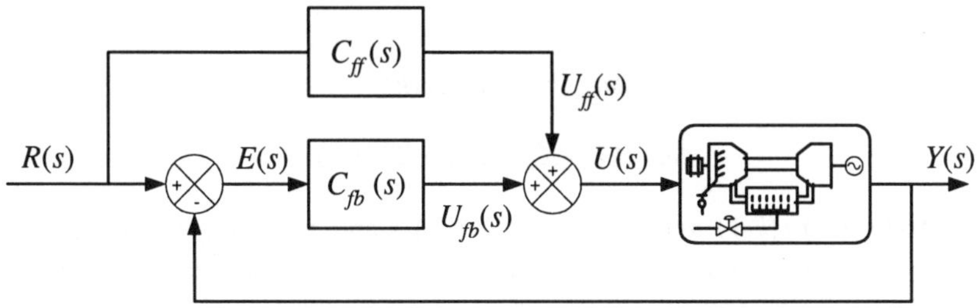

Figure 3. PI-2DF controller.

In the PI-2DF controller, the feedforward controller $C_{ff}(s)$ solves the speed reference tracking problem and the feedback controller $C_{fb}(s)$ regulates speed and compensates for disturbances. The feedback controller $C_{fb}(s)$ is a PI controller:

$$C_{fb}(s) = \frac{U_{fb}(s)}{E(s)} = K_{fb} + \frac{K_i}{s} \tag{1}$$

where K_{fb} is the proportional gain, K_i is the integral gain, $E(s)$ is the error signal and $U_{fb}(s)$ is the corresponding feedback control action. On the other hand, the feedforward controller $C_{ff}(s)$ is a P controller:

$$C_{ff}(s) = \frac{U_{ff}(s)}{R(s)} = K_{ff} \tag{2}$$

where K_{ff} is the proportional gain and $U_{ff}(s)$ is the feedforward control action. The final control action generated by PI-2DF controller is given by:

$$U(s) = U_{ff}(s) + U_{fb}(s) \tag{3}$$

Realization of the PI-2DF as a digital controller requires discrete-time versions of (1) and (2) that can be obtained, for a sampling period T and discrete time index k, using the following approximations:

$$t \approx kT \tag{4}$$

$$\int_0^t e(t)dt \approx T\sum_{j=1}^{k} e(j) \tag{5}$$

and taking into account that:

$$\Delta u_{ff}(k) = u_{ff}(k) - u_{ff}(k-1) \tag{6}$$

$$\Delta u_{fb}(k) = u_{fb}(k) - u_{fb}(k-1) \tag{7}$$

yields:

$$\Delta u_{ff}(k) = K_{ff}\Delta r(k) \tag{8}$$

$$\Delta u_{fb}(k) = K_p\Delta e(k) + K_i Te(k) \tag{9}$$

where $\Delta r(k)=r(k)-r(k-1)$ and $\Delta e(k)=e(k)-e(k-1)$ stand for changes in the speed reference and error signals, respectively. The discrete-time version of the PI-2DF controller is shown in Figure 4, where z^{-1} is a sampling period time delay.

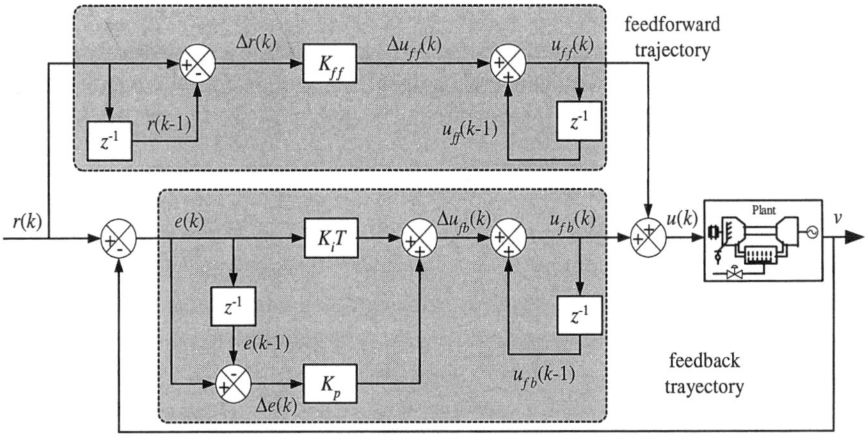

Figure 4. Discrete-time PI-2DF controller for a CTG.

3. Neurofuzzy 2 Degrees-of-freedom PI Controller

Although the two-degrees-of-freedom structure allows solution of the tracking and regulation control problems, the K_{ff}, K_i and K_p fixed gains in the PI-2DF controller are good for a single point of operation. Nevertheless, the advantages of the PI-2DF control strategy may be extended throughout the CTG range of operation by replacing these gains with wide-range mappings, as shown in Figure 5. In this work, such mappings are to be realized by means of one-input-one-output neurofuzzy systems to be able to design any arbitrary relationships between inputs and outputs, as required at different points of operation during startup of CTG (Figure 6).

Neurofuzzy systems allow design of fuzzy systems using the automatic learning methods of neural networks. Initially, the neurofuzzy systems are trained to reproduce the input-output relationship defined by linear mappings with slopes equal to the K_{ff}, K_i and K_p gains. This approach allows placement of the PI-2DF controller with neurofuzzy systems in the speed control loop without disturbing the current CTG startup response. Figures 7, 8 and 9 define the input triangular and output singleton membership functions of the neurofuzzy systems for the case with seven partitions of the operating space. Likewise, Table 1 lists the IF-THEN rules of the three neurofuzzy systems. Once in the loop, a few parameters of the neurofuzzy systems may be changed to improve CTG response, as will be explained shortly.

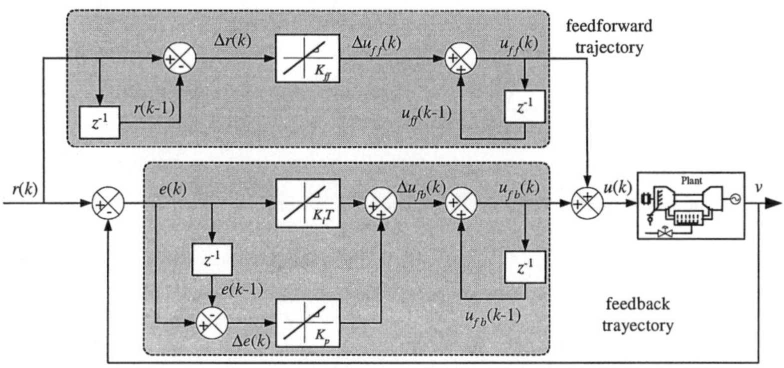

Figure 5. Discrete-time PI-2DF controller with wide-range mappings.

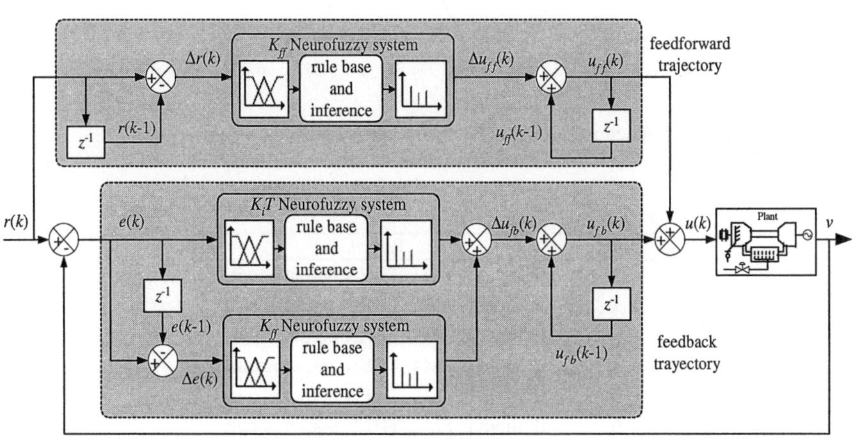

Figure 6. Neurofuzzy systems in the PI-NF2DF controller.

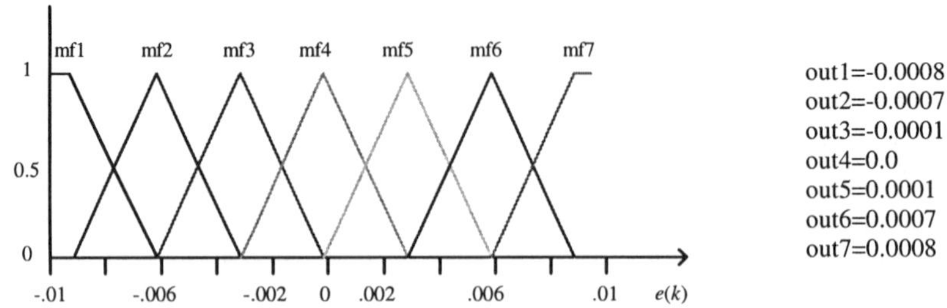

Figure 7. Membership functions for linear K_i neurofuzzy system.

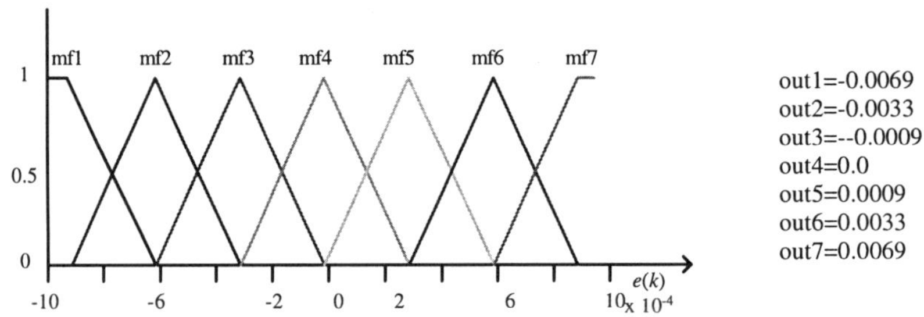

Figure 8. Membership functions for linear K_p neurofuzzy system.

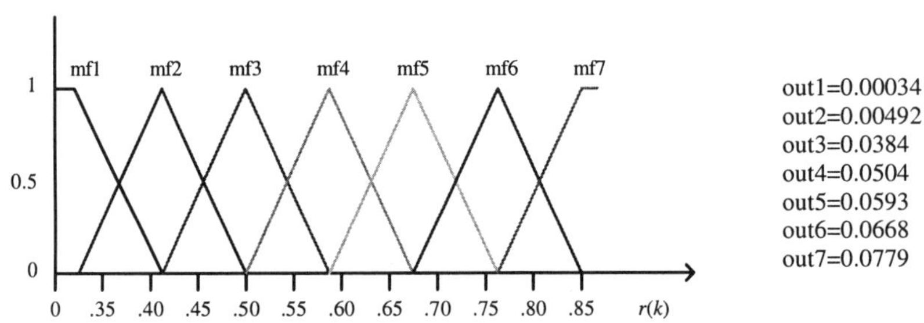

Figure 9. Membership functions for linear K_{ff} neurofuzzy system.

Table 1. Rule base of neurofuzzy system

Num	Rule
1	IF input is FM1, THEN output is out1
2	IF input is FM2, THEN output is out2
3	IF input is FM3, THEN output is out3
4	IF input is FM4, THEN output is out4
5	IF input is FM5, THEN output is out5
6	IF input is FM6, THEN output is out6
7	IF input is FM7, THEN output is out7

4. Tuning of PI-NF2DF Controller

4.1. *Knowledge-Based Manual Tuning*

Neurofuzzy systems are known to be universal approximators. The input-output mapping can be directly modified by changing the output singletons. In this work, changes are made in the K_{ff} neurofuzzy system to improve speed tracking response and in K_p and K_i neurofuzzy systems to improve disturbance rejection to major operation events. Changes are made at partitions (rules) where the speed response needs improvement. The amount of change is determined by the operator based on his experience and performance of current CTG startup response. To this aim, in this work the control effort (CE) and IAE indexes were found to be helpful. Figure 10 shows the tuning procedure followed to improve performance of PI-NF2DF control over PI control. As an example, Figure 11 shows the K_i mapping before (linear) and after (nonlinear) tuning.

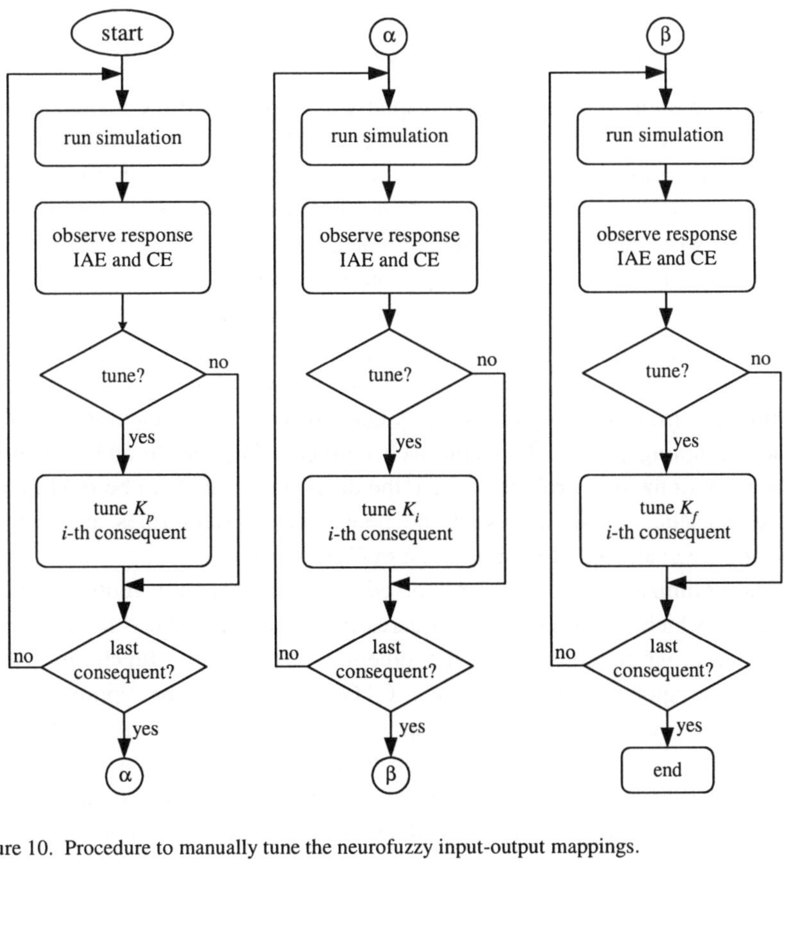

Figure 10. Procedure to manually tune the neurofuzzy input-output mappings.

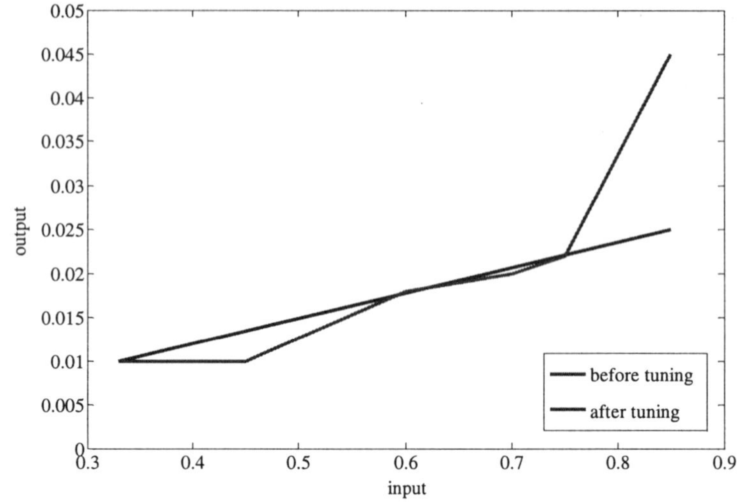

Figure 11. Input-output mapping of *Ki* neurofuzzy system, before and after tuning.

4.2. *Automatic Tuning of PI-NF2DF Controller*

Performance improvement of CTG response at startup with the manually-tuned nonlinear PI-NF2DF controller may provide excellent results. However, there is always uncertainty about how good the improvement is, or in other words, how good the knowledge-based tuning could be. To provide an answer, tuning is also carried out with a numerical optimization program.

Tuning of the neurofuzzy input-output mappings is formulated as a constrained multiobjective optimization problem, where the goal is for the CTG to track the speed reference as close as possible from ignition speed through synchronization speed. Hence, the speed tracking error has to be minimized for all sampling instants throughout startup. One objective (cost function) is defined at each sampling period. This approach defines a large-scale and computing-intensive optimization problem, where the decision variables to be modified are all the consequent parameters of the neurofuzzy system rules and the cost functions are evaluated after simulation of CTG startup.

The optimization problem was solved using the lsqnonlin routine available in Matlab, which performs a least-squares fit to minimize speed tracking error. The main program consists of the tunepi.m and fcost.m functions. Tunepi declares the global variables of the CTG model, sets the options for the optimization routine and calls the lsqnonlin routine to perform multiobjective optimization. Fcost gets the parameters calculated by the optimization routine, reads and writes to the neurofuzzy data structures, performs simulation of CTG startup and returns the vector of absolute values of speed tracking errors to the optimization function. The lsqnonlin function performs optimization with constraints in the decision variables by setting upper and lower limits to each one of them; this guarantees mappings with positive slope and direct action PI-NF2DF controllers. Stopping the optimization search is carried out setting up a tolerance variable to a given value.

5. Simulation Experiments and Results

Feasibility demonstration of PI-NF2DF controller is carried out by means of simulation experiments with the mathematical model of a 24 MW CTG in a graphical simulation environment in a personal computer [5]. Experiments consist in performing CTG startup simulations with both the conventional PI and the PI-NF2DF controllers in the already presented discrete-time versions. Speed tracking performance is evaluated with the IAE and CE indexes:

$$IAE = \int_0^{\infty} \left| e(t) \right| \, dt \tag{10}$$

$$EC = \int_0^{\infty} u^2(t) \, dt \tag{11}$$

Figure 12 shows CTG startup with PI control tuned at K_p=3.5 y K_i=0.7. Five operation events at startup are considered. Point 1 corresponds to turning on the starting device. Point 2 indicates the beginning of the acceleration ramp or speed reference pattern at 1920 rpm with a change in set point and activation of the speed closed loop control. Point 3 indicates turning off the starting device. This event is seen as an external disturbance by the speed control loop. Point 4 at 4920 rpm corresponds to closing the bleeding valves and opening the inlet guide vanes. At this point all air mass flow from the compressor enters the combustion chamber and the turbine causing a sudden decay in temperature and speed increase, such that overspeed may occur. The control system has to bring the speed back to the speed reference value. Point 5 at 5100 rpm is known as synchronization speed with a big change in the speed reference slope. Here, the speed control loop has to manage the large CTG inertia to keep speed constant in preparation for synchronization to the power grid.

Figure 13 shows the CTG startup with both the PI and nonlinear PI-NF2DF controllers. After tuning, performance with the PI control reported IAE=2968.8 and with the PI-ND2GL IAE=2136. These results essentially demonstrate that performance may be significantly enhanced using the knowledge-based manual tuning procedure. Results also show that manual tuning yields a performance close to that obtained with the numerical optimization tuning. Figure 14 shows the control signals provided by the same controllers. The CE indexes are very close one to another, which means that the improved performance with the PI-NF2DF controller does not necessarily implies larger control effort.

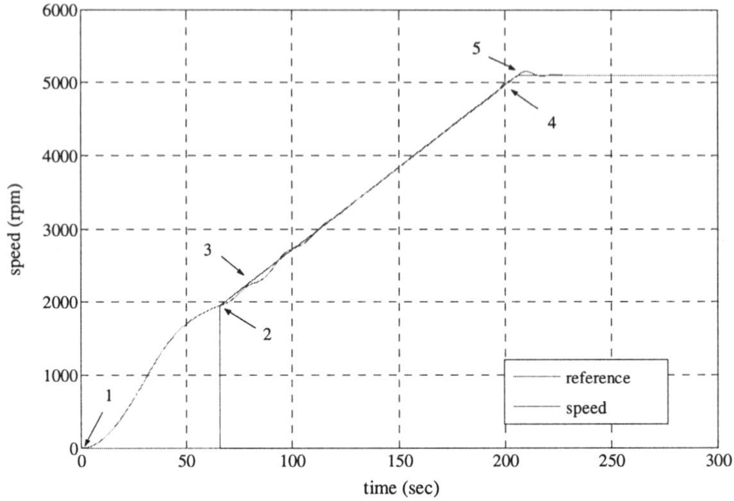

Figure 12. CTG startup with conventional PI control.

6. Conclusions

This paper introduced a manual tuning procedure for the PI-NF2DF speed control for a CTG startup. The tuning procedure modifies the neurofuzzy system input-output mappings by changing the rule consequents based on operator experience and knowledge. The obtained nonlinear PI-NF2DF controller may outperform the conventional PI control where required. The proposed tuning procedure may be used on-site in actual CTG, in which case fewer iterations would be required.

Acknowledgements

Authors acknowledge Dr. Salvador Gonzalez and Mr. Rafael Chavez from the Electrical Research Institute for their support to research activities.

References

1. Woodward Governor Co., Manual 26144B. (2002).
2. L. Castelo, R. Garduño and E. Quintero, IEEE 17th. RVP-AI/2004. GEN-13 (2004).
3. L. Castelo and R. Garduño, IEEE 18th. RVP-AI/2005. CSA-22 (2005).
4. L. Castelo, M.Sc. Thesis. Cenidet. (2004).

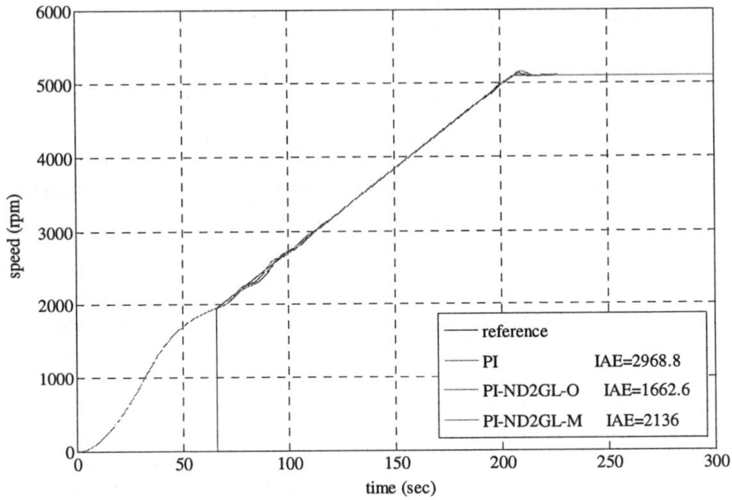

Fig 13. CTG startup with PI and PI-NF2DF controller.

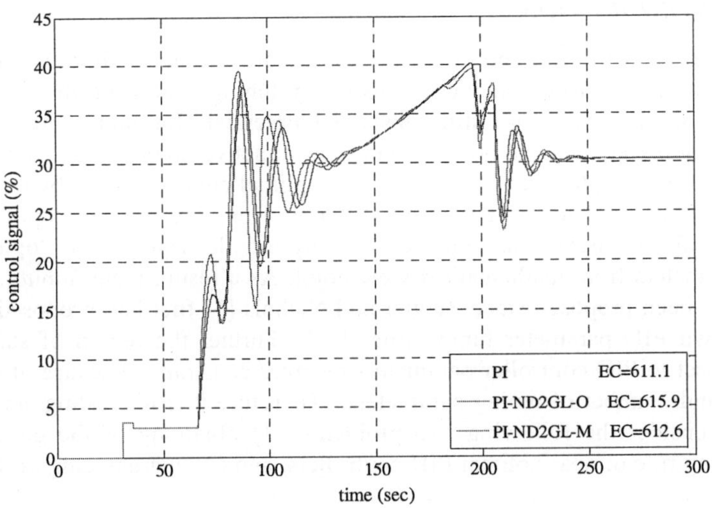

Fig 14. CTG startup with PI and PI-NF2DF controller.

STABLE PID CONTROLLER DESIGN VIA PARALLEL FEEDFORWARD COMPENSATOR

ZENTA IWAI, IKURO MIZUMOTO,YUICHI NAKASHIMA[†]

Department of Mechanical Systems Engineering, Kumamoto University, 2-39-1,Kurokami, Kumamoto, Japan 860-8555

MASANORI TAKAHASHI

Department of Mechanical Systems Engineering, Kyushu Tokai University, 9-1-1 ,Toroku, Kumamoto, Japan 862-8652

In this paper, a new design method of PID controller is proposed. The method utilizes the so- called almost strict positive realness (ASPR) of the plant so that the stability of the PID control system is guaranteed by use of K-Y lemma and Lyapunov's stability theorem. An application of the proposed basic design concept to a practical design of tracking PID control system is also discussed. The result is applied to the design of non ASPR PID control system by introducing the parallel feedforward compensator. The effectiveness and robustness of the proposed method is examined through simulations by comparing it with other conventional PID controller parameter tuning rules based on the first order with time delay process model.

1. *INTRODUCTION*

Most of instrument and process engineers are familiar with PID control. There is a well-established practice of installing, tuning, and using the controllers. In spite of this, there are still substantial potentials for improving PID control among researchers and practitioners. The attractive point of PID control is that it offers the simple means to realize near optimal control system by tuning only 3 controller parameters. However, different from the ordinary optimal control algorithms, there exists no exact way of determining 3 "optimal" PID parameters from mathematical view point. As a result, many tuning algorithms have been proposed since Zeigler and Nichols [1] first had proposed their well known PID parameter tuning rules [1,2]. Further the region of stability with respect to PID controller parameters receives constraints because of the luck of enough number of tuning parameters. Therefore it is interesting and important to consider the following two problems: (1) obtaining of the general simple tuning rule of near optimal PID controllers, and (2) guaranteeing of the stability

[†] Work partially supported by grant 17560398 of the Japan Society for the Promotion of Science

for the control system with 3 adjustable PID controller parameters. In fact, tuning and stabilizing of PID controllers have over the years been objects for a great amount of research. It is noted that tuned PID controller does not necessarily behave to cover the stability of any type of controlled plant. As to the stability analysis and synthesis of PID control system, it was shown that Hermite-Biehler Theorem can be utilized not only to derive conditions for the existence of the set of stabilizing controller parameters but also as a convenient analytical method to design compensators[3,4]. However these conditions do not contain the clear information concerning the improvement of the control performance at present stage.

In this paper, a new approach is proposed concerning the design of PID control system. This approach utilizes the special process characteristics called ASPRness (almost strict positive realness) [5,6]. Here it is shown that the control system can be stable by a PID feedback controller if the controlled plant is ASPR. The specific features of this approach are as follows. It always gives stable PID control system so that it does not need to consider the constraints for stability region as to PID controller parameters. The obtained control system includes the internal model coping with the tracking of reference input and the elimination of disturbance, and the so called parallel feedforward compensator (PFC) which realizes the ASPRness of the augmented control system[6,7,8]. Here it is given a very simple design procedure of PFC compared to the known method. The effectiveness of the proposed method is examined and compared with conventional PID tuning methods by using the benchmark test example shown in the reference[11].

2. *BASIC CONCEPT OF STABLE PID CONTROLLER FOR ASPR PLANTS*

Let us consider the n-th order controllable and observable SISO plant:

$$\begin{aligned} \dot{x} &= Ax + bu \\ y &= c^T x \end{aligned} \tag{2.1}$$

We assume that eq. (2.1) satisfies the following assumption.

[*Assumption 1*]

Eq. (2.1) is ASPR. That is, there exist positive definite matrices $P = P^T > 0$ and $Q = Q^T > 0$ such that

$$(A - k_p^* bc^T)^T P + P(A - k_p^* bc^T) = -Q \ , \ Pb = c \tag{2.2}$$

for all $k_p^* \geq {}^{\exists}k_{10} > 0$ [6].

Then we have the following lemma.

[*Lemma 1*]

Suppose that the assumption 1 holds. Then plant (2.1) can be stabilized by the following PID controller:

$$u^* = -k_P^* y - k_D^* \dot{y} - k_i^* w \ , \ \dot{w} = y \tag{2.3}$$

where

$$k_p^* \geq {}^\exists k_{10} > 0, k_D^* > 0, k_i^* > 0 \tag{2.4}$$

(Proof of Lemma 1)

Substituting eq. (2.3) into eq. (2.1) leads to the following equation:

$$\begin{aligned} \dot{x} &= A^* x - k_D^* b\dot{y} - k_i^* bw \\ y &= c^T x \end{aligned} \tag{2.5}$$

where

$$A^* = A - k_p^* bc^T$$

Let

$$V = x^T Px + k_p^* y^2 + k_i^* w^2 \tag{2.6}$$

be the candidate of Lyapunov function. Then, from the assumption 1, we can evaluate its derivative along the trajectory of eq. (2.5) as follows.

$$\dot{V} \leq -x^T Qx \leq 0 \tag{2.7}$$

It means $\lim_{t\to\infty} x(t) = 0$. From this, we can conclude $\lim_{t\to\infty} y(t) = 0$ and the boundedness of $w(t)$. Q.E.D.

3. *APPLICATION TO THE DESIGN OF STABLE TRACKING PID CONTROL SYSTEMS*

In this section, the basic design concept of stable adaptive PID Controller is applied to the design of stable adaptive tracking PID Control system. Let us consider the following n-th SISO order plant:

$$\begin{aligned} \dot{x}(t) &= Ax(t) + bu(t) + b_1 d(t) \\ y(t) &= c^T x(t) \end{aligned} \tag{3.1}$$

where $d(t)$ denotes the input disturbance. The problem to be considered here is a construction of stable adaptive tracking PID control system which can achieve the tracking of the output $y(t)$ to the reference input $r(t)$. Suppose that $r(t)$ satisfies the following differential equation which is known as the internal model:

$$D(p)r(t) = 0 \ , \ D(p) = p^p + d_1 p^{p-1} + \cdots + d_p \tag{3.2}$$

where " p " denotes the differential operator. Define $z(t), v(t)$ and the tracking error $e(t)$ as follows.

$$z(t) = D(p)x(t) \ , \ v(t) = D(p)u(t) \ , \ e(t) = y(t) - r(t) \tag{3.3}$$

Then, operating $D(p)$ from both sides of eq.(3.3) and taking into consideration eqs.(3.1) and (3.2) lead to the following equation:

$$D(p)e(t) = c^T z(t) \tag{3.4}$$

Further we assume that the disturbance $d(t)$ also satisfies the following disturbance model:

$$D(p)d(t) = 0 \tag{3.5}$$

From eqs.(3.1)-(3.5), we can obtain the following equation:

$$\begin{aligned} \frac{d}{dt}\bar{x}(t) &= \bar{A}\bar{x}(t) + \bar{b}v(t) \\ \bar{y}(t) &= \bar{c}^T\bar{x}(t) \end{aligned} \tag{3.6}$$

where

$$\bar{x} = \begin{bmatrix} z \\ e \\ \vdots \\ e^{(p-1)} \end{bmatrix}, \bar{A} = \begin{bmatrix} A & 0 & \cdots & 0 \\ 0^T & 1 & \ddots & \vdots \\ \vdots & \ddots & \ddots & 0 \\ 0^T & \cdots & 0 & 1 \\ c^T & -d_p & \cdots & -d_1 \end{bmatrix}, \bar{b} = \begin{bmatrix} b \\ 0 \\ \vdots \\ 0 \end{bmatrix}, \bar{c} = \begin{bmatrix} 0 \\ 1 \\ 0 \\ \vdots \\ 0 \end{bmatrix} \qquad \bar{y} = e(t)$$

The input and output of this system are $v(t)$ and $\bar{y}(t)$, respectively. However

$$\bar{c}^T\bar{b} = 0 \tag{3.7}$$

It means that the relative degree of the system is equal or greater that 2 so that the system (3.6) is not ASPR. To improve the situation, we introduce the following $n_f - th$ order parallel feedforward compensator (PFC)[7,8].

$$\begin{aligned} \frac{d}{dt}x_f(t) &= A_f x_f(t) + b_f v(t) \\ y_f(t) &= c_f^T x(t)\ ,\ c_f b_f > 0 \end{aligned} \tag{3.8}$$

By combining eq. (3.6) and eq. (3.8), the following extended system is obtained.

$$\begin{aligned} \frac{d}{dt}x_a(t) &= A_a x_a(t) + b_a v(t) \\ y_a(t) &= \bar{y}(t) + y_f(t) = c_a^T x(t) \end{aligned} \tag{3.9}$$

where

$$x_a = \begin{bmatrix} \bar{x} \\ x_f \end{bmatrix}, A_a = \begin{bmatrix} \bar{A} & 0 \\ 0 & A_f \end{bmatrix}, b_a = \begin{bmatrix} \bar{b} \\ b_f \end{bmatrix}, c_a = \begin{bmatrix} \bar{c} \\ c_f \end{bmatrix}$$

[*Assumption 2*]

Extended system (3.9) is ASPR. That is, eq. (3.9) satisfies the assumption 1. In other words, PFC (3.8) can be designed so as to the resultant extended system (3.9) becomes ASPR. Then we have the following theorem.

[*Theorem 1*]

Assume that the assumption 2 holds. Then, the following PID controller:

$$v = -k_p y_a - k_d \dot{y}_a - k_i w \ , \ \dot{w} = y_a \tag{3.10}$$

stabilizes the closed loop system.

(Proof of theorem 1) It is apparent from the proof of theorem 1. (Q.E.D.)

It is noted that in this case $\lim_{t\to\infty} x_a(t) = 0$ holds. This relation includes $\lim_{t\to\infty} e(t) = 0$. It means that the output tracking to reference input $r(t)$ is attained. A schematic block diagram of the control system is shown in Fig.1.

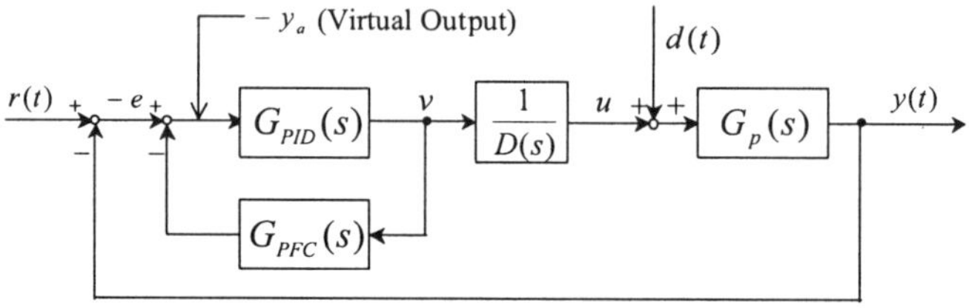

Figure.1 Schematic Diagram of PID Control System with PFC.

4. *A CONCRETE DESIGN SCHEME OF PFC*

It is necessary to construct a PFC which realizes the ASPRness of the extended system (3.9) to design the above stated tracking PID control system. The plant expressed by the transfer function $G(s)$ is called as ASPR if $G(s)$ satisfies the following conditions[6]:

[*ASPR Condition*]

(1) The relative degree of the plant is 0 or 1.
(2) The leading coefficient is positive.
(3) Plant is minimal phase.

Let

$$G_a(s) = c_a{}^T (sI - A_a)^{-1} b_a = \frac{1}{D(s)} G_p(s) + G_f(s) \tag{4.1}$$

$$G_p(s) = c^T (sI - A)^{-1} b, G_f(s) = c_f{}^T (sI - A_f)^{-1} b_f \tag{4.2}$$

If the plant $G_p(s)$ is minimal phase, then we have known a general procedure concerning the construction of a PFC $G_f(s)$ which makes extended plant $G_a(s)$ be ASPR[8]. This procedure is well known in the design of the simple adaptive control systems. However it becomes rather complicated when the relative degree is greater than 2. Here a new simple construction method of PFC is proposed. Let

$$G_p(s)=\frac{N_p(s)}{D_p(s)},G_p^*(s)=\frac{N_p^*(s)}{D_p^*(s)} \tag{4.3}$$

be the transfer function of the plant and its nominal known plant model, respectively. It is noted that there always exist some differences between real plant and its nominal plant model. Now let us assume the following assumption.

[*Assumption 3*]

(1) $D_p(s), D_p^*(s)$ are stable polynomial.

(2) Let

$$\begin{aligned} T(s) &= D_p^*(s)N_p(s)-D_p(s)N_p^*(s) \\ &= d_m s^m+\cdots+d_0 \end{aligned} \tag{4.4}$$

where m is an appropriate integer. Then there exists a small positive constant δ such that

$$|d_i|\le\delta, i=0,\cdots,m. \tag{4.5}$$

(3) $D(s)$ is a stable polynomial.

Then the following theorem holds.

[*Theorem 2*]

Let

$$G_f(s)=G_{PFC}(s)=G_{AS}(s)-\frac{1}{D(s)}G_P^*(s) \tag{4.6}$$

be a PFC of the control system where

$$G_{AS}(s)=\frac{N_{AS}(s)}{D_{AS}(s)} \tag{4.7}$$

is an ASPR transfer function which is given by the designers. Further we assume the assumption 3 holds. Then the extended system $G_a(s)$ becomes ASPR.

(Proof of Theorem 2)

$$\begin{aligned} G_a(s) &= \frac{1}{D(s)}G_p(s)+G_{PFC}(s) \\ &= G_{AS}(s)+\frac{1}{D(s)}\{G_p(s)-G_p^*(s)\} \\ &= \frac{N_{AS}(s)D(s)D_P(s)D_p^*(s)+D_{AS}(s)\{N_P(s)D_P^*(s)-D_P(s)N_P^*(s)\}}{D_{AS}(s)D(s)D_P(s)D_p^*(s)} \\ &= \frac{N_{AS}(s)D(s)D_P(s)D_p^*(s)+D_{AS}(s)T(s)}{D_{AS}(s)D(s)D_P(s)D_p^*(s)} \end{aligned} \tag{4.8}$$

Since $G_{AS}(s)$ is ASPR, $N_{AS}(s)$ is Hurwitz polynomial and its relative degree does not exceed 1. Further the plant and its nominal models are proper or strictly proper. Thus the relative degree of $G_a(s)$ is 0 or 1. From the assumption 3(2), the characteristic roots of the numerator of (4.7) are dominated by the characteristic roots of

$$N_{AS}(s)D(s)D_P(s)D_P^*(s) = 0$$

for sufficiently small δ because of the continuity of the roots of algebraic equation concerning coefficients. Thus they remain in the left half plane according to the assumption 3(1 and 3). Hence $G_a(s)$ is ASPR. Q.E.D.

It is noted that the proposed method is very easy to give ASPRness to the controlled system. However it requires more knowledge and constraints concerning the nominal model and the stability of the plant and internal model beforehand compared to the usual method.

5. *CONSIDERATION CONCERNING STEADY STATE OF THE CONTROL SYSTEM*

Let us consider the PID control system shown in Figure.1, where

$$C(s) = G_{PFC}(s) = \frac{C_1(s)}{s}, \; C_1(s) = k_D s^2 + k_P s + k_I \tag{5.1}$$

is a PID controller.

[*Theorem 3*]

Suppose that the assumption 2 holds and PFC has the form shown in eq.(4.6). Further suppose that the reference input and disturbance input are step inputs and

$$G_{PFC}(0) = 0\,. \tag{5.2}$$

in eq.(4.6). Then, we have

$$\lim_{t\to\infty}(y(t) - r(t)) = \lim_{t\to\infty} e(t) = 0\,. \tag{5.3}$$

(Proof of theorem 3)

Define $e(s) = r(s) - y(s)$.

Then, from Figure.1, we have

$$e(s) = \frac{A_2(s)}{A_1(s)} r(s) - \frac{A_2(s)G_P(s)}{A_1(s)} d(s) \tag{5.4}$$

where

$$A_1(s) = sD(s) + D(s)C_1(s)G_{PFC}(s) + C_1(s)G_P(s)$$
$$A_2(s) = sD(s) + D(s)C_1(s)G_{PFC}(s)$$

We assume that $r(t)$ and $d(t)$ are unit step inputs. Then

$$\lim_{t\to\infty} e(t) = \lim_{s\to 0} se(s)$$
$$= \frac{D(0)C_1(0)G_{PFC}(0)}{D(0)C_1(0)G_{PFC}(0)+C_1(0)G_P(0)}$$
$$+\frac{D(0)C_1(0)G_{PFC}(0)G_P(0)}{D(0)C_1(0)G_{PFC}(0)+C_1(0)G_P(0)} \tag{5.5}$$
$$=0$$

holds from eq.(5.2). Q.E.D.

It is also apparent from the process of above stated proof that the tracking error vanishes
as far as $r(t)$ and $d(t)$ satisfy eqs.(3.2) and (3.5).

6. *Example*

Consider the fourth-order process with a wide range of time constants [11].

$$G_P(s) = \frac{K}{(T_1 s+1)(T_2 s+1)(T_3 s+1)(T_4 s+1)} \tag{6.1}$$
$$K=1, T_1=1, T_2=0.2, T_3=0.04, T_4=0.008$$

This plant can be approximated such as

$$\widetilde{G}_P(s) = \frac{Ke^{-Ls}}{(Ts+1)}, T=1, K=1, L=0.248 \tag{6.2}$$

By using first-order Pade approximation, we have the following approximated model of eq.(6.1)[10].

$$G_P{}^*(s) = \frac{K^*}{(T^* s+1)(L^* s+!)}, T^*=1, K^*=1, L^*=0.248 \tag{6.3}$$

The purpose of the simulation is to examine the effect of proposed method by using the step reference input and the step disturbance. Hence we can set $D(s)=1$ in this case.

(1) Reference input and disturbance input:

$$r(t)=1.0, 0\le t, d(t)=1.0, 5\le t \tag{6.4}$$

(2) PFC:

$$G_{PFC}(s) = G_{AS}(s) - G_P{}^*(s)$$
$$G_{AS}(s) = \frac{K_{AS}}{T_{AS}s+1}, T_{AS}=1, K_{AS}=1 \tag{6.5}$$

(3) Matching condition (5.2):

$$K_{AS} = K^* \tag{6.6}$$

(4) PID Controller parameters:

$$v = -k_p y_a - k_d \dot{y}_a - k_i w \ , \ \dot{w} = y_a$$
$$k_P = 15, k_I = 21, k_D = 3 \tag{6.7}$$

Figure.2 shows an output response for step reference input. 100% change of disturbance is added after 5 sec. Its affect gradually decreases to zero. The result is compared to the results based on other well known PID parameter tuning rules: Zeigler-Nichols(Z-N), Chien, Hrones and Reswick(CHR) and Internal Model Control(IMC). Tuned values of PID parameters are selected based on the approximated model (6.2) [3(IMC, CHR),11(Z-N)] and these values are shown in Table 1 and the results are shown in Figure.3. Further, to examine the robustness of the algorithm, we give parameter value changes to plant parameters : T_2, T_3, T_4 in eq.(6.1) as shown in Table 2. These parameter variations correspond to the change of the ratio τ between the apparent time delay and apparent time constant from 0.248 to 0.6 approximately in the approximated model (6.2). We use the same PID controller parameters which are already calculated in the case of $\tau = 0.248$ [3,11]. The simulation results are shown in Figures.4-6, respectively.

Throughout these simulations, it was shown that the proposed method gave a good response as far as the plant model is known. Further, the robustness against the plant parameter variation and the addition of the disturbance were kept equal or much better than other conventional method

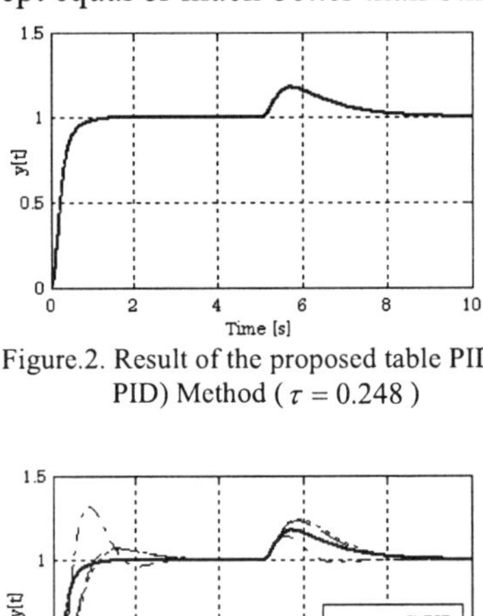

Figure.2. Result of the proposed table PID(S-PID) Method ($\tau = 0.248$)

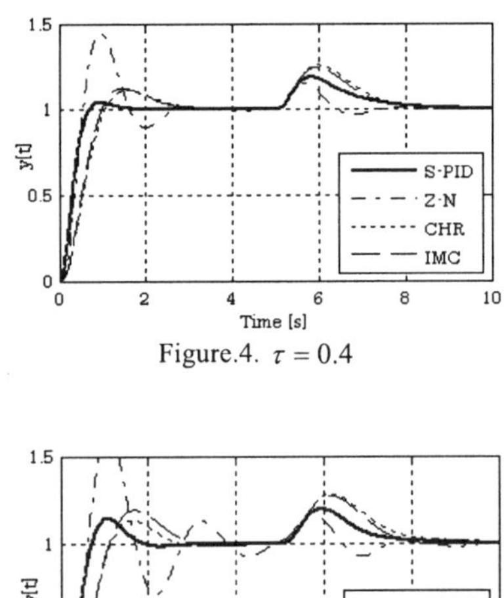

Figure.4. $\tau = 0.4$

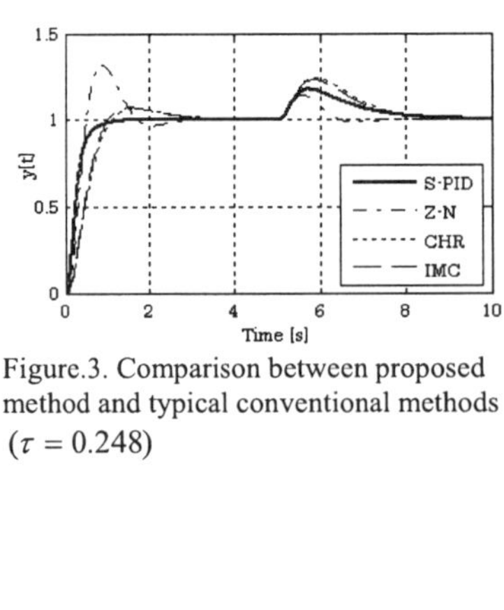

Figure.3. Comparison between proposed method and typical conventional methods ($\tau = 0.248$)

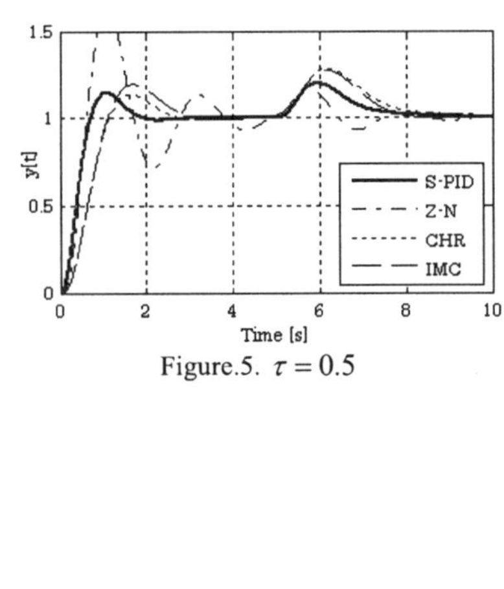

Figure.5. $\tau = 0.5$

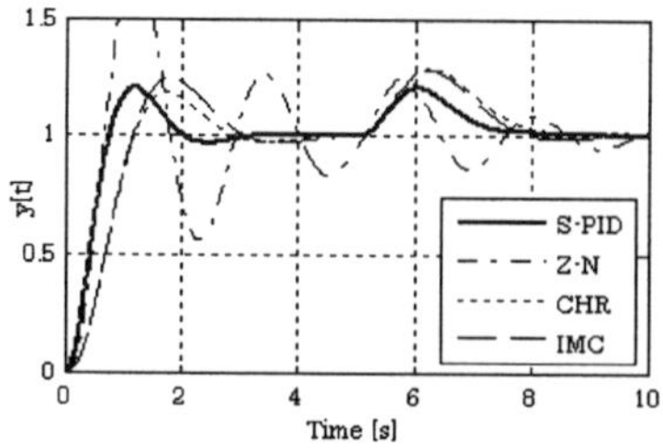

Figure.6. $\tau = 0.6$

Table 1. PID parameters in conventional method

	k_P	k_I	k_D
Z-N	4.84	9.86	0.61
CHR	2.42	2.42	0.30
IMC	2.65	2.36	0.29

Table 2. Plant parameter variations in the simulation.

	T_2	T_3	T_4
Fig.4	0.28	0.056	0.011
Fig.5	0.40	0.080	0.016
Fig.6	0.48	0.096	0.019

7. *Conclusion*

A new design method of stable PID control system utilizing the ASPRness of the plant was proposed. It is also shown that if the plant does not have ASPR characteristics, then PID control system can be realized by using the parallel feedforward compensator (PFC). A very simple and practical design method of PFC was also derived for stable plant and the offset zero design condition for set point change and step disturbance were given concerning the design of PFC.

References

1. J.G.Ziegler and N.B.Nichols, *Trans.ASME*. **64**, 759 (1942).
2. K. J.Astrom and T. Hagglund, *PID Control, Theory, Design and Tuning,* ISA, USA, second edition (1995).
3. G. J.Silva, A. Datta and S. P. Bhattacharyya, *PID Controllers for Time-Delay Systems,* Birkhauser, USA (2005).
4. A.Roy and K. Iqbal. *Proc. of IFAC World Congress*. Tu-M21-TO/6,Praha (2005).
5. I.Barkana , J. of Guidance. Control and Dynamics. **Vol.28**. No.4, 631 (2005).
6. H.Kaufman, I. Bar-Kana and K. Sobel *Direct Adaptive Control Algorithms, Theory and Applications.* Springer-Verlag, USA (1994).
7. Z.Iwai and I. Mizumoto, *Int. J. Control*, **Vol.59**. No.6, 1543 (1994).
8. Z.Iwai, I.Mizumoto and M.Deng, *Proc. of 33rd IEEE CDC*, 2827 (1994).
9. H.Ohtsuka, Z.Iwai and I.Mizumoto, *Proc. of IFAC World Congress*.Th-MO4-TO/5, Praha (2005).
10. S. L.Shah, Z. Iwai, I. Mizumoto and M. Deng, *J.of Process Control*. **Vol.7**. No.6, 439 (1997).
11. D.E.Seborg, T.F. Edgar and D.A. Mellichamp, *Process Dynamics and Control (2nd Edition)*. John Wiley & Sons,Inc (2004).
12. Astrom,K, H. Panagopoulos and T. Hagglund, *Automatica*. **34**, 571 (1998).

THE PARTIAL LINEARIZATION METHOD FOR TRACKING THE TIME-VARIANT REFERENCE FUNCTION

TOMOHIRO HENMI

Department of Electro-Mechanical Engineering
Takamatsu National College of Technology
355 Chokushicho, Takamatsu, Kagawa 761-8058, Japan
E-mail: henmi@takamatsu-nct.ac.jp

MINGCONG DENG and AKIRA INOUE

Department of Systems Engineering
Faculty of Engineering
Okayama University
3-1-1 Tsushima-naka, Okayama 700-8530, Japan
E-mail: {deng, inoue}@suri.sys.okayama-u.ac.jp

Feedback linearization methods are well known techniques for the control of nonlinear systems. Among them, the input-output feedback linearization method is input-output map is linearized by changing variables of the systems and using nonlinear feedback constructed by structure theories of the systems. A partial linearization method, which sets arbitrary state variable of systems as output of systems and sub-systems concerned in this state are partially linearized by input-output feedback linearization method, is widely applied to the control of underactuated systems. In general method, the linearized subsystems can track a constant reference value by a *PD* controller which is included in a nonlinear feedback. However, the *PD* controller is not guaranteed for tracking control of the time-variant reference function. In this paper, the new partial linearization method that guarantees that the linearized sub-systems track the time-variant reference function is proposed. The proposed method is based on a Lyapunov theorem and back-stepping control method. Numerical simulations are given to show the effectiveness of the proposed method.

Keywords: Partial linearization; Back-stepping control; Tracking control; Nonlinear system

1. Introduction

Control methods based on the Lyapunov theorem are well known for the control of nonlinear systems, and many methods have been proposed. Also, the feedback linearization methods based on the structure theory of systems are well known techniques for the control of nonlinear systems.[1–4] The feedback linearization methods have two types, the full-state linearization, where the state equation can be completely linearized,[3] and the input-output map is linearized while the state equation may be only partially linearized. Simplicity of this method in addition to the advantage of the ability to use linear control theory on the linearized system has fostered its application in many systems. On the other hand, a partial linearization method is one of the nonlinear control methods that sets an arbitrary state variable of systems as output of systems and sub-systems concerned in this state are partially linearized by input-output feedback linearization method. It is widely applied to the control of underactuated systems.[3,5–7]

The underactuated system means a system that controls all degree of freedom using less actuators in number than the number of generalized coordinates. Thus, the design method of controllers of underactuated systems gives some reductions of numbers of necessary actuators, of the cost and of the weight of systems, hence this control system design problem is an interesting research topic in industry. Especially, in the space industry, the control of underactuated systems is applicable to not only realizing reduction of numbers of necessary actuators, of the cost and of the weight of the space robot, but also dealing with the emergency, e.g., in case of controlling the space robot with the failure of partial actuators.

So far, using partial linearization method, the linearized subsystems can track a constant reference value by a *PD* controller which is included in a nonlinear feedback. However, the *PD* controller is not guaranteed to track the time-variant reference function. Then in this paper, a new partial linearization method that guarantees the linearized sub-systems to track the time-variant reference function is proposed. The proposed method is designed based on Lyapunov theorem and back-stepping control method. In order to show the proposed method, the numerical simulation is performed using the cat-type inverted pendulum, which is one of the typical examples of underactuated systems.

In Section 2, a problem discussed in this paper is defined. The purpose of this problem is for tracking control to the time-variant reference function using the new partial linearization control method.

In Section 3, the new partial linearization method designed based on a Lyapunov theorem and back-stepping control method is proposed. The proposed method guarantees the linearized sub-systems to track the time-variant reference function.

In Section 4, in order to show the effectiveness of the proposed method, numerical simulations using cart-type single inverted pendulum are shown.

2. Problem setup

In this paper, a tracking control problem of the following single-input-single-output n-dimensional affine system is discussed.

$$\dot{x} = f(x) + g(x)u \tag{1}$$

$$y = h(x) \tag{2}$$

where f, g, and h are sufficiently smooth in a domain $D \subset R^n$.The purpose of this problem is tracking control of y to the time-variant reference function $r_1(t)$, which is assumed continuous and n-th differentiable function.

First, the system (1),(2) is linearized by the conventional partial linearization method.

If the relative degree of system is $k(1 \leq k \leq n)$ and using the following nonlinear feedback,

$$u = (L_g L_f^{k-1} h(x))^{-1}(-L_f^k h(x) + v) \tag{3}$$

Then the input-output map reduces to

$$y^{(k)} = v \tag{4}$$

A change of variables to transform (1),(2) is considered.

$$z = \begin{bmatrix} z_1 \\ z_2 \\ \vdots \\ z_k \\ --- \\ z_{k+1} \\ \vdots \\ z_n \end{bmatrix} = \Phi(x) = \begin{bmatrix} h \\ L_f h \\ \vdots \\ L_f^{k-1} h \\ --- \\ w_{k+1}(x) \\ \vdots \\ w_n(x) \end{bmatrix} \tag{5}$$

Using (3) and (5), the system (1),(2) are transformed to

$$\begin{bmatrix} \dot{z}_1 \\ \dot{z}_2 \\ \vdots \\ \dot{z}_{k-1} \\ \dot{z}_k \\ -- \\ \dot{z}_{k+1} \\ \vdots \\ \dot{z}_n \end{bmatrix} = \begin{bmatrix} \begin{bmatrix} 0 & 1 & 0 & \cdots & 0 \\ 0 & 0 & 1 & \cdots & 0 \\ \vdots & \vdots & \vdots & \ddots & \vdots \\ 0 & 0 & 0 & \cdots & 1 \\ 0 & 0 & 0 & \cdots & 0 \end{bmatrix} \begin{bmatrix} z_1 \\ z_2 \\ \vdots \\ z_{k-1} \\ z_k \end{bmatrix} \\ -------- \\ a_{k+1}(z) \\ \vdots \\ a_n(z) \end{bmatrix} + \begin{bmatrix} 0 \\ 0 \\ \vdots \\ 0 \\ 1 \\ --- \\ b_{k+1}(z) \\ \vdots \\ b_n(z) \end{bmatrix} v \tag{6}$$

$$y = \begin{bmatrix} 1 & 0 & \cdots & 0 \end{bmatrix} z \tag{7}$$

where

$$a_i(z) = (L_f w_i) - (L_g w_i)(L_g L_f^2 h)^{-1} L_f^2 h \big|_{x=\Phi^{-1}(z)} \tag{8}$$

$$b_i(z) = (L_g w_i)(L_g L_f^2 h)^{-1} \big|_{x=\Phi^{-1}(z)} \tag{9}$$

This system can be divided into a linear subsystem (upper part) and a nonlinear subsystem (lower part).

In the previous method,[1,2,5] the linearized subsystems can track a constant reference value r by a *PD* controller v shown as follow,

$$v = -k_1(z_1 - r) - k_2 z_2 \cdots - k_k z_k \tag{10}$$

However, the *PD* controller is not guaranteed for tracking the time-variant reference function $r_1(t)$. Then, the new partial linearization method which guarantees that the linearized sub-systems to track the time-variant reference function is proposed.

3. The partial linearization method to tracking the time-variant reference function

The partial linearization method to tracking the time-variant reference function is proposed. Proposed method is based on Lyapunov theorem and back-stepping control. The Algorithm of proposed method and proof of it are shown as follow.

The Algorithm of the proposed method

- Step 1 : At first, define the error function e_1 as follow:

$$e_1 = y - r_1(t) \tag{11}$$

If $r_1(t)$ has a element of $y^{(k-1)}$ i.e. $\dot{r}_1(t)$ can be shown as $\dot{r}_1(t) = y^{(k)}r_{11}(t)+r_{12}(t)$, where the function $r_{11}(t)$ is element of $y^{(k)}$ and the function $r_{12}(t)$ is other element, the v designed with assuming the system has a condition $r_{11}^{-1}(t) \neq 0$

$$v = (\dot{y} + r_{12} + \alpha_1 e_1)r_{11}^{-1}(t) \qquad (\alpha_1 > 0) \tag{12}$$

Else define the new target function r_2 of $\dot{y}$ as follow

$$r_2(t) := -\alpha_1 e_1 + \dot{r}_1 \tag{13}$$

and go to Step2.

- Step 2 : Define the error function $e_2 = \dot{y} - r_2(t)$,

  If $r_2(t)$ has a element of $y^{(k-1)}$ the v designed with assuming the system has a condition $r_{21}^{-1}(t) \neq 0$

$$v = (\ddot{y} + r_{22} + \alpha_2 e_2)r_{21}^{-1}(t) \qquad (\alpha_2 > 0) \tag{14}$$

  Else define the new target function r_3 of $\ddot{y}$ as follow

$$r_3(t) := -\alpha_2 e_2 + \dot{r}_2 \tag{15}$$

  and go to Step3

  ⋮

- Step k : Define the error function $e_k = \dot{y}^{(k-1)} - r_k(t)$ and v is designed as follow

$$v = -\alpha_k e_k + \dot{r}_k \qquad (\alpha_k > 0) \tag{16}$$

Proof.

- **In the case of Step 1 :** Define the Lyapunov candidate be $V_1 = e_1^2/2$.

  In condition of satisfying $\dot{r}_1(t) = y^{(k)}r_{11}(t) + r_{12}(t)$, the derivative V_1 is obtained as follow:

$$\begin{aligned}\dot{V}_1 &= e_1\dot{e}_1 = e_1(\dot{y} - \dot{r}_1) \\ &= e_1(\dot{y} - y^{(k)}r_{11}(t) + r_{12}(t)) \end{aligned} \tag{17}$$

  Now, using nonlinear feedback (3), the input-output map reduces to $y^{(k)} = v$. Therefore if v is designed as (12), then the function V_1 is the Lyapunov stability as follow

$$\dot{V}_1 = e_1(\dot{y} - vr_{11}(t) + r_{12}(t)) \tag{18}$$

$$= -\alpha_1 e_1^2 \tag{19}$$

and we obtain,

$$\lim_{t\to\infty} e_1 = 0, \qquad \lim_{t\to\infty} y = r_1(t).$$

- **In the case of Step 2 :** Define the Lyapunov candidate be $V_2 = e_2^2/2$.
In condition of satisfying $\dot{r}_2(t) = y^{(k)} r_{21}(t) + r_{22}(t)$, the derivative V_2 is obtained as follow:

$$\begin{aligned} \dot{V}_2 &= e_1 \dot{e}_2 = e_2(\dot{y} - \dot{r}_2) \\ &= e_2(\dot{y} - y^{(k)} r_{21}(t) + r_{22}(t)) \end{aligned} \tag{20}$$

Now, using nonlinear feedback (3), the input-output map reduces to $y^{(k)} = v$. Therefore if v is designed as (14), then the function V_2 is the Lyapunov stability as follow

$$\dot{V}_2 = e_2(\dot{y} - v r_{21}(t) + r_{22}(t)) \tag{21}$$

$$= -\alpha_2 e_2^2 \tag{22}$$

and we obtain,

$$\lim_{t\to\infty} e_2 = 0, \qquad \lim_{t\to\infty} \dot{y} = r_2(t).$$

And since $r_2 = -\alpha_1 e_1 + \dot{r}$, $\dot{y} - \dot{r}_1 = -\alpha_1 e_1$ i.e., $\dot{V}_1 = -\alpha_1 e_1^2$ is given. Therefore, we obtain,

$$\lim_{t\to\infty} e_1 = 0, \qquad \lim_{t\to\infty} y = r_1(t).$$

- **In the case of Step k :** Define the Lyapunov candidate be $V_k = e_k^2/2$.
If v is designed as (14), then, the derivative $V_k \sim V_1$ are obtained as follow:

$$\dot{V}_k = -\alpha_k e_k^2, \qquad \dot{V}_{k-1} = -\alpha_{k-1} e_{k-1}^2, \qquad \cdots \qquad \dot{V}_1 = -\alpha_1 e_1^2$$

That is, the function $V_k \sim V_1$ is the Lyapunov stability and we obtain,

$$\begin{aligned} &\lim_{t\to\infty} e_k = 0, && \lim_{t\to\infty} y^{(k-1)} = r_k(t) \\ &\lim_{t\to\infty} e_{k-1} = 0, && \lim_{t\to\infty} y^{(k-2)} = r_{k-1}(t) \\ &\qquad \vdots && \qquad \vdots \\ &\lim_{t\to\infty} e_1 = 0, && \lim_{t\to\infty} y = r_1(t). \end{aligned}$$

Then, using proposed method, y can track to time-variant reference function $r_1(t)$. □

4. Numerical simulations

In order to show the performance of the proposed method, the simulation using a single cart-type inverted pendulum is performed.

The controlled system is shown as follows

$$\ddot{\theta} = \frac{mgl}{I}\sin\theta - \frac{ml}{I}\cos\theta \cdot \ddot{z} \tag{23}$$

where, θ is angular position, z position of the cart, m is mass of the pendulum, l is length from joint to the center of mass of the first pendulum, I is inertia of the pendulum around the joint, and g is gravity acceleration. Now control input is defined as $u := \ddot{z}$ and output is defined as $y := \theta$.

The purpose of this control is tracking control of $y = \theta$ to time-varinat reference function $r_1(t)$ as shown as

$$r(t) = 0.3\sin(10\pi t^2). \tag{24}$$

Using the nonlinear feedback u as shown as

$$u = \frac{A\sin\theta - v}{B\cos\theta} \quad (A = mgl/I, B = ml/I) \tag{25}$$

we obtain $\ddot{\theta} = v$. where, v is input for partially linearized subsystem.

Now, designing the v by using proposed algorithm, we obtain

$$v = -\alpha_2(\dot{\theta} + \alpha_1(\theta - r_1) - \dot{r}_1) - \alpha_1(\dot{\theta} - \dot{r}_1) + \ddot{r}_1. \tag{26}$$

where, $\alpha_1 = \alpha_2 = 20$.

In order to compare the proposed method and previous method, numerical simulation of each method are performed. The result of simulation are shown in Fig.1 and 2. These graphs show the angle of the pendulum [rad], the angler velocity of the pendulum [rad/sec], the control input [m/sec^2], the reference angle of the pendulum rad and the error function.

The graph of Fig.2 shows the simulation result by using proposed controller (25) and (26), and the graph of Fig.1 shows the simulation result by using previous controller (25) and (10) with $k_1 = k_2 = 5$.

From Fig.1 and 2, despite the error function e_1 of the previous method is away from 0 when the frequency of the reference function is large, the angle of pendulum θ in proposed method always tracking to the time-variant reference function and $e_1 = 0$ is achieved.

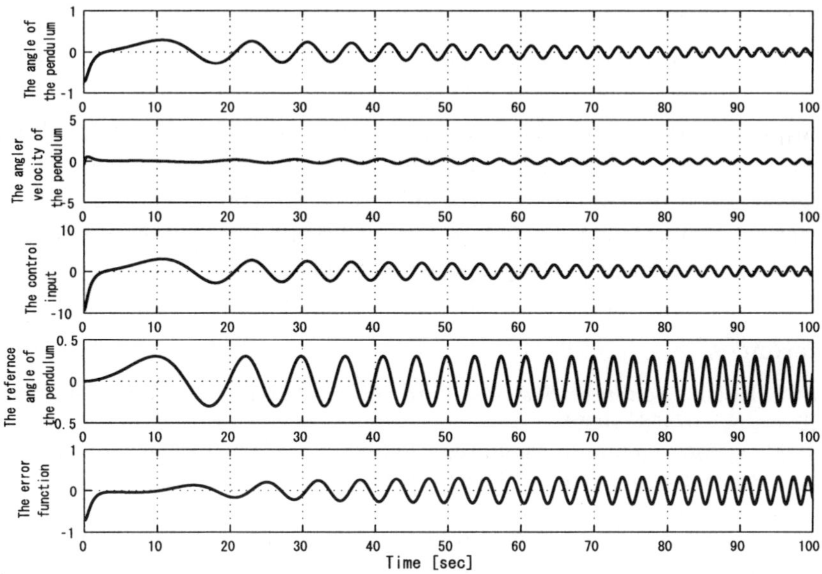

Fig. 1. Tracking control of the cart-type single inerted pendulum via the method in [1]

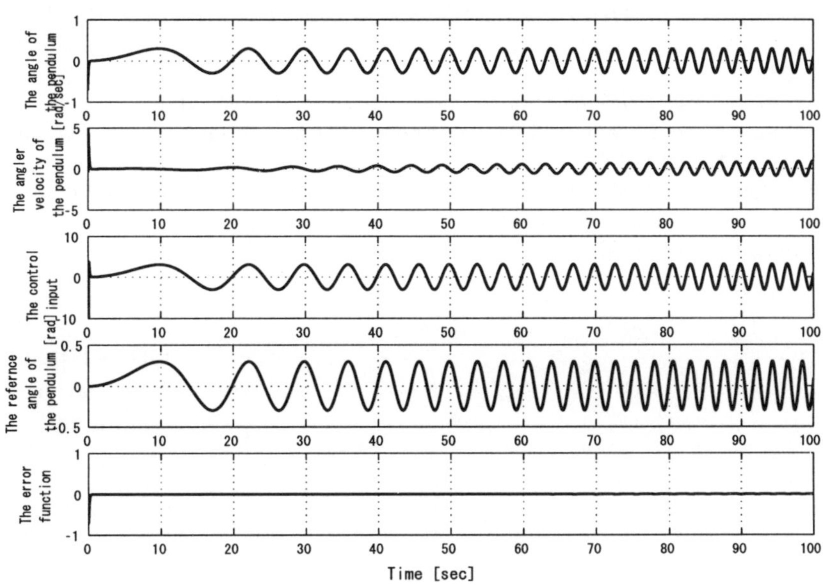

Fig. 2. Tracking control of the cart-type single inerted pendulum via the proposed method

5. Conclusions

In this paper, a new partial linearization method that guarantees the linearized sub-systems to track the time-variant reference function is proposed. In the proposed method, the designed function of v which is virtual input for the linearized subsystem. The Algorithm of the proposed method is designed based on Lyapunov theorem and back-stepping control method. Numerical simulations for cart-type inverted pendulum system are given to show the effectiveness of the proposed method.

References

1. H. K.Khalil, *Nonlinear Systems -third edition-* (Prentice Hall, Upper Saddle River, NJ, 07458, 1996).
2. T. Mita, *Introduction to Nonlinear Control Theory -Skill Control of Underactuated Robots-(In Japanese)* (SHOKODO Co., Ltd., Tokyo, 2000).
3. O. J. Rojas and G. C. Goodwin, Preliminary analysis of a nonlinear control scheme related to feedback linearization, in *Proceedings of 40th IEEE Conference on Decision and Control*, 2001.
4. T. Henmi, M. Deng, A. Inoue, N. Ueki and Y. Hirashima, *A partial linearization method to compensate input disturbances of nonlinear systems*, in *ADVANCES IN THE DYNAMICS INSTRUMENTATION AND CONTROL*, eds. C. Y. Su, S. Rakheja, R. Bhat and E. Wang (The World Scientific Press, 2004).
5. M. W. Spong, *IEEE Control Systems Magazine* **45**, 725 (2000).
6. I. Fantoni, R. Lozano and M. W. Spong, *IEEE Transactions on Automatic Control* **15**, 49 (1995).
7. A. Ohsumi and M. Matsumura, wing-up control of an inverted pendulum via partial feedback linearization, in *Proceedings of 26th SICE Symposium on Control theory*, 1997.

FAULT DIAGNOSIS AND IDENTIFICATION FOR DC MOTORS

D. R. ESPINOZA-TREJO*

Facultad de Ingeniería, CIEP, UASLP,
E-mail: espinoza_trejo_dr@uaslp.mx

D. U. CAMPOS-DELGADO

Facultad de Ciencias, UASLP, Av. Salvador Nava s/n, Zona Univ.,
C.P. 78290, S.L.P., México, E-mail: ducd@fciencias.uaslp.mx

The conditions for fault detection and isolation (FDI) based on differential geometry are analyzed for both linear and nonlinear DC motor configurations. Hence, the sets of faults that can be detected and isolated are deduced, and it is derived that the presence of an unknown perturbation in the mechanical equation of the motor limits the capability of detecting mechanical faults. Based on the geometric approach, observers are developed to generate fault residuals. Simulation and experimental results are shown to verify the analysis.

1. Introduction

Nowadays, the DC motors are still extensively used in the industry, since they are easier to control for variable speed and torque conditions than induction motors. However, they require a continuous maintenance schedule, due to the mechanical wear or aging. In this way, it is interesting to evaluate the typical fault conditions in this electrical machine. Furthermore, it is appealing to take advantage of the well-known dynamical characterizations of the motor (analytical models) in order to detect these faulty scenarios. So far in the literature, techniques based on parameters estimation and neural networks [7], fuzzy logic [9], and signal processing [5] have been applied for the DC motor in a linear configuration. Nevertheless, a detailed study of the fault detection and isolation of the multiples configurations of the DC motor have not been carried out so far. The first step in the fault detection

*D.R. Espinoza-Trejo acknowledges the financial aid provided by CONACYT through a doctoral scholarship (# 166718).

and isolation (FDI) design needs to determine the set of faults that can be identified and isolated within the motor, in order to generate next the indicative fault signals (**residuals**). One approach for FDI relies on using model-based observers to generate residuals [6], [2]. But in order to achieve this objective, the mathematical model of the motor has to present certain dynamic conditions among the state variables, possible perturbations, and fault signals to achieve successfully the FDI process. Note that the DC motor faces a practically unknown perturbation, *load torque*, that could be constant or time-varying. Hence, this paper looks to study the structural conditions in terms of differential geometry [8],[4], in order to achieve a perfect fault detection and isolation for the linear and nonlinear configurations of the DC motor.

2. DC Motor Modeling

There are three different configurations to operate a DC motor. These configurations are classified according to the connection of the field and armature windings. These are: separately excited configuration, parallel and series connections. For brevity, only the first two configurations will be analyzed here. In the following derivations, the parameters of the DC motor are defined as: R_a armature resistance, L_a armature inductance, R_f field resistance, L_f field inductance, M mutual inductance, B mechanical friction, J inertia, k_b induced emf constant, and T_l load torque.

2.1. *Separately Excited Configuration*

This configuration needs two voltage sources for its operation, one for the field (stator) winding, and another for the armature (rotor) winding. Usually, a constant voltage source is applied to the field (a constant field is assumed), then the electric torque is proportional only to the armature current. Hence, controlling this variable, the dynamics of the motor drive system can be adjusted. The mathematical model of this configuration is linear, and it is shown in Eq. (1), (assuming the field current constant, $I_f = constant \Rightarrow k_b = MI_f$). In this model, consider $u = V_a$ as the control input, the states of the system as $x_1 = i_a$ the armature current, and $x_2 = \omega$ the angular velocity. It is further assumed that the output of the system is the complete state (which is common in practical applications), $y_1 = i_a$ and $y_2 = \omega$.

$$\begin{aligned} \dot{x}_1 &= -\tfrac{R_a}{L_a} x_1 - \tfrac{k_b}{L_a} x_2 + \tfrac{1}{L_a} u, \\ \dot{x}_2 &= \tfrac{k_b}{J} x_1 - \tfrac{B}{J} x_2 - \tfrac{1}{J} T_l \end{aligned} \qquad (1)$$

2.2. *Shunt Configuration (Parallel)*

Contrary to the separately excited configuration, this connection only needs one power supply for its operation. The mathematical model of this configuration is nonlinear, and it is shown in Eq. (2). It is assumed that there is available a variable resistor R_{adj} to adjust the maximum velocity in the motor (*field weakening*) [3], i.e. $\omega_{max} = (R_f + R_{adj})/M$. Note that the nonlinearity of the shunt configuration complicates its control algorithm for variable speed applications. In this model, it is considered $u = V$ as the control input, the states of the system as $x_1 = i_a$ the armature current, $x_2 = i_f$ the field current, and $x_3 = \omega$ the angular velocity. It is assumed that the output of the system is also the complete state, $y_1 = i_a$, $y_2 = i_f$ and $y_3 = \omega$.

$$\begin{aligned} \dot{x}_1 &= -\tfrac{R_a}{L_a}x_1 - \tfrac{M}{L_a}x_2x_3 + \tfrac{1}{L_a}u \\ \dot{x}_2 &= -\tfrac{R_f+R_{adj}}{L_f}x_2 + \tfrac{1}{L_f}u \\ \dot{x}_3 &= -\tfrac{B}{J}x_3 + \tfrac{M}{J}x_1x_2 - \tfrac{1}{J}T_l \end{aligned} \quad (2)$$

3. FPRG for Linear Systems

The fundamental problem of residual generation (FPRG) for linear systems is considered assuming the following class of linear systems:

$$\dot{x} = Ax + Bu + \sum_{i=1}^{k} L_i\nu_i, \quad y = Cx \quad (3)$$

where $x \in \mathbb{R}^n$ describes the states vector, $u \in \mathbb{R}^m$ the known control inputs, and $y \in \mathbb{R}^q$ the measured system outputs. $\nu_i \in \mathbb{R}^{k_i}$ with $i \in \mathbf{k}$ describes the behavior (concerning time and magnitud) of the ith fault and is denoted as the fault signal. $\mathbf{k}$ denotes the finite set $\{1, 2, ..., k\}$. Assuming the case of multiple failures (EFPRG) as declared in [8]. It is assumed that k failure events are present. The objective is then to design an observer that generates k residuals, r_i, such that they are affected only for the ith failure mode ν_i. It has been shown in [8] that the existence of a solution for the FPRG of linear systems is established by the following theorem:

Theorem 3.1. *FPRG has a solution if and only if*

$$\mathcal{S}_i^* \cap \mathcal{L}_i = 0; \quad \forall i \in \mathbf{k} \quad (4)$$

where $\mathcal{L}_i = Im\{L_i\}$ and $\mathcal{S}_i^* := \inf \underline{\mathcal{S}}(\sum_{j\neq i}\mathcal{L}_j)$, which denotes the infimal unobservability subspace which contains all the contributions $\mathcal{L}_j$ from the faults ν_j, $j \neq i$, i.e., the subspace $\mathcal{P}_i = \sum_{j\neq i}\mathcal{L}_j$.

4. FPRG for Input-Affine Nonlinear Systems

The FPRG for nonlinear systems is considered assuming an input-affine representation:

$$\dot{x} = f(x) + \sum_{k=1}^{m} g_k(x)u_k + \sum_{i=1}^{s} l_i(x)f_i + \sum_{j=1}^{d} n_j(x)w_j$$
$$y = h(x) \tag{5}$$

where $x \in \mathbb{R}^n$ represents the state vector, u_k the known control inputs, $k = 1, ..., m$, f_i the fault modes $i = 1, ..., s$, w_j the disturbances $j = 1, ..., d$, and $y \in \mathbb{R}^q$ the output vector. Moreover, $f, g_1, ..., g_m, l_1, ..., l_s, n_1, ..., n_d$, and h are smooth vector fields. It is assumed that $\mathbf{s} = \{1, ..., s\}$ failure events could be present. Again, the objective is to design an observer that generates s residuals, r_κ, such that they are affected only by the κth failure mode f_κ. Necessary conditions for the solvability of the FPRG for input-affine nonlinear systems (lNLFPRG) have been provided in [4], and they are briefly recalled here. Let P_κ be the distribution generated by the vector fields l_i, $i \neq \kappa$, and n_j, $j = 1, ..., d$, i.e.,

$$P_\kappa = span\{l_1, ..., l_{\kappa-1}, l_{\kappa+1}, ..., l_s, n_1, ...n_d\} \tag{6}$$

Following the differential geometric approach of [4], a solution to the FPRG for input-affine nonlinear systems exists only if:

$$l_\kappa \notin (\Omega_\mathcal{O})^\perp \tag{7}$$

where $\Omega_\mathcal{O}$ denote the largest observability codistribution contained in $(P_\kappa)^\perp$, and $(\Omega_\mathcal{O})^\perp$ is an unobservability distribution.

5. FDI for a DC Motor

This section presents the fault detection and isolation results obtained by applying the geometric approach to the linear and nonlinear configurations of the DC motor (separately excited and parallel connection). Various faults can occur on these systems. In [1], it is illustrated the different fault scenarios that can be observed in applications where DC motors are used. In this paper due to space limitations, only sensor and actuator faults are studied (additive faults).

5.1. *Fault Scenarios for a Separately Excited DC Motor*

The three different faults listed in Table 1 are considered as generic faults, and they are studied here with the geometric FDI analysis.

Table 1. Faults Analyzed in a Separately Excited Config.

Fault	Symbol	Type
speed sensor fault	Δ_ω	additive-abrupt
armature current fault	Δ_{ia}	additive-abrupt
DC bus offset fault	Δ_u	additive-abrupt

5.1.1. *Geometric FDI Analysis for a Separately Excited DC Motor*

The linear DC motor system in Eq. (1), has to be rewritten as in the form shown in Eq. (3) to be able to apply the geometric approach as stated in [8]. One way to obtain the required form is to include sensor faults as *pseudo-actuator faults.* A procedure to achieve this goal is described in [11] and used here in the following analysis. On other hand, in accordance with Theorem 5 given in [11], the number of failure modes (i.e. the number of the actuators and sensors which may fail) should be less than or equal to the number of outputs of system, so that the FPRG for linear systems has a solution. Therefore, only a limited subset of faults can be considered in the geometric FDI analysis. In this case the system only has two outputs, then, only subsets formed at most for two arbitrary fault signals should be considered. In any other condition, the FPRG is not solvable for arbitrary fault signals. In Table 2, three different FPRGs are considered assuming that the **load torque T_l is known**. The geometric FDI analysis is used in order to check the possibility of isolating the two faults from each other. Hence, it is obtained that $FPRG_1$ to $FPRG_3$ are considered as a strongly identifiable family, as declared in [8], because they fulfilled the conditions given in [8] and recalled here in Sec. 3. On other hand, when the **load**

Table 2. Faults Sets in a Separately Excited Config.

FPRG	Fault1	Fault2
$FPRG_1$	Δ_ω	Δ_u
$FPRG_2$	Δ_ω	Δ_{ia}
$FPRG_3$	Δ_u	Δ_{ia}

torque T_l information is unknown (which is the most common situation in any application), it can only be considered subsets with one fault and the load torque, such that the FPRG could have a solution. In Table 3, three different $FPRGs$ are considered assuming that the load torque is an arbitrary signal. The geometric FDI analysis was applied to obtain that $FPRG_4$ to $FPGR_6$ are considered also a strongly identifiable family.

Table 3. Fault Sets for a Separately Excited Config. with Unknown Disturbance.

FPRG	Fault	Disturbance
$FPRG_4$	Δ_ω	T_l
$FPRG_5$	Δ_{ia}	T_l
$FPRG_6$	Δ_u	T_l

5.2. *Fault Scenario for a DC Motor in Parallel Connection*

The four different faults listed in Table 4 are now considered as generic faults, and they are studied in this subsection for the geometric FDI analysis.

Table 4. Faults Analyzed in a Parallel Config.

Fault	Symbol	Type
speed sensor fault	Δ_ω	additive-abrupt
armature current sensor fault	Δ_{ia}	additive-abrupt
field current sensor fault	Δ_{if}	additive-abrupt
DC bus offset fault	Δ_u	additive-abrupt

5.2.1. *Geometric FDI Analysis for a DC Motor in Parallel Connection*

The nonlinear DC motor system in Eq. (2) has to be rewritten in the form shown in Eq. (5) to be able to apply the geometric approach as stated in [4]. Once more, the method to obtain the required structure is to include sensor faults as *pseudo-actuator faults*. On the other hand, in agreement with [4], the number of residual should be less or equal that the number of outputs of system, such that the solution for the FPRG in nonlinear systems could exist. At most groups of three faults could be studied. Considering any three faults (see Table 4) and (T_l) like an **unknown input**, applying the geometric FDI analysis it was proven that the corresponding FPRGs are not solvable for arbitrary signals. Only with the assumption that the load torque is known, then some sets will have a solution. Furthermore, when a speed sensor fault is considered, only under the assumption that the load torque is an unknown input but constant, then a solution is obtained, (see Table 5). Now, when mechanical faults occur such as: static or dynamic air-gap irregularities, bearing failures, shaft misalignment, brush wear, etc. then, the mechanical equation parameters B, J and M are affected by the faults as a variation of their nominal value (Δ_B, Δ_J, Δ_M), respectively. Hence, these faults are modeled like multiplicative-incipient. When Δ_B, Δ_J faults are considered in the geometric FDI analysis, it is found that the vector field associated to the faults Δ_B or Δ_J, e.g. $\Delta_B \Rightarrow (0, 0, \Delta_B x_3/J)^T$ for each point $x \in \mathbb{R}^3$ with ($x_3 \neq 0$) generates the same vector space that the one associated with the disturbance $T_l \Rightarrow (0, 0, 1/J)^T$. Hence, the isolation problem in this case does not have a solution, i.e., if it is required to place the image of the vector associated with the disturbance T_l in the unobservability subspace, such that it does not affect the output (residual) r, then automatically the image of the vector field associated to the mechanical faults (Δ_B or Δ_J) will also belong to the unobservability subspace. Next, it is summarized in Table 5 for brevity only some FPRGs that are solvable.

Table 5. Feasible Faults Sets for the DC Motor in Parallel Config.

FPRG	Faults	Disturbance
$FPRG_7$	Δ_ω	$T_l \rightarrow$ (constant input)
$FPRG_8$	Δ_u, Δ_ω	$T_l \rightarrow$ (constant input)
$FPRG_9$	Δ_u, Δ_{ia}	$T_l \rightarrow$ (arbitrary input)
$FPRG_{10}$	Δ_u, Δ_ω, Δ_{if}	$T_l \rightarrow$ (known input)

5.2.2. *Simulation Results for a DC Motor in Parallel Connection*

In this subsection, the $FPRG_8$ is considered for illustration purposes. It is assumed a DC bus offset fault (f_3) at $t = 40$ s with a 30% drop of its nominal value, a speed sensor fault (f_4) at $t = 70$ s with a 50% drop of its real value, and a load torque step of 2 n.m. at $t = 20$ s. In order to design the residual generator, in [4], it is shown that the fulfillment of the necessary condition in Eq. (6) implies the existence of a change of coordinates in the output and state spaces. Once these changes of coordinates have been performed, the new resulting subsystem is not affected by all signals but one (fault of interest). Hence, in this case a Luenberger observer is designed [2] (Cap. 9.2 (3)) for the resulting subsystem to generate the residual. In Figure 1, it is shown the obtained results that verify the theoretical derivation.

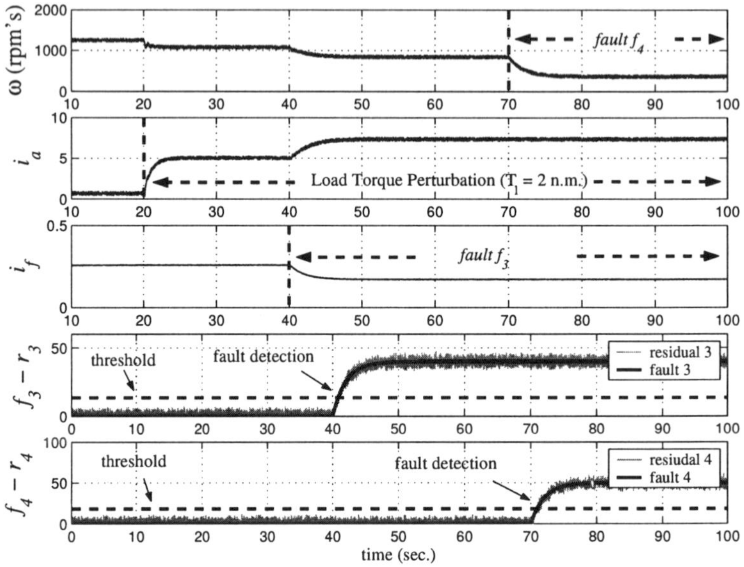

Figure 1. Simulation Results for Parallel Config.: (TOP) Velocity Measurement, (MIDDLE 1) Armature Current Measurement, (MIDDLE 2) Field Current Measurement, (MIDDLE 3) Fault (f_3) and Residual (r_3), (BOTTOM) Fault (f_4) and Residual (r_4).

5.2.3. *Experimental Results for a DC Motor in Parallel Connection*

The observers (residual generators) were implemented experimentally in a dSPACE DS1103 system running at a sampling frequency of 10 kHz. The test-bed consists of a 2 HP Shunt DC Motor that is connected to a 2 HP Permanent Magnet DC Motor utilized as a load. A tacogenerator measures the angular velocity of the shaft, at a proportion of 50 V/1000 RPM with an error of ±10 %. There are measurements of the armature and field currents through hall-effect sensors. It is important to mention that the three measurements are noisy during the experiments, as it will be observed in the implementation plots, and this issue presents a challenge for the observers to show good robustness. The motor voltage V is supplied by a DC-DC chopper working under a PWM scheme (switching frequency $10kHz$). One test was carried out, and $FPRG_8$ was considered (see Table 5). It is assumed a speed sensor fault (f_1) at approximately $t = 28$ s with a 50% drop of its real value, a DC bus offset fault (f_2) at $t = 55$ s with a 25% drop of its nominal value, and a load torque ($T_l = 1$ n.m.) applied at $t = 0$ s. By the experimental results in Figure 2, it can be observed that both residuals are not zero without faults due to parametric uncertainty. However, when a fault occurs the residuals are notably affected by the fault associated to each one of them. This consideration (model uncertainties) represents a strong limitation to carry out accurately the detection task in variable speed and load torque applications.

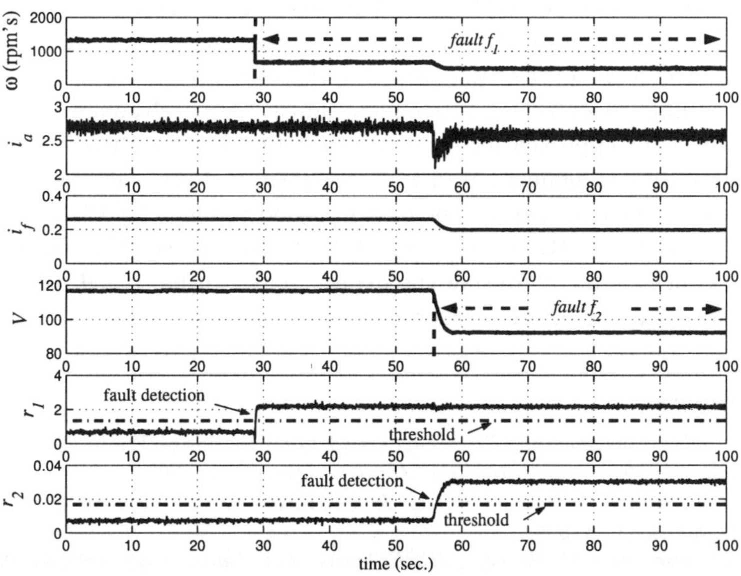

Figure 2. Experimental Results: (TOP) Velocity Measurement, (MIDDLE 1) Armature Current Measurement, (MIDDLE 2) Field Current Measurement, (MIDDLE 3) Motor voltage, (MIDDLE 4) Residual (r_1), (BOTTOM) Residual (r_2).

6. Conclusions and Final Remarks

In accordance with the restrictions of the system (number of outputs, inputs and unknown perturbations), only limited faults subsets can be considered in the geometric FDI analysis. For the separately excited DC motor, if it is measured the load torque, then it can be considered at most groups of two faults for FDI. If not, then at least all faults shown in Table 1 can be discriminated from the disturbance T_l. Meanwhile, for the DC motor in parallel connection, if it is assumed the load torque to be known, then it should be considered at most groups of three faults for the geometric FDI analysis. If not, then all faults shown in Table 4 at least can be discriminated from the disturbance T_l (speed sensor fault only with $T_l = constant$). Hence, it is concluded that the geometric FDI analysis can be very restrictive under the supposition of concurrent faults. Therefore, it is proposed as future work to carry out the geometric FDI analysis relaxing conditions for FDI assuming non-concurrent faults, as declared in [8] in Sec. V for linear systems, or in [10] in Sec. III-C for nonlinear systems. On other hand, mechanical faults that have their effect on mechanical parameters as B or J, cannot be discriminated from the disturbance T_l, since their dynamical effects are in the same direction than the disturbance, which limits the capability of detecting mechanical faults. Now, although the model-based observer approach for residual generation has the disadvantage of a lack of robustness against parametric uncertainty, if this uncertainty can be quantified then adaptive thresholds are a good solution for FDI in variable speed applications. By the experimental results, it is seen that the problem of noisy measurements during the experiments can be solved using signal filters for each one of the measurements.

References

1. D.U. Campos-Delgado, E. Palacios and D.R. Espinoza-Trejo, *Int. Conf. on Dynamics, Instrumentation and Control*, Queretaro, (2006).
2. J. Chen and R.J. Patton, *Robust-Model Based Fault Diagnosis for Dynamic Systems*, KAP, (1999).
3. J. Chiasson and M. Bodson, *IEEE Trans. Aut. Cont.*, **V38**, 1662 (1993).
4. C. De Persis, A. Isidori, *IEEE Trans. Aut. Cont.* **V46**, 853 (2001).
5. M. Hajiaghajani, A. Toliyat, and I. M. Panahi, *IEEE Trans. Energy Conv.*, **V19**, 60 (2004).
6. R. Isermann, *Fault-Diagnosis Systems*, Springer, (2005).
7. X. Q. Liu, H. Y. Zhang, J. Liu and J. Yang, *IEEE Trans. Ind. Electr.*, **V47**, 1021 (2000).
8. M. Massoumnia, G. C. Verghese and A. S. Willsky, *IEEE Trans. Aut. Cont.* **V34**, 1729 (1989).
9. L. J. Miguel, and L. F. Blázquez, *Eng. Appl. Art. Intel.*, **V18**, 423 (2005).
10. R. Mattone and A. De Luca, *Proc. of the 44th IEEE Conf. on Decision and Cont.*, 1005 (2005).
11. S. H. Zad and M. Massoumnia, *Automatica*, **V35**, 887 (1999).

DYNAMIC PRINCIPAL COMPONENTS ANALYSIS WITH ADAPTIVE STANDARDIZATION FOR FAULT DETECTION IN MIMO SYSTEMS

JESÚS MINA and CRISTINA VERDE

Instituto de Ingeniería-UNAM
verde@servidor.unam.mx
Coyoacan DF 04510, México
Fax: (52)-55-56233600 ext 8052

In the present work the problem of false alarms in a DPCA-based supervision system is tackled. In order to reduce the false alarms rate, an extension to the DPCA-based monitoring is proposed which take into account the non-stationary property in data due to changes in the operation point of multivariate linear dynamic systems. The idea is to include on-line for each new multivariate observation an adaptation in the standardization stage according to estimated means of the data. The inputs means estimation is carried out through single moving average and the outputs means estimation is carried out through identified nominal inputs-outputs relations and estimated input means. The proposed methodology is evaluated in a three interconnected tanks system.

Keywords: MIMO Linear Systems, Fault Detection, Dynamic Principal Component Analysis, Adaptive Standardization.

1. Introduction

Automatic fault diagnosis in industrial processes is becoming a field of engineering knowledge of growing importance, that is related to product quality, process availability, process safety. This fact has motivated the design of methodologies with diverse points of view during the last decades. In the framework of automatic control theory, different model-based methods for dynamic systems have been proposed which assume the existence of an explicit model in normal condition obtained from primary physical principles e.g. [1]. In this case a model is used as a reference and must contain analytical redundancy to distinguish between fault and normal condition.

However, for complex systems the lack of adequate models is recognized and the process behavior is available from historical data of multiple measurements associated to control, outputs, process indicators, etc. In this case the analytical model-based fault detection and isolation, FDI, approaches are not powerful; however, it is possible to resort to data-driven based techniques which get an implicit model of the process from historical data. Methods based on qualitative models and artificial intelligence tools have been proposed in the Safe Process community [2], [3].

The traditional method used in the industry to detect a fault is a univariate monitoring using a threshold, where an alarm signal is activated if a variable exceeds its threshold. However in a complex system with hundreds of variables, the univariate monitoring can become in an unmanageable problem, additionally it could happen that more than one alarm be turned on even when only one fault occurs or the process is in normal condition. This last phenomenon is due to the latent correlation among variables.

One of the multivariate statistical techniques used to data compression is the Principal Component Analysis (PCA) [4]. Based in a correlation analysis, the key of the PCA is the description of multivariate process data in a lower dimension orthogonal space, called *principal components space* which corresponds to a linear combinations of the original variables. The dimension reduction in the new space depends on the degree of correlation among variables.

If the data are from variables of dynamic systems, to take into account the inherent auto-correlation, the PCA must be carried out considering time lags of variables, this extension is called Dynamic PCA [5]. The Safe Process community has used the DPCA methodology as a tool to get an implicit model from historical data of a multivariate process operating under nominal conditions and using this implicit model to carry out fault detection tasks [6,7].

Here it is important to note that the DPCA based modeling is obtained from multivariate data of the process under stationary conditions, this is, around an specific operation point, therefore, the changes in the operation point are interpreted as faults by the fault detection algorithm.

The above described problem has been tackled with adaptive versions of PCA [8,9]; however the adaptation is based in the variations of actual multivariate observations without distinguish the real causes of changes in the variables, driving, by the other side, that a fault detection algorithm based in this reasoning does not detect faults.

In order to make a DPCA based fault detection algorithm robust to changes in the operation point and sensitive to faults, in the present work is proposed for MIMO linear systems, an adaptation in the standardization stage according to estimated means of the data. The inputs means estimation is carried out through single moving average and the outputs means estimation is carried out through identified nominal inputs-outputs relations and estimated input means.

In the following, the DPCA based fault detection will be briefly reviewed, next, the proposed extension of DPCA based fault detection for changes in the operation point will be described. Finally the proposed methodology will be evaluated for the three tanks system.

2. Fault Detection via DPCA

Let the matrix $\mathbf{X}$ be a set of historical data composed of n_t observations from r input variables and s output variables, taken from a dynamic process operating in nominal conditions and around an operation point, this is

$$\mathbf{X} = \begin{bmatrix} \mathbf{u}_1 \cdots \mathbf{u}_r \, \mathbf{y}_1 \cdots \mathbf{y}_s \end{bmatrix}_{(n_t \times p)} \tag{1}$$

To take into account the auto-correlation of each variable or time series in the DPCA based modeling it is necessary to express them in w time lags, so, for example in the case of the input variable $\mathbf{u}_1$ it is obtained the following matrix

$$\vec{\mathbf{U}}_1 = \begin{bmatrix} \mathbf{u}_1(n_t) & \mathbf{u}_1(n_t-1) & \cdots & \mathbf{u}_1(n_t-w) \\ \mathbf{u}_1(n_t-1) & \mathbf{u}_1(n_t-2) & \cdots & \mathbf{u}_1(n_t-w+1) \\ \vdots & \vdots & \ddots & \vdots \\ \mathbf{u}_1(w+1) & \mathbf{u}_1(w+1-1) & \cdots & \mathbf{u}_1(1) \end{bmatrix}_{(n \times w)}$$

similarly for the other time series finally it is obtained the following matrix denotated *trajectory matrix*

$$\vec{\mathbf{X}} = \begin{bmatrix} \vec{\mathbf{U}}_1 \cdots \vec{\mathbf{U}}_r \, \vec{\mathbf{Y}}_1 \cdots \vec{\mathbf{Y}}_s \end{bmatrix}_{(n \times m)} \tag{2}$$

where $n = n_t - w$ and $m = p\,(w+1)$. So, the implicit modeling with DPCA from $\vec{\mathbf{X}}$ can be summarized as follows:

(1) Standardize $\overrightarrow{\mathbf{X}}$ in relation to its means and standard deviations, this is

$$\widetilde{\mathbf{X}}(i,j) = \frac{\overrightarrow{\mathbf{X}}(i,j) - \mu_{\overrightarrow{\mathbf{X}}}(j)}{\sigma_{\overrightarrow{\mathbf{X}}}(j)}$$

for $i = 1, \ldots, n$ and $j = 1, \ldots, m$. Where

$$\mu_{\overrightarrow{\mathbf{X}}} = \left[\mu_{\overrightarrow{\mathbf{U}}}\ \mu_{\overrightarrow{\mathbf{Y}}}\right]_{(1\times m)} \tag{3}$$

$$\sigma_{\overrightarrow{\mathbf{X}}} = \left[\sigma_{\overrightarrow{\mathbf{U}}}\ \sigma_{\overrightarrow{\mathbf{Y}}}\right]_{(1\times m)} \tag{4}$$

(2) Transform the standardized data in the principal components space $\mathbf{Z} = \widetilde{\mathbf{X}}\mathbf{V}_t$ where $\mathbf{V}_t \in \Re^{m\times l}$ is composed of an appropriate selection of l eigenvectors associated to the correlation matrix $\mathbf{R} = \frac{1}{n-1}\widetilde{\mathbf{X}}^T\widetilde{\mathbf{X}}$.

(3) Due to the orthogonal property of $\mathbf{Z}$ it is possible to describe each observation in a univariate parameter, in this case it is used the Hotelling parameter defined as

$$T^2_{\mathbf{Z}_i} = \mathbf{Z}_i \mathbf{S}_{\mathbf{Z}}^{-1} \mathbf{Z}_i^T$$

where $\mathbf{S}_{\mathbf{Z}}$ is the covariance matrix of $\mathbf{Z}$.

(4) Calculate the normal condition threshold UCL from the probability density function of the set of parameters $T^2_{\mathbf{Z}_i}$ [10]

$$UCL = \frac{(n-1)^2\left(\frac{l}{n-l-1}\right)F\left(\frac{\alpha}{2}; l, n-l-1\right)}{n\left(1 + \left(\frac{l}{n-l-1}\right)F\left(\frac{\alpha}{2}; l, n-l-1\right)\right)} \tag{5}$$

where n and l are the dimensions of $\mathbf{Y}^l$ and α is a level of significance.

On the base of this modeling the procedure to evaluate and classify an actual observation $\overrightarrow{\mathbf{x}}_a \in \Re^{1\times m}$ is summarized as follows:

(1) Standardize with respect to means (3) and standard deviations (4).
(2) Map to the principal components space, $\mathbf{z}_a$.
(3) Reduce to the univariate parameter, $T^2_{\mathbf{z}_a}$.
(4) Compare with UCL. A deviation over the threshold indicates a fault.

3. DPCA based Fault Detection with Adaptation

The false alarm of a new multivariate observation $\overrightarrow{\mathbf{x}}_a$ when the system is in steady state around a new operation point (new means), occurs because the standardization of $\overrightarrow{\mathbf{x}}_a$ is calculated with the fixed means (3) of the training operation point instead of the new mean. By the other hand, an important fact is that any nominal actual observation $\overrightarrow{\mathbf{x}}_a$ in steady state around other mean values set is affine to any observation of the historical data matrix $\widetilde{\mathbf{X}}$, and as consequence the correlation matrix between the m variables is invariant, this is analized below. This preservation of correlation does not happen if the actual observation $\overrightarrow{\mathbf{x}}_a$ is captured under fault conditions, this property can be used to distinguish between normal changes in the operating point of the process and variations due to faults.

3.1. *Preservation of the Correlation Structure in a MIMO Linear System under Changes in the Operation Point*

Lets consider a linear MIMO system with random inputs vector $\mathbf{u} \in \Re^r$ and outputs vector $\mathbf{y} \in \Re^s$, which under steady state conditions can be expressed as

$$\mathbf{y} = \mathbf{A}\mathbf{u} \tag{6}$$

this is, the outputs are expressed as a linear combination of the inputs. By the other side, the statistical parameters of mean and covariance of the input and output variables are

$$\mu_\mathbf{u} = E\left[\mathbf{u}\right] \; ; \Gamma_\mathbf{u} = E\left[\mathbf{u}\mathbf{u}^T\right] - \mu_\mathbf{u}\mu_\mathbf{u}^T$$
$$\mu_\mathbf{y} = E\left[\mathbf{y}\right] \; ; \Gamma_\mathbf{y} = E\left[\mathbf{y}\mathbf{y}^T\right] - \mu_\mathbf{y}\mu_\mathbf{y}^T$$

which satisfy the following relations

$$\mu_\mathbf{y} = \mathbf{A}\mu_\mathbf{u} \; ; \Gamma_\mathbf{y} = \mathbf{A}\Gamma_\mathbf{u}\mathbf{A}^T$$

Now, lets define the vector $\mathbf{x} = \left[\, \mathbf{u}^T \; \mathbf{y}^T \,\right]^T$ of input and output variables with the following statistical parameters

$$\mu_\mathbf{x} = \begin{bmatrix} \mu_\mathbf{u} \\ \mu_\mathbf{y} \end{bmatrix} = \begin{bmatrix} \mu_\mathbf{u} \\ \mathbf{A}\mu_\mathbf{u} \end{bmatrix}$$

$$\Gamma_\mathbf{x} = E\left[\mathbf{x}\mathbf{x}^T\right] - \mu_\mathbf{x}\mu_\mathbf{x}^T = \begin{bmatrix} \Gamma_\mathbf{u} & \Gamma_\mathbf{u}\mathbf{A}^T \\ \mathbf{A}\Gamma_\mathbf{u} & \Gamma_\mathbf{y} \end{bmatrix} = \begin{bmatrix} \Gamma_\mathbf{u} & \Gamma_\mathbf{u}\mathbf{A}^T \\ \mathbf{A}\Gamma_\mathbf{u} & \mathbf{A}\Gamma_\mathbf{u}\mathbf{A}^T \end{bmatrix} \tag{7}$$

The relations in (7) describe the structural correlation between the input and output variables. In the case of a change in the operation point the structural correlation in (7) does not change. To verify the later statement it is only necessary to evaluate in $\Gamma_\mathbf{x}$ the preservation of $\Gamma_\mathbf{u}$ under changes in $\mu_\mathbf{u}$. Adding to $\mathbf{u}$ a vector of constant values $\mathbf{c}$ results in $\mathbf{u}_C = \mathbf{u} + \mathbf{c}$ with the following statistical parameters

$$\mu_{\mathbf{u}_C} = E\left[\mathbf{u}_C\right] = E\left[\mathbf{u} + \mathbf{c}\right] = \mu_\mathbf{u} + \mathbf{c} \tag{8}$$

$$\begin{aligned}\Gamma_{\mathbf{u}_C} &= E\left[\mathbf{u}_C\mathbf{u}_C^T\right] - \mu_{\mathbf{u}_C}\mu_{\mathbf{u}_C}^T \\ &= E\left[(\mathbf{u}+\mathbf{c})(\mathbf{u}+\mathbf{c})^T\right] - (\mu_{\mathbf{u}}+\mathbf{c})(\mu_{\mathbf{u}}+\mathbf{c})^T\end{aligned}$$

$$\Gamma_{\mathbf{u}_C} = E\left[\mathbf{u}\mathbf{u}^T\right] - \mu_{\mathbf{u}}\mu_{\mathbf{u}}^T = \Gamma_{\mathbf{u}} \tag{9}$$

as it is seen $\Gamma_{\mathbf{u}_C} = \Gamma_{\mathbf{u}}$ indicating that the correlation structure is preserved under changes in the means of the input variables.

3.2. *Estimating the Inputs-Output Relations*

In the proposed fault detection algorithm the objective is to carry out an adaptive standardization process from on-line estimation of the nominal means of input and output variables achieving the standardized components $\widetilde{\mathbf{x}}_a$ of a nominal observation vector $\overrightarrow{\mathbf{x}}_a$, around any operation point, keep the same correlation structure given in $\mathbf{R}$. To achieve this adaptation, it is assumed the existence of nominal inputs-output linear relations, which have to be identified in the modeling stage additionally to the DPCA based statistical model. Here it is proposed to identify Moving Average models MA(q) for each one of the output variables. Then, in the case of a MIMO linear system with r inputs and s outputs, each one of the outputs can be expressed as a linear combination of the inputs at different time lags $q_{1i}, q_{2i}, \cdots, q_{ri}$

$$y_i(t) = \sum_{k=0}^{q_{1i}} a_{k1i}u_1(t-k) + \cdots + \sum_{k=0}^{q_{ri}} a_{kri}u_r(t-k) \tag{10}$$

for $i = 1, \cdots, s$.

The $a_{k1i}, \cdots, a_{kri}$ parameters are obtained through correlation analysis for system identification [11]. By the other side, according to (10) the output mean estimation will be given by

$$\widehat{y}_i(t) = \sum_{k=0}^{q_{1i}} \widehat{a}_{k1i}u_1(t-k) + \cdots + \sum_{k=0}^{q_{ri}} \widehat{a}_{kri}u_r(t-k) \tag{11}$$

$$E\left[\widehat{y}_i(t)\right] = \widehat{\mu}_{y_i}(t) = \sum_{k=0}^{q_{1i}} \widehat{a}_{k1i}E\left[u_1(t-k)\right] + \cdots + \sum_{k=0}^{q_{ri}} \widehat{a}_{kri}E\left[u_r(t-k)\right] \tag{12}$$

See from (12) that the mean in the output $y_i(t)$ at time t depends of the $q_{1i}+1, \cdots, q_{ri}+1$ means of $u_1, \cdots, u_r$ respectively $E\left[u_1(.)\right], \cdots, E\left[u_r(.)\right]$, which here are estimated through SMA for a window of data for the corresponding input u_j.

3.3. *Fault Detection with Adaptive Standardization*

For any actual observation vector, with input and output variables, expressed in w time lags

$$\overrightarrow{\mathbf{x}}_a = \left[\overrightarrow{\mathbf{u}}_{a1} \cdots \overrightarrow{\mathbf{u}}_{ar} \; \overrightarrow{\mathbf{y}}_{a1} \cdots \overrightarrow{\mathbf{y}}_{as} \right] \tag{13}$$

the procedure to evaluate and classify such actual observation is summarized as follows:

(1) Estimate through SMA the the means of the actual input data, $\widehat{\mu}_{\overrightarrow{\mathbf{u}}_a}$, and through (12) the nominal means of the output variables, $\widehat{\mu}_{\overrightarrow{\mathbf{y}}}$; next construct the vector

$$\widehat{\mu}_{\overrightarrow{\mathbf{x}}_a} = \left[\widehat{\mu}_{\overrightarrow{\mathbf{u}}_a} \, \widehat{\mu}_{\overrightarrow{\mathbf{y}}} \right]_{(1 \times m)} \tag{14}$$

(2) Standardize the m terms in (13) using the estimated means vector (14) and the historical standard deviations in (4), this is

$$\widetilde{\mathbf{x}}_a(j) = \frac{\overrightarrow{\mathbf{x}}_a(j) - \widehat{\mu}_{\overrightarrow{\mathbf{x}}_a}(j)}{\sigma_{\overrightarrow{\mathbf{X}}}(j)}$$

for $j = 1, \ldots, m$.

(3) Transform the vector $\widetilde{\mathbf{x}}_a$ in the principal component subspace $\mathbf{z}_a$ through $\mathbf{V}_t$

$$\mathbf{z}_a = \widetilde{\mathbf{x}}_a \mathbf{V}_t$$

(4) Map $\mathbf{z}_a$ in the behavior symptom $T^2_{\mathbf{z}_a}$ through

$$T^2_{\mathbf{z}_a} = \mathbf{z}_a \mathbf{S}_{\mathbf{Z}}^{-1} \mathbf{z}_a^T$$

(5) If the resulting value deviates from the normal condition threshold a fault is present in the system.

The key of the proposed methodology is the continuous estimation of (3) using the nominal linear relations (12) to carry out an appropiate standardization.

4. Application Example: Three Tanks System

The tanks system is composed of three cylindrical tanks, interconected at the bottom by pipes and with valves V_1 in the link between tanks 2 and 3, and V_2 in the link between tank 2 and the outside, which aperture can be manipulated in order to emulate faults e.g. pipe blockage. The tank dimensions are: $h_T = 0.63m$, $A_T = 0.01539m^2$. The system is feed by two inputs Q_1 to the tank 1 and Q_2 to the tank 2 which are measured just as the ouput variables h_1 and h_2 with are the levels of tanks 1 and 2 respectively. The matemathical model is the following

$$\begin{array}{l} A_T \frac{dh_1}{dt} = Q_1 + Q_{31} - Q_{10} \\ A_T \frac{dh_3}{dt} = Q_{23} - Q_{31} \\ A_T \frac{dh_2}{dt} = Q_2 - Q_{23} - Q_{20} \end{array} ; \begin{array}{l} Q_{10} = K_1 \sqrt{h_1} \\ Q_{31} = K_{31} \rho \left(h_3 - h_1 \right) \\ Q_{23} = K_{23} \rho \left(h_2 - h_3 \right) \\ Q_{20} = K_2 \sqrt{h_2} \end{array} \tag{15}$$

where $\rho(x) \triangleq sgn(x)\sqrt{|x|}$. For the experiments the system was simulated under the following operation point: $h_1^0 = 0.147m$, $h_2^0 = 0.276m$, $h_3^0 = 0.195m$, $Q_1^0 = 4.75 \times 10^{-5} m^3/s$, $Q_2^0 = 7.35 \times 10^{-5} m^3/s$, $K_1 = 1.8165 \times 10^{-4}$, $K_{31} = 1.0055 \times 10^{-4}$, $K_2^0 = 9.804 \times 10^{-5}$ and $K_{23}^0 = 7.8047 \times 10^{-5}$.

Taking a set of 400 nominal observations measured every 10s the principal components space resulted in dimensions 301×81, so, for an $\alpha = 0.01$ the resulting threshold is $UCL = 110.0976$. By the other side, the inputs-output relations identified were $h_1 = f(Q_1, Q_2, q_1)$ and $h_2 = f(Q_1, Q_2, q_2)$ with time lags of order $q_1 = 44$ and $q_2 = 59$ respectively. The following evaluation conditions were tested:

(1) Fault condition, blockage in the pipe which links tanks 2 and 3, the time occurrence fault is $9000s$.

(2) Normal operation of the system during $14000s$, with changes in U_1 of $+20\%$ in $3000s < t < 6000s$; -20% in $9000s < t < 12000s$ and in U_2 of $+20\%$ in $4500s < t < 7500s$; -20% in $10500s < t < 13500s$.

The first evaluation has as goal to compare the behavior of the DPCA-based fault detection and the fault detection of the proposed algorithm. The monitoring results are given in Fig. 1 which shows that both algorithms develop a similar response to the fault detection.

The second test is in normal operating conditions but before changes in the operation point, the monitoring results are given in Fig. 2, where it is cleared observed that the DPCA-based fault detection (Mon1) interprets

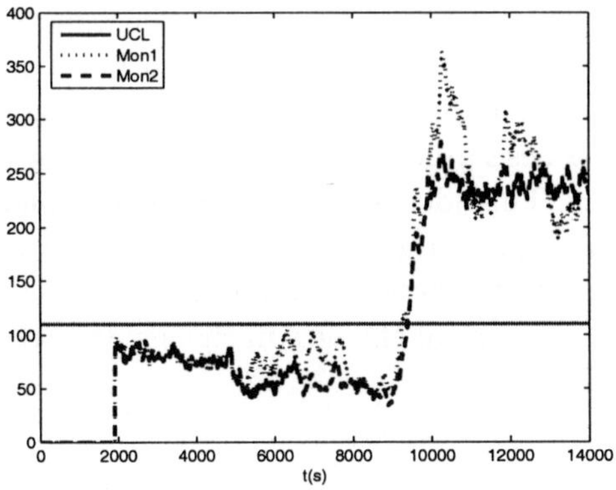

Fig. 1. Fault condition: UCL - Threshold of normal condition; Mon1 - DPCA-based monitoring; Mon2 - DPCA with adaptation.

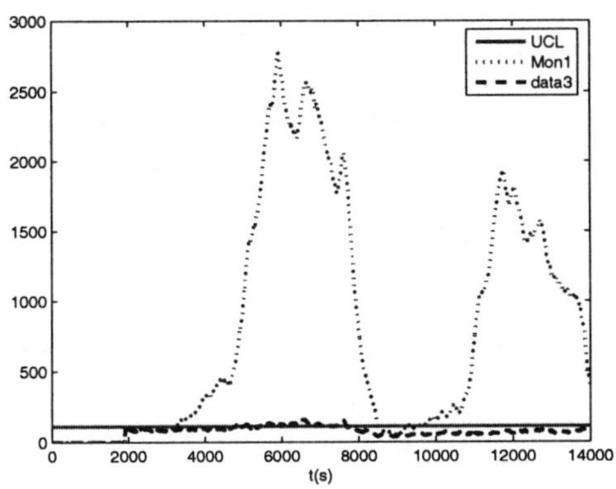

Fig. 2. Normal condition: UCL - Threshold of normal condition; Mon1 - DPCA-based monitoring; Mon2 - DPCA with adaptation.

the normal changes in the operation point as faults, however, the proposed algorithm (Mon2) is robust before these changes.

5. Conclusions

Here, a modification to the DPCA algorithm for fault detection has been proposed, in which an adaptive standardization with respect to on-line estimated statistical parameters is carried out. The key of the proposed algorithm is that the adaptation is from statistical parameters of the variables which are estimated through identified nominal linear inputs-output relations. Additionally, it is important to note that these nominal relations do not looks for precise description of the output variables but only for mean estimation purposes. This idea allows to deal with non-stationary signals and to reduce significatively the rate of false alarms. It was shown through a series of tests the effectiveness of the proposed fault detection algorithm to distinguish between normal changes in signals and the variations due to the presence of faults.

Acknowledgment

Supported by the EOLI Project of the European Community INCO program contract ICA4-CT-2002-10012 and IN102306-2-DGPA-UNAM

References

1. R. J. Patton, P. M. Frank and R. N. Clark, *Issues of Fault Diagnosis for Dynamic Systems* (Springer-Verlag, London, 2000).
2. L. H. Chiang, E. L. Russell and R. D. Braatz, *Fault Detection and Diagnosis in Industrial Systems* (Advanced Textbooks in Control and Signal Processing, Springer-Verlag, 2001).
3. V. Venkatasubramanian, R. Rengaswamy, S. N. Kavuri and K. Yin, *Computers and Chemical Engineering* **27**, 327 (2003).
4. J. E. Jackson, *A Users Guide to Principal Components* (John Wiley, New york, 1991).
5. W. Ku, R. H. Storer and C. Georgakis, *Chemometrics and Intelligent Laboratory Systems* **30**, 179(November 1995).
6. E. L. Russell, L. H. Chiang and R. D. Braatz, *Chemometrics and Intelligent Laboratory Systems* **51**, 81(May 2000).
7. J. Chen and K.-C. Liu, *Chemical Engineering Science* **57**, 63(January 2002).
8. N. B. Gallagher, B. M. Wise, S. W. Butler, D. D. White and G. G. Barna, Development and Benchmarking of Multivariate Statistical Process Control Tools for a Semiconductor Etch Process: Improving Robustness Through Model Updating, in *ADCHEM'97*, (Banff, Canada, 1997).
9. W. Li, H. H. Yue, S. Valle-Cervantes and S. J. Qin, *Journal of Process Control* **10**, 471(October 2000).
10. N. D. Tracy, J. C. Young and R. L. Mason, *Journal of Quality Technology* **24**, 88(April 1992).
11. G. E. P. Box, G. M. Jenkins and G. C. Reinsei, *Time Series Analysis: Forecasting and Control* (Prentice Hall, New Jersey, 1994).

ADAPTIVE OBSERVER-BASED FAULT DETECTION TO A PROCESS CONTROL EXPERIMENTAL SYSTEM

AKIRA INOUE, MINGCONG DENG, TOMOHARU OGITA AND SHINICHI YOSHINAGA*

Department of System Engineering, Okayama University
3-1-1 Tsushima-Naka,
Okayama, 700-8530, Japan
E-mail: {inoue, deng }@suri.sys.okayama-u.ac.jp

**Department of Mechanical Engineering, Takamatsu National College of Technology*
335 Chokushicho, Takamatsu, 761-8058, Japan

A fault detection system is designed to a process experimental system by using an adaptive observer. The adaptive observer consists of the unknown-input state observer, and a parameter adjusting law. The observer is used to identify the uncertainties in the method. A simulation result obtained to a water level control system is given to show the effectiveness of the proposed system.

1. Introduction

In some cases, control system deals with a disaster or an accident. For the safety and reliability of the control system, fault diagnosis or faults detection technology is studied by many researchers [1]. There are two ways to detect faults in a plant. One method is a mechanical method that uses increased number of the sensors. Another one is an analytical method that uses process information of the plant [2], [3], [5]. Mainly the analytical methods are studied because that the mechanical method costs more than the analytical method. In this paper, by extending the design method in [5], we use the analytical method to detect fault in a process control experimental system. That is, we propose a method to estimate the fault signal of a process experimental system by using an unknown-input state observer. In this case, the considered process includes uncertainties, such as process parameter deviations. The proposed observer estimates the uncertainties and also identifies the plant parameters while fault detection works.

The outline of this paper is as follows. In Section 2 we show the model of the experimental system and state the problem setup. Section 3 gives the fault detection scheme for the case of plant with uncertainties by using an adaptive observer. Simulation example on the experimental system with uncertainties is illustrated in Section 4.

2. Modelling and Problem Formulation

The diagram of the experimental system is given in Fig. 1. Definition of

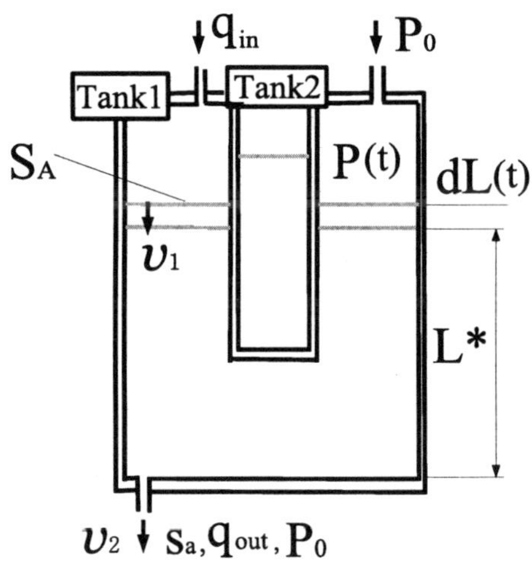

Figure 1. The diagram of the experimental system

system parameters is given as follows.

L: Water level(m) ,
v_1: Velocity of water level variation(m/s),
P_0: The atmospheric pressure (N/m^2),
S_A: Sectional square of Tank 1(m^2),
ρ : Density of water(kg/m^3)

dL: Variation of water level(m)
v_2 : Velocity of outflow(m/s)
$P(t)$: Pressure in tank(N/m^2)
s_a : Square of drain pipe(m^2)
g :Gravity acceleration(m/s^2)

Based on Bernoulli theorem, we have

$$\frac{v_1^2}{2g} + \frac{P(t)}{\rho g} + L(t) = \frac{v_2^2}{2g} + \frac{P_0}{\rho g} \tag{1}$$

$$S_A v_1 = s_a v_2 \tag{2}$$

If $s_a/S_A \ll 1$ is satisfied, we obtain that $(s_a/S_A)^2 \simeq 0$. Then,

$$v_2 = \sqrt{\frac{2}{\rho}(P(t) - P_0) + 2gL(t)} \tag{3}$$

Using the relationship of outflow water $q_{out}(\mathrm{m}^3/\mathrm{s}) = s_a v_2$ and defining inflow water as $q_{in}(\mathrm{m}^3/\mathrm{s})$, we have $S_A \dot{L}(t) = q_{in} - q_{out}$, and

$$\dot{L}(t) = \frac{1}{S_A} q_{in} - \frac{s_a}{S_A}\sqrt{\frac{2}{\rho}(P(t) - P_0) + 2gL(t)} \tag{4}$$

Using linearization at $P(t) = P_0$, L^*(water level of state of equilibrium), $q_{in}^* = q_{out}^*$, we have

$$\dot{x}(t) = \frac{1}{S_A} u(t) - \frac{s_a}{S_A}\frac{\sqrt{2g}}{2\sqrt{L^*}} x(t) + \frac{1}{S_A}\bar{g}(x, u, t, \theta) \tag{5}$$

Where, $x(t), u(t)$ are variables, and $L(t) = L^* + x(t)$, $q_{in} = q_{in}^* + u(t)$. $\bar{g}(x, u, t, \theta)$ is uncertainty which satisfies the following relationship.

$$\begin{aligned} \bar{g}(x, u, t, \theta) &= \sum_{i=1}^{\ell} \varphi_i(u, t)\theta_i \\ &= \varphi^T(u, t)\theta \end{aligned} \tag{6}$$

$$\varphi(u, t) = [\varphi_1(u, t), \ldots, \varphi_\ell(u, t)]^T \tag{7}$$

$$\theta = [\theta_1, \ldots, \theta_\ell]^T \tag{8}$$

Where, $\varphi(u, t) = (\Delta_1 + \Delta_2/s)\tilde{\varphi}(u, t)$, θ is unknown parameter vector, s is the Laplace operator, and Δ_1, Δ_2 are constants.

For the above process model, we rewrite it by a general description as follows.

$$\dot{\boldsymbol{x}}(t) = A\boldsymbol{x}(t) + B\boldsymbol{u}(t) + F\boldsymbol{f}(t) + G\boldsymbol{g}(\boldsymbol{u}, t, \boldsymbol{\theta}) \tag{9}$$

$$\boldsymbol{y}(t) = C\boldsymbol{x}(t) \tag{10}$$

where,$\boldsymbol{x} \in R^n$ denotes the state vector and is unobservable. $\boldsymbol{u} \in R^r$ and $\boldsymbol{y} \in R^m$ are the measurable control input vector and the measurable output vector, respectively. A, B, C, F and G are known matrices with appropriate dimensions. $\boldsymbol{f}(t) \in R^p$ represents the fault vector which is considered as an unknown time function. In this paper, if there exists fault, then $\boldsymbol{f}(t) \neq 0$, otherwise $\boldsymbol{f}(t) = 0$. $\boldsymbol{g}(\boldsymbol{u}, t, \boldsymbol{\theta}) \in R^l$ is the unknown disturbance vector satisfied the following relation when it is scalar.

$$\begin{aligned} \boldsymbol{g}(\boldsymbol{u}, t, \boldsymbol{\theta}) &= \sum_{i=1}^{l} \varphi_i(\boldsymbol{u}, t)\theta_i \\ &= \boldsymbol{\varphi}^T(\boldsymbol{u}, t)\boldsymbol{\theta} \qquad (11) \\ \boldsymbol{\varphi}(u, t) &= [\varphi_1(u, t), \cdots, \varphi_l(u, t)]^T \qquad (12) \\ \boldsymbol{\theta} &= [\theta_1, \cdots, \theta_l]^T \qquad (13) \end{aligned}$$

where, $\boldsymbol{\varphi}(\boldsymbol{u}, t)$ is known and $\boldsymbol{\theta}$ in unknown vector. In the case of $\boldsymbol{g}(\boldsymbol{u}, t, \boldsymbol{\theta}) \in R^l$, we have $\boldsymbol{\varphi}(\boldsymbol{u}, t) \in R^{l \times l}$. The objective is to estimate $\boldsymbol{\theta}$ and $\boldsymbol{f}(t)$ by using an adaptive observer.

3. Structure of Fault Detection Observer

In this section, a fault detection scheme for the case of plant with uncertainty by using an adaptive observer is shown.

Suppose that the fault vector is constant, combining the fault signal model to the process model (9), (10), augmented model is obtained as follows.

$$\dot{\boldsymbol{x}}_e(t) = A_e \boldsymbol{x}_e(t) + B_e \boldsymbol{u}(t) + \bar{G}\bar{\boldsymbol{g}}(\boldsymbol{u}, t, \boldsymbol{\theta}) \qquad (14)$$

$$y_a(t) = C_e \boldsymbol{x}_e(t), \boldsymbol{x}_e(t) = [\boldsymbol{x}(t) f(t)]^T \qquad (15)$$

where

$$\bar{\boldsymbol{g}}(\boldsymbol{u}, t, \boldsymbol{\theta}) \triangleq \begin{bmatrix} \boldsymbol{g} \\ 0 \end{bmatrix} \qquad (16)$$

$$\bar{G} \triangleq \begin{bmatrix} G & 0 \\ 0 & 0 \end{bmatrix} \qquad (17)$$

$$A_e \triangleq \begin{bmatrix} A & F \\ 0 & 0 \end{bmatrix} \qquad (18)$$

$$B_e \triangleq \begin{bmatrix} B \\ 0 \end{bmatrix} \qquad (19)$$

$$C_e \triangleq \begin{bmatrix} C & 0 \\ 0 & 0 \end{bmatrix} \qquad (20)$$

and $\bar{\boldsymbol{g}} \in R^{(l+n)}$ and $\bar{G} \in R^{(l+n)}$. Define the estimate of $\boldsymbol{\theta}$ as $\widehat{\boldsymbol{\theta}}$, using the estimate we can design an adaptive observer as follows.

$$\begin{aligned}\dot{\widehat{\boldsymbol{x}}}_e(t) &= A_e\widehat{\boldsymbol{x}}_e(t) + B_e\bar{\boldsymbol{u}}(t) - K_e\boldsymbol{\varepsilon}(t)\\ \widehat{\boldsymbol{y}}_1(t) &= C_e\widehat{\boldsymbol{x}}_e(t)\\ \boldsymbol{\varepsilon}(t) &\triangleq \widehat{\boldsymbol{y}}_a(t) - \boldsymbol{y}_a(t)\end{aligned} \tag{21}$$

where

$$\begin{aligned}\widehat{\boldsymbol{u}} &= \widehat{g}\\ \widehat{g} &= \boldsymbol{g}(\boldsymbol{x}, \boldsymbol{u}, t, \widehat{\boldsymbol{\theta}}) = \boldsymbol{\varphi}(\boldsymbol{u}, t)\widehat{\boldsymbol{\theta}}\\ \boldsymbol{u}(t) &= \bar{\boldsymbol{u}}(t) - \widehat{\boldsymbol{u}}(t)\end{aligned}$$

In this paper, for the simplicity it is assumed that $B = G$. The following parameter adjusting law is given to estimate the unknown parameter $\boldsymbol{\theta}$.

$$\widehat{\boldsymbol{\theta}} = \widehat{\boldsymbol{\theta}}_I + \widehat{\boldsymbol{\theta}}_P \tag{22}$$

$$\dot{\widehat{\boldsymbol{\theta}}}_I(t) = -\Gamma_I\boldsymbol{\varphi}(\boldsymbol{u}, t)(\widehat{\boldsymbol{\theta}} - \boldsymbol{\theta}) \tag{23}$$

$$\widehat{\boldsymbol{\theta}}_P(t) = -\Gamma_P\boldsymbol{\varphi}(\boldsymbol{u}, t)(\widehat{\boldsymbol{\theta}} - \boldsymbol{\theta}) \tag{24}$$

where $\Gamma_I \geq 0$ and $\Gamma_P > 0$ are adjusting gain matrix. The augmented error system is shown as follows.

$$\dot{\boldsymbol{e}}(t) = A_0\boldsymbol{e}(t) + G(g(u, t, \widehat{\boldsymbol{\theta}}) - g(u, t, \boldsymbol{\theta})) \tag{25}$$

$$\boldsymbol{e}(t) = \widehat{\boldsymbol{x}}_e(t) - \boldsymbol{x}_e(t) \tag{26}$$

$$\boldsymbol{\varepsilon}(t) \triangleq C_e\boldsymbol{e}(t) \tag{27}$$

provided that the transfer function between $(g(u, t, \widehat{\boldsymbol{\theta}}) - g(u, t, \boldsymbol{\theta}))$ and $\varepsilon(t)$ is strictly positive real and the error system is observable. As a result, we have $\boldsymbol{\varepsilon}(t) \to 0$ [4] and $\boldsymbol{e}(t) \to 0$. Further, the fault signal $f(t)$ can be estimated by the above adaptive observer.

4. Simulation

As an example of fault detection, a process control experimental system for water level control (see Fig.1, L^*) which has a fault at valve ("in" for Tank1) is considered. As mention above, this paper considers the system with fault signal and uncertainties caused by approximation on real system. That is, we propose a method to estimate the fault signal by using an unknown-input state observer. The proposed scheme estimates the uncertainties and also identifies the system parameters while fault detection works. We show a simulation result to confirm the effectiveness (Figs. 2, 3), where the following fault signal $f(t)(\mathrm{m}^3/\mathrm{s})$ is used.

$$f(t) = \begin{cases} 0.0 & t < 2000 \text{ (s)} \\ f_1 * (t-2000)/500 & 2000 \leq t \leq 2500 \text{ (s)} \\ f_1 & 2500 < t \leq 6000 \text{ (s)} \\ -f_1 * (t-6000)/500 + f1 & 6000 < t \leq 6500 \text{ (s)} \\ 0.0 & 6500 < t \leq 9000 \text{ (s)} \end{cases}$$

$$f_1 = 8.33 * 10^{-6}$$

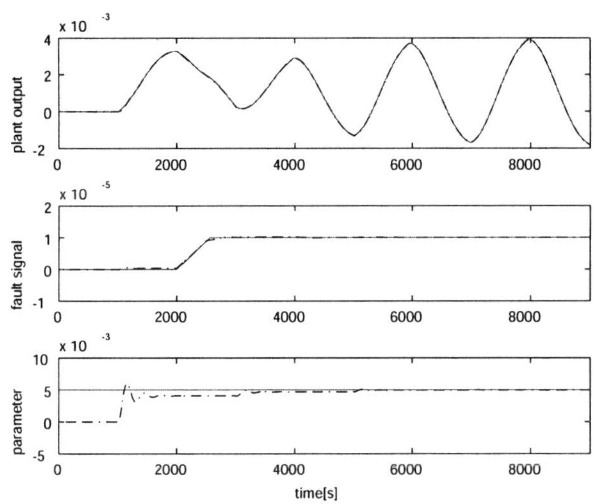

Figure 2. The responses of plant output, fault signal and the estimated parameters. Upper figure is for the water level change, the solid line is the real water level and the dashed line is the estimated one; Middle figure is for fault signal detection, the solid line is $f(t)$ and the dashed line describes the estimated $f(t)$; Lower figure is for parameter estimation of uncertainty, the solid line is the real uncertain parameter and the dashed line is the estimated one

5. Conclusion

In this paper, a design method to detect faults in process experimental system with uncertainties is proposed. The method is based on adaptive

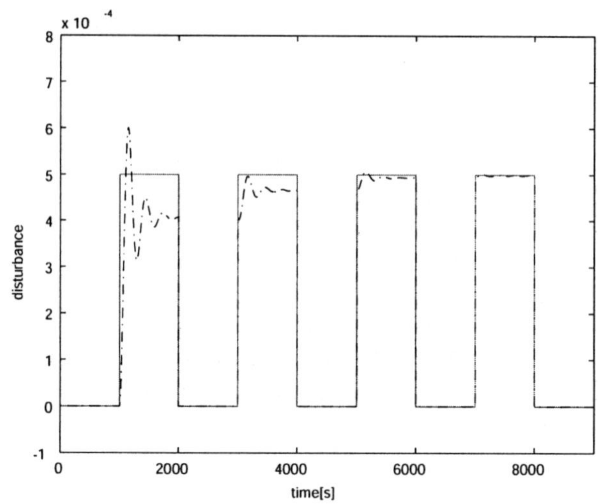

Figure 3. The responses of the uncertainty, the step-wise line is the real uncertainty and another is the estimated one

observer approach. The proposed design scheme estimates the uncertainties and also estimates the process states while fault detection works. Simulation example is illustrated to show the effectiveness of the proposed method.

Acknowledgments

The authors would like to thank JSPS for their financial aid (Grant in aid No. 16101005).

References

1. J. Korbicz, J. M. Koscielny, Z. Kowalczuk and W. Cholewa, *Fault Diagnosis, Models, Artificial Intelligence, Applications.* Springer, 2004.
2. D. W. Gu and F. W. Poon, "A robust fault-detection approach with application in a rolling-mill process," *IEEE Trans. on Control Systems Technology*, Vol.11, No.3, pp.408-414, 2003.
3. P. M. Frank, "Fault diagnosis in dynamic systems using analytical and knowledge-based redundancy -A survey and some new results," *Automatica*, Vol.26, No. 3, pp.459-474, 1990.
4. Z. Iwai, A. Inoue, and S. Kawaji, *Observer*, Corona Publishing Co., Ltd., Tokyo, 1988 (in Japanese).
5. A.Inoue, M.Deng and S.Yoshinaga: Fault detection by using an adaptive observer, Proc. of ICCAS, pp. 710-713, KINTEX, Gyeonggi-Do, Korea, June 2005.

VIBRATION SUPPRESSION OF SYSTEMS WITH UNKNOWN PARAMETERS

MASANORI TAKAHASHI

Kyushu Tokai University,
9-1-1 Toroku, Kumamoto 862-8652 Japan
E-mail: mtakahasi@ktmail.ktokai-u.ac.jp

SHOSAKU KINOSHITA, MASATAKA GOROMARU AND YOSHINORI KAWASAKI

Ariake National College of Technology,
150 Higashi-hagio, Omuta, Fukuoka 836-8585 Japan

ZENTA IWAI

Kumamoto University,
2-39-1 Kurokami, Kumamoto 860-8555

This paper deals with vibration suppression of systems with unknown parameters. Fundamentally, free-vibration can be suppressed by reducing dynamic energy. In this paper, we propose the two novel vibration suppression methods based on inelastic collision of masses. Also the proposed methods will be applied to the problem on reduction of residual vibration in so-called *stacker crane*. To confirm the effectiveness of the proposal, several experiments will be explored.

1. Introduction

According to *law of conservation of momentum*, inelastic collision of masses reduces the dynamic energy stored in conservative systems. By using this fundamental physical phenomenon, two simple vibration suppression methods are developed for systems with unknown parameters. One is asymptotic suppression by iterative collision and separation of masses. The other is perfect suppression by *two-shots* collision. The advantages of these methods are that they do not require any *a priori* information about the vibration systems.

Furthermore, the proposed methods are applied to vibration reduction problem on so-called *stacker cranes*. Stacker cranes have been widely utilized in various warehouses. The difficulty of the vibration reduction problem is that the frequency of the vibration changes with the weight and the hight of the platform. To cope with this problem, we utilize the proposed vibration suppression methods for stacker cranes. Several experiments are explored in order to confirm the effectiveness of the proposals.

2. Vibration Suppression Methods

2.1. *Problem statement*

Consider a vibration system on a smooth plane surface as shown in Fig. 1, which consists of a main mass M, an auxiliary mass m and a spring k. The masses M and m obey the following equations of motion.

$$\begin{aligned} M\ddot{X} &= -kX - T, \quad X(t_0) \neq 0 \\ m\ddot{x} &= T, \quad x(t_0) = X(t_0) \end{aligned} \tag{1}$$

where X and x are the displacements of the masses M and m. T is the internal force between the main mass M and the auxiliary mass m. The parameters, M, m and k are supposed to be unknown. Furthermore, assume that we can control an internal force T in order to realize collision ($T > 0$) and separation ($T = 0$) of the masses.

The main task is to completely suppress the undamped free-vibration of the main mass M by using the auxiliary mass m.

2.2. *Method I*

The vibration can be suppressed by the following procedure, which is also shown in Fig. 1.
Step 0. The masses M and m stick and move together ($X = x$).
Step 1. Separate the masses M and m, when their velocities become zero ($\dot{X} = \dot{x} = 0$).
Step 2. Collide the main mass M with the additional mass m and stick them[a], when the main mass M returns to the equilibrium point ($X = 0$).
Step 3. Go back to Step 1.

Now, let E_n and A_n be the dynamic energy in the system and the amplitude of the main mass M in the nth separation (Step 1) respectively. Furthermore, let V_n^- denote the velocity of the main mass M just before

[a]The collision in Step 2 is the perfectly inelastic collision.

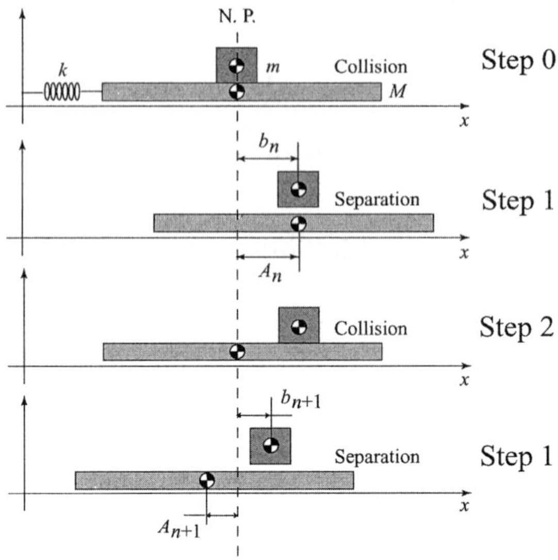

Figure 1. Asymptotic suppression by iterative collision (Method I)

the nth collision. Then, because of *conservation of energy*, we can have

$$E_n = \frac{1}{2}kA_n^2 = \frac{1}{2}MV_n^{-2} \tag{2}$$

After the nth collision (Step 2), the velocity V_n^+ of the combined mass $M + m$ becomes,

$$V_n^+ = \frac{M}{M+m}V_n^- \tag{3}$$

Hence, after the nth collision, the dynamic energy E_{n+1} is stored in the system as,

$$E_{n+1} = \frac{1}{2}(M+m){V_n^+}^2 = \frac{1}{2}kA_{n+1}{}^2 \tag{4}$$

where A_{n+1} is the amplitude of the main mass M in the $n+1$th separation (Step 1). Hence, from (2) and (4), the following relationships hold.

$$E_{n+1} = S^2 E_n, \quad A_{n+1} = SA_n \tag{5}$$

$$S := \left(\frac{M}{M+m}\right)^{\frac{1}{2}} \tag{6}$$

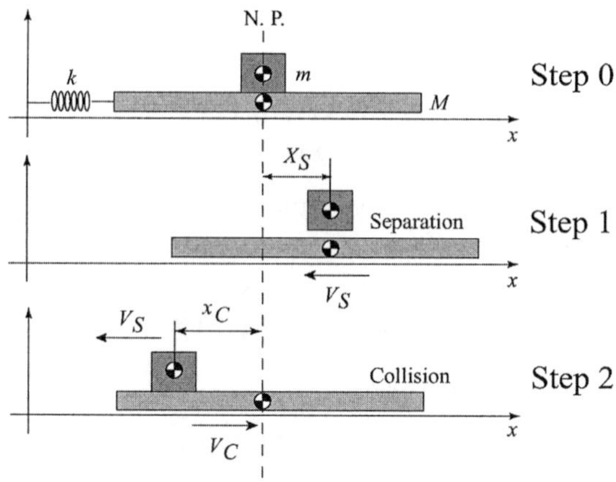

Figure 2. Perfect suppression by 'two-shots' collision (Method II)

Therefore, from the fact that $0 < S < 1$, we can have

$$\lim_{n\to\infty} E_n = 0, \quad \lim_{n\to\infty} A_n = 0 \tag{7}$$

This means that the vibration can be suppressed after the infinite number of collisions.

2.3. *Method II*

Here, we consider the case where $M < m$. In this case, two inelastic collisions suppress the vibration completely.

Step 0. To specify the suppression ratio S given in (6), we execute Step 0 $\sim$ 3 in Method I once as follows (do not separate the masses in Step 3).

Step 0 $\to$ Step 1 ($X = A_1$) $\to$ Step 2 $\to$ Step 3 (Step 1, $X = A_2$).

Measure the amplitudes A_1 and A_2, and calculate $S = A_2/A_1$.

Step 1. Separate the masses M and m, when the combined mass $M + m$ reaches the following point,

$$X = X_s := A_2\left(1 - \frac{M}{m}\right)^{\frac{1}{2}} = A_2\left(\frac{1-2S^2}{1-S^2}\right)^{\frac{1}{2}} \tag{8}$$

Step 2. Collide the main mass M with the auxiliary mass m and stick them, when the main mass M returns to the equilibrium point ($X = 0$) and the directions of two velocities $\dot{X}$ and $\dot{x}$ differ mutually ($\dot{X} \cdot \dot{x} < 0$).

The dynamic energy E stored in Step 1 is given by

$$E = \frac{1}{2}kA_2{}^2 = \frac{1}{2}kX_s{}^2 + \frac{1}{2}MV_S{}^2 + \frac{1}{2}mV_S{}^2 \tag{9}$$

where V_S is the velocity of the main mass M (also the auxiliary mass). From (8) we have

$$E = \frac{m(M+m)}{2M}{V_S}^2 \tag{10}$$

Now, let V_C be the velocity of the main mass M just before the collision. Then, the dynamic energy (9) can be represented as

$$E = \frac{1}{2}M{V_C}^2 + \frac{1}{2}m{V_S}^2 \tag{11}$$

Thus, from (10) and (11) we have

$$m^2{V_S}^2 - M^2{V_C}^2 = (mV_S + MV_C)(mV_S - MV_C) = 0 \tag{12}$$

Taking $V_S \cdot V_C < 0$ into consideration, we have that $mV_S - MV_c \neq 0$. Hence, after the collision, the momentum of the masses becomes zero as,

$$mV_S + MV_C = 0 \tag{13}$$

This implies that the dynamic energy becomes zero ($E = 0$) because the velocities of the masses become zero at the equilibrium point. Thus, the vibration can be suppressed completely by three collisions.

Remark: If $M < m$ then from (6) we have

$$S < (0.5)^{\frac{1}{2}}$$

Hence, in Step 1, if the suppression ratio S does not satisfy the above inequality then we utilize Method I.

3. Residual Vibration Reduction in a Stacker Crane

3.1. *Vibration reduction problem*

The stacker crane consists of the cart and the platform fundamentally. The cart can move forward and backward, and the platform can be lifted up and down (Fig. 1). Thus, the combination of their movements makes it possible to realize 2D transportation in a warehouse [1].

Recently, stacker cranes tend to lack rigidity due to the light-weight structures. Hence, the residual vibration in the body is often induced after the cart stops. In addition, the frequency of the vibration always changes with respect to the hight of the platform and the load.

To cope with these problems, we apply the proposed vibration methods I and II to the vibration reduction problem on stacker cranes.

3.2. *Modification of the structure of the platform*

To exploit the proposed vibration suppression methods, we modify the structure of the platform as shown in Fig. 4. The conventional platform is settled on the frame of the crane. In contrast, the proposed one has the wheels and so it can moves independently of the crane. Furthermore, the wheels have the mechanical brakes which generate the internal force T seen in (1). By controlling the internal force T, we can collide the platform (as the auxiliary mass m) with the frame (as the main mass M). From (6), the heavier load m provides the higher suppression effect S in these methods. This is the reason why the proposed method is useful and suitable for stacker cranes.

Thus, we can apply the proposed vibration suppression methods and suppress the residual vibration in the stacker cranes.

4. Experimental results

4.1. *An experimental model of the stacker crane*

To confirm the effectiveness of the proposed methods, we explore the experiments by using a small-scaled model of the stacker crane, which is also shown in Fig. 5. The modified platform has the four wheels and the electromagnet which generates the internal force for collision and separation. The parameters of the experimental model are listed in Table 1.

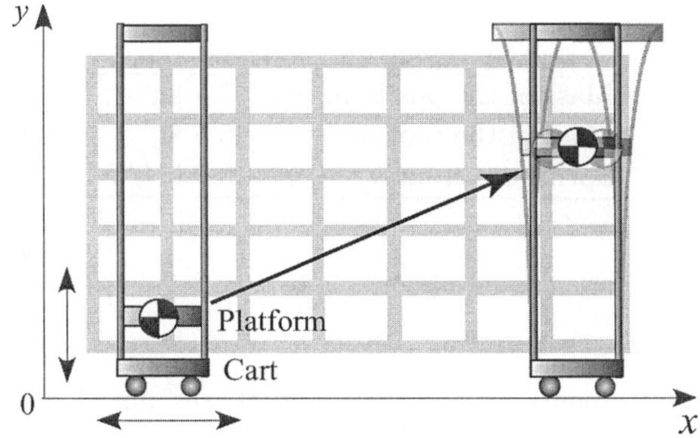

Figure 3. Illustration of a stacker crane

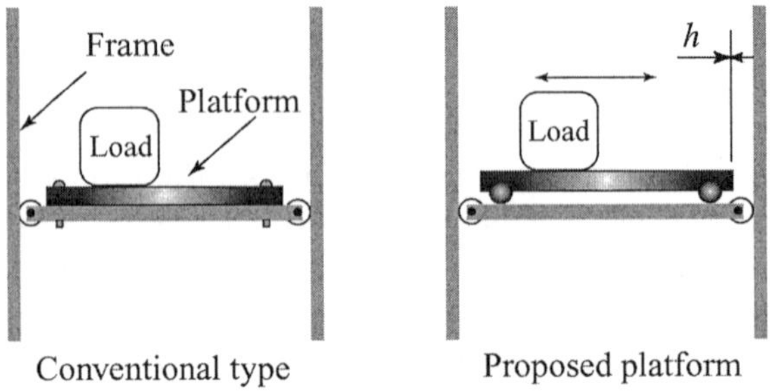

Figure 4. A modified structure of the platform for vibration reduction

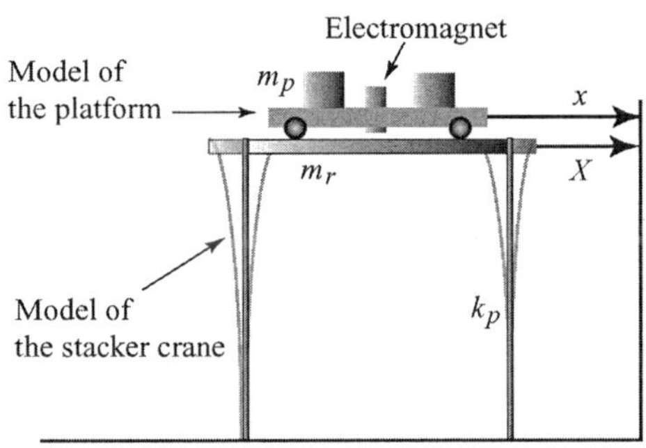

Figure 5. Overview of the experimental apparatus

Table 1 Parameters of the model

Mass of the platform m_p	7.5 kg
Mass of the roof m_r	5.5 kg
Spring constant k_p	0.2 kN/m

4.2. *Experimental results*

In both the experiments by Method I and II, the initial displacement X of the crane (roof) are set as $|X(0)| \simeq 100$ [mm]. Fig. 6 and 7 show the results. The dashed lines indicate the free-vibrations with no control. Both methods can suppress the vibration efficiently.

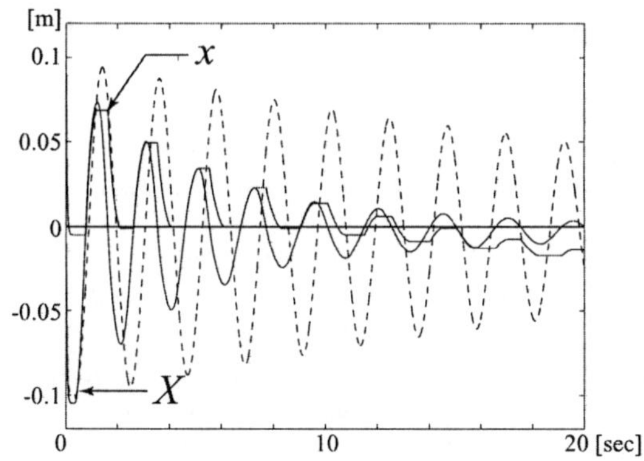

Figure 6. Results by Method I

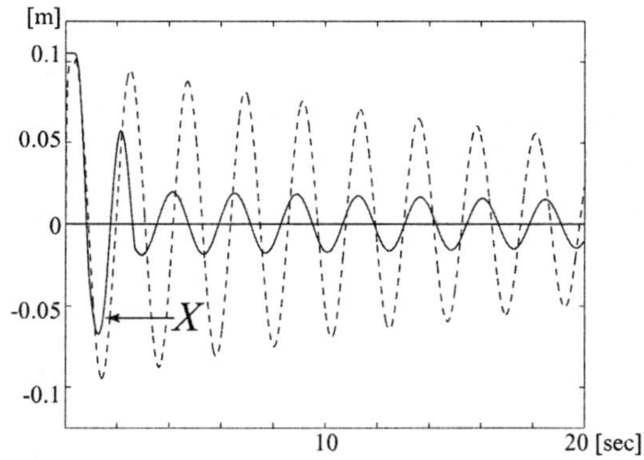

Figure 7. Results by Method II

5. Conclusions

This paper has presented the new vibration suppression methods, and shown the applications to vibration reduction problem on stacker cranes. It has been shown by theoretical analysis in dynamic energy and the experiments that free-vibration can be suppressed successfully.

References

1. M. Takahashi, S. Kinoshita, K. Tsutsumi, Y. Kawasaki and Z. Iwai, *Proceedings of the 10th Asia-Pacific Vibration Conference*, Vol. 2, 680 (2003)

A ROBUST STATE OBSERVER FOR nDOF LAGRANGIAN SYSTEMS

David I. Rosas Almeida

Autonomous University of Baja California, Engineering Faculty, Blvd. Benito Juarez s/n, Mexical, B.C. Mexico
E-mail: drosas@uabc.mx

Joaquin Alvarez

Scientific Research and Advanced Studies Center of Ensenada, Km 107 Carr. Tij.-Ens., Ensenada, B.C., Mexico,
E-mail: jqalvar@cicese.mx

We present a design methodology for a robust Luenberger-type state observer for nDOF-Lagrangian systems. The observer maintains its characteristic of exponential convergence to the steady-state in spite of the presence of bounded parameter variations and bounded external disturbances. The observer incorporates a discontinuous term in the feedback signal, so the stability proof is based on analysis tools for variable structure systems. The design methodology is experimentally illustrated.

Keywords: Lagrangian systems, robust observer, variable structure systems.

1. Introduction

One of the main problems in the implementation of control techniques based on state feedback is that frequently the state vector is not available. In some situations the state vector can be measured through sensors; however, in some other cases its measurement is not possible or turns out to be very expensive. This problem can be solved using observers; these systems calculate the state variables needed to implement a controller.

One of the most used scheme to solve this problem is the so-called Luenberger observer that, in the beginning, it was used only for linear systems [1]. At present, there exist many versions of this scheme for their application to nonlinear systems, see for example [2].

The classic Luenberger scheme performs well when the plant dynamics is more or less exactly known. Ideally, we need an exact model of the plant to apply this scheme; however, in the presence of perturbations the estimation of the states may not be sufficiently accurate, degrading the controller performance.

To solve this problem some schemes for robust observers have been proposed. In [2] it is proposed an approach to robust state observer based on Luenberger observer that requires the solution of an algebraic Riccati equation; this scheme is applied to perturbed linear systems. Other schemes use the Kalman filter, see for example [3] and [4].

On the other hand, in recent years several results have appeared on the stability of systems with variable structure, see for example [5] and [6], where it is shown that this class of systems can have good convergence properties and, at the same time, they can display good performance in face to bounded external disturbances [7].

In this sense, several observers have been proposed based on sliding mode techniques [8]. Such observers show good characteristics of robustness to bounded external perturbations and converge to the steady-state in finite time. However, for MIMO systems the design procedure can be a difficult task.

One important class of nonlinear systems are the so called Lagrangian systems. Mechanisms and some circuits are some examples of systems that belong to this class.

In this paper we present a new design of a robust observer for nDOF Lagrangian systems. The design is based on the Luenberger observer for non linear systems, where we add a discontinuous term in the feedback, such that a variable structure system is produced, and the tools developed for this kind of systems can be applied. The equilibrium point of the error system between the plant and the observer maintains its characteristic of exponential stability in spite of parameter variations and nonvanishing bounded external disturbances.

The organization of the paper is as follows. In the second section a theorem showing the stability properties of a class of discontinuous second order systems is discussed. In the third section we present the observer design. In the fourth section, the design technique is experimentally illustrated with the design and implementation of an observer for a simple pendulum. In the final section we present some conclusions.

2. Stability of a class of perturbed second order systems

We present in this section a preliminary result that will be useful to design the observer.

Consider the following second order system

$$\dot{v}_1 = v_2, \qquad (1)$$
$$\dot{v}_2 = -av_1 - bv_2 + \epsilon(t) - c\,\mathrm{sgn}(v_1),$$

where a and b are positive constants, $\epsilon(t)$ is an external perturbation with the bound $|\epsilon(t)| \leq \rho$, where ρ is a constant, c is a control parameter, and $\mathrm{sgn}(\cdot)$ is the signum function. Define the matrix A as

$$A = \begin{bmatrix} 0 & 1 \\ -a & -b \end{bmatrix}, \qquad (2)$$

and the matrix P, which is the solution of the Lyapunov equation for the (Hurwitz) matrix A, as

$$P = \begin{bmatrix} p_{11} & p_{12} \\ p_{12} & p_{22} \end{bmatrix}. \qquad (3)$$

The stability properties of (1) are given by the following theorem.

Theorem 2.1. *For system (1), if*

$$c > 2\lambda_{\max}(P)\sqrt{\frac{\lambda_{\max}(P)}{\lambda_{\min}(P)}}\left(\frac{a\rho}{\theta}\right),$$

for some $0 < \theta < 1$, *then the origin of the state space will be a global exponentially stable equilibrium point in Lyapunov sense.*

Proof. The proof is divided in two parts; first we define the nominal system as (1) with $\epsilon(t) = 0$, and prove the stability of the origin using tools from variable structure systems. After that, we find the condition on c such that the stability properties are maintained for the perturbed system.

The nominal system of (1) is defined as

$$\dot{v}_1 = v_2, \qquad (4)$$
$$\dot{v}_2 = -av_1 - bv_2 - c\,\mathrm{sgn}(v_1).$$

System (4) has two structures: S_1 for $v_1 > 0$,

$$S_1 : \begin{bmatrix} \dot{v}_1 \\ \dot{v}_2 \end{bmatrix} = \begin{bmatrix} v_2 \\ -av_1 - bv_2 - c \end{bmatrix},$$

and S_2 for $v_1 < 0$

$$S_2 : \begin{bmatrix} \dot{v}_1 \\ \dot{v}_2 \end{bmatrix} = \begin{bmatrix} v_2 \\ -av_1 - bv_2 + c \end{bmatrix}.$$

Each structure has a different equilibrium point; for S_1 it is $\overline{v}_{S_1} = (-c/a, 0)$, and for S_2 we have $\overline{v}_{S_2} = (c/a, 0)$. Note that these equilibria are placed in the region where the system dynamics is given by the other structure (S_2 for $\overline{v}_{S_1}$, S_1 for $\overline{v}_{S_2}$).

Each equilibrium point is global exponentially stable with the following Lyapunov functions; for S_1

$$V_{S_1}(v) = v^T P v + 2v^T P\gamma + \left(\frac{c}{a}\right)^2 p_{11}, \tag{5}$$

and for S_2

$$V_{S_2}(v) = v^T P v - 2v^T P\gamma + \left(\frac{c}{a}\right)^2 p_{11}, \tag{6}$$

where $\gamma = \left[c/a \; 0\right]^T$.

A direct application of the criterion given in [9] to prove the existence of sliding modes allows to conclude that the discontinuity surface given by $\sigma = v_1 = 0$ is not a sliding surface.

Note also that the solutions cross the line $v_1 = 0$ from quadrant II to quadrant I, and from quadrant IV to quadrant III.

Now consider the functions $V_{S_1}(v)$, $V_{S_2}(v)$. These functions intersect at the origin with a value $V_{S_i}(0) = (c/a)^2 p_{11}$, for $i = 1, 2$. Define two neighborhoods of the origin, Ω_ε with radio $\varepsilon > 0$, and Ω_β defined in the following form,

$$\begin{aligned} \Omega_\beta &= \Omega_1 \cup \Omega_2, \\ \Omega_1 &= \left\{v \in \Re^2 \mid v_1 \geq 0, V_{S_1}(v) \leq \beta\right\}, \\ \Omega_2 &= \left\{v \in \Re^2 \mid v_1 < 0, V_{S_2}(v) \leq \beta\right\}, \end{aligned}$$

where $\beta > (c/a)^2 p_{11}$. Finally, define a neighborhood Ω_δ with radio $\delta < \varepsilon$ (δ can depend on ε and β; $\delta(\varepsilon, \beta)$) such that $\Omega_\delta \subset \Omega_\beta$.

Define a set of times $T = \{t_1, t_2, \ldots, t_i, \ldots\}$, at these moments the structure commutations appear. We assume that $t_0 < t_1 < t_2 < \ldots$

If $\|v(t_0)\| < \delta$ and $v(t_0) \in \Omega_k \subset \Omega_\beta$ for some $k = 1, 2$ (the $k-$th structure is active), then the first change of structure appears at time t_1, and because $\dot{V}_{S_k} < 0$, we have $\|v(t_0)\| > \|v(t_1)\|$, then $V_{S_k}(v(t_0)) > V_{S_k}(v(t_1))$.

Now $v(t_1)$ is the initial condition for the next structure and, by construction of V_{S_i}, $V_{S_k}(v(t_1)) > V_{S_{k+1}}(v(t_1))$ by a factor $4|v_2(t_1)| p_{12}(c/a)$. The second commutation appears at time t_2; the system goes from Ω_{k+1} to Ω_k, $\|v(t_1)\| > \|v(t_2)\|$, $V_{S_{k+1}}(v(t_1)) > V_{S_{k+1}}(v(t_2))$ and $V_{S_{k+1}}(v(t_2)) > V_{S_k}(v(t_2))$ and so on for all $t_i \in T$.

Then we see that the sequences $W_1 = \{V_{S_k}(t_1), V_{S_k}(t_3), ...\}$ and $W_2 = \{V_{S_{k+1}}(t_2), V_{S_{k+1}}(t_4), ...\}$ are strictly decreasing and lower bounded and converge to $(c/a)\, p_{11}$, and also it is satisfied that $\|v(t_{i+1})\| < \|v(t_i)\| < \cdots < \|v(t_0)\| < \delta < \varepsilon \ \forall t > t_0, \forall i$.

For all $\varepsilon > 0$ and $\beta > (c/a)^2 p_{11}$ we can find a number δ so that the trajectories initiating in Ω_δ will remain within the neighborhood Ω_ε for all $t \geq t_0$. Therefore, the origin is stable in the Lyapunov sense.

To demonstrate asymptotic stability it is enough to notice that

$$\lim_{i \to \infty} V_{S_k}(t_i) = \lim_{i \to \infty} V_{S_{k+1}}(t_i) = \frac{c}{a} p_{11},$$

this is the value that takes both Lyapunov functions at the origin; then

$$\lim_{t \to \infty} v(t) = 0.$$

To demonstrate exponential stability it is enough to notice that the solution in the interval time $[t_0, t_1)$ decreases in exponential form due to the exponential stability of the equilibrium point of that structure.

Finally, to demonstrate that this result is global it is enough to notice that each structure has a global exponentially stable equilibrium; therefore, all properties mentioned before will remain for all initial condition.

Now we analyze the perturbed system (1). Consider the structure S_1 of the system (the analysis of the structure S_2 is similar), and make the following change of variables $z_1 = v_1 + c/a$ and $z_2 = v_2$. The dynamics of system (1), in the new state space, is given by

$$\begin{aligned} \dot{z}_1 &= z_2, \\ \dot{z}_2 &= -a z_1 - b z_2 + \epsilon(t), \end{aligned}$$

or in simplified form, $\dot{z} = Az + g$, where $g = \left[0 \ \epsilon(t)\right]^T$.

Propose a Lyapunov function

$$V(z) = z^T P z,$$

where the matrices A and P are defined by equations (2) and (3), respectively. The derivative of V is given by

$$\dot{V}(z) = -z^T z + 2 z^T P g \leq -\|z\|^2 + 2\lambda_{\max}(P)\|z\|\rho.$$

Because $a > 0$ and $b > 0$, we can apply lemma 9.2 given in [10], and conclude that, for all $\|z(t_0)\| > \mu$ the solution $z(t)$ satisfies

$$\|z(t)\| \leq k \exp(-\gamma(t - t_0))\|z(t_0)\| \qquad \forall t_0 \leq t < t_0 + t_f,$$

and

$$\|z(t)\| \leq \mu \qquad \forall t \geq t_0 + t_f,$$

where t_f is a finite time, and

$$k = \sqrt{\frac{\lambda_{\max}(P)}{\lambda_{\min}(P)}}, \gamma = \frac{(1-\theta)}{2\lambda_{\max}(P)}, \mu = 2\lambda_{\max}(P)\sqrt{\frac{\lambda_{\max}(P)}{\lambda_{\min}(P)}}\frac{\rho}{\theta},$$

for some θ, $0 < \theta < 1$. This part shows that the ball of radius μ, with center located at $(-c/a, 0)$, is an attractor for structure S_1, denoted as B_{S_1}.

The equilibrium point of the structure S_2

$$\begin{aligned} \dot{v}_1 &= v_2, \\ \dot{v}_2 &= -av_1 - bv_2 + \epsilon(t) + c, \end{aligned}$$

has the same stability properties that the equilibrium point of the structure S_1. Then the ball B_{S_2} of radius μ and centered at $(c/a, 0)$ is an attractor for S_2.

Then each structure of the nominal system has an equilibrium point symmetrically located on the v_1-axis at a distance $r = c/a$ from the origin. If this distance is grater than μ, i. e. if

$$c > 2\lambda_{\max}(P)\sqrt{\frac{\lambda_{\max}(P)}{\lambda_{\min}(P)}}\left(\frac{a\rho}{\theta}\right), \tag{7}$$

then the two attractor B_{S_1} and B_{S_2} of the perturbed system (1) do not intersect each other, and the behavior of the solution of the perturbed system will be qualitatively similar to the behavior of the nominal system, for which these attractors correspond when $\mu = 0$. Therefore, the perturbed system converges to the origin in the same way than the nominal system□

3. State observer design

Consider now a nDOF Lagrangian system described by the following equation

$$M(q)\ddot{q} + C(q,\dot{q})\dot{q} + D\dot{q} + g(q) + \varepsilon(t) = \tau, \tag{8}$$

where $q \in \Re^n$ is the generalized position vector, $M(q)$ is the inertia matrix, $C(q,\dot{q})$ is the centrifuge and Coriolis matrix, D is a positive definite diagonal matrix that contains the viscous friction coefficients of each link, $g(q)$ is the gravitational force vector, $\varepsilon(t)$ is a perturbation term that may encloses parametric variations and external perturbations; we assume that this term is bounded, $\|\varepsilon(t)\| \leq \rho$, and τ is the input torque. All matrices and vectors are defined with the suitable dimensions.

Defining the state variables $x_1 = q$, $x_2 = \dot{q}$, the state space representation of system (8) is

$$\begin{bmatrix} \dot{x}_1 \\ \dot{x}_2 \end{bmatrix} = \begin{bmatrix} x_2 \\ M^{-1}(\cdot)\left(-C(\cdot)x_2 - Dx_2 - g(\cdot) - \varepsilon(t) + \tau\right) \end{bmatrix}, \tag{9}$$

$$y = x_1, \tag{10}$$

where $y \in \Re^n$ is the output equation given by the position vector x_1. We assume that the behavior of system (9-10) is bounded for any bounded input τ and any bounded perturbation $\varepsilon(t)$.

We propose an observer for system (9 - 10) as

$$\begin{bmatrix} \dot{\hat{x}}_1 \\ \dot{\hat{x}}_2 \end{bmatrix} = \begin{bmatrix} \hat{x}_2 \\ M^{-1}(\cdot)\tau \end{bmatrix} + f(y - \hat{y}), \tag{11}$$

$$\hat{y} = \hat{x}_1, \tag{12}$$

where the vector $f(y - \hat{y})$ has the form

$$f(y - \hat{y}) = \begin{bmatrix} C_1(y - \hat{y}) \\ C_2(y - \hat{y}) + C_3 \mathrm{sgn}(y - \hat{y}) - M^{-1}(\cdot)g(\cdot) \end{bmatrix}, \tag{13}$$

where C_1, C_2 and C_3 are definite positive diagonal matrices. Define the error variables $e_1 = x_1 - \hat{x}_1$, $e_2 = x_2 - \hat{x}_2$, then the error dynamics is described by

$$\begin{bmatrix} \dot{e}_1 \\ \dot{e}_2 \end{bmatrix} = \begin{bmatrix} e_2 - C_1 e_1 \\ \Psi(t) - C_2 e_1 - C_3 \mathrm{sgn}(e_1) \end{bmatrix}. \tag{14}$$

where $\Psi(t) = M^{-1}(x_1)\left(-C(x_1, x_2)x_2 - Dx_2 - \varepsilon(t)\right)$.

Proposition 3.1. *For system (14) it is possible to find a set of matrices C_1, C_2 and C_3 such that the origin of the error space will be a global exponentially stable equilibrium point. Then the system defined by (11) and (12) is an observer for the system defined by (9) and (10).*

Proof. Because x_1 and x_2 are bounded, the term $\Psi(t)$ is bounded, then each term $\psi_i(t)$, $i = 1, ..., n$ is bounded too. From (14) we can see that this system is a set of n subsystems given by

$$\begin{bmatrix} \dot{e}_{1,i} \\ \dot{e}_{2,i} \end{bmatrix} = \begin{bmatrix} e_{2,i} - c_{1,i}e_{1,i} \\ -c_{2,i}e_{1,i} - c_{3,i}sgn(e_{1,i}) + \psi_i(t) \end{bmatrix}$$

for $i = 1, ..., n$. Now make a change of variables

$$v_{1,i} = e_{1,i}, v_{2,i} = e_{2,i} - c_{1,i}e_{1,i}.$$

The dynamics of system (14) in the new state space is given by

$$\begin{aligned} \dot{v}_{1,i} &= v_{2,i}, \\ \dot{v}_{2,i} &= -c_{2,i}v_{1,i} - c_{1,i}v_{2,i} + \psi_i(t) - c_{3,i}sgn(v_{1,i}), \end{aligned} \tag{15}$$

for $i = 1, ..., n$. This system has the form of system (1), and by applying theorem 1 we can find the conditions on $c_{1,i}$, $c_{2,i}$, and $c_{3,i}$ □

We can see that the proof is very easy taking the result presented in the last section; now we present the conditions on the constants to design an observer

4. A simple pendulum example

Consider a simple pendulum model given by

$$\begin{aligned} \dot{x}_1 &= x_2, \\ \dot{x}_2 &= -k_1x_2 - k_3\sin(x_1) + k_4\tau + \varepsilon(t), \\ y &= x_1 \end{aligned} \tag{16}$$

where $k_1 = 2.9996^{-2}$, $k_3 = 67.912$, $k_4 = 55.549$ and $\varepsilon(t)$ is a perturbation term that satisfies the following bound

$$|\varepsilon(t)| \leq \rho,$$

where ρ is a constant. In this case we suppose that $\rho = 0.5$. Now, the state observer for system (16) is proposed to be

$$\begin{aligned} \dot{\widehat{x}}_1 &= \widehat{x}_2 + v_1, \\ \dot{\widehat{x}}_2 &= -k_1\widehat{x}_2 - k_3\sin(\widehat{x}_1) + k_4\tau + v_2, \\ \hat{y} &= \hat{x}_1, \end{aligned}$$

where v_1 and v_2 are given by

$$\begin{aligned} v_1 &= c_1e_1, \\ v_2 &= c_2e_1 + c_3\text{sgn}(e_1) + k_3(\sin(\widehat{x}_1) - \sin(x_1)). \end{aligned}$$

We select the constants $c_1 = 2$, $c_2 = 10$, $\theta = 0.9$ and $c_3 = 10$.

The experimental results are shown in the following figures. For $c_3 = 0$ (classic Luenberger observer) the error between the real angle and the observed angle is very large, see Figure 1. In this and the following figures the vertical line indicates the time where the torque was applied to the pendulum.

Figure 2 shows the experimental results for the proposed observer, with $c_3 = 10$. As we can see, after a transient due to the initial condition, the error is almost zero.

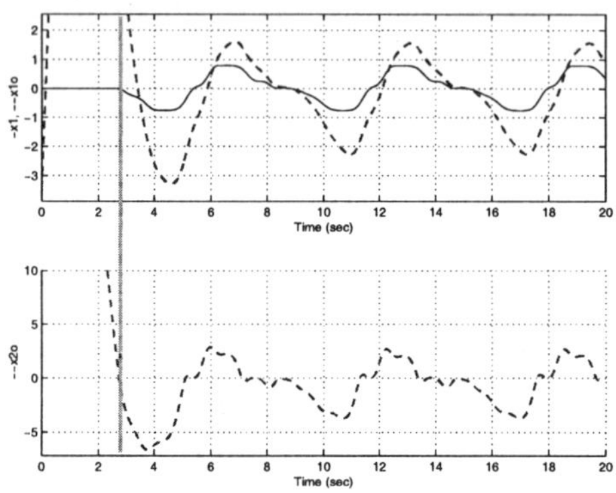

Fig. 1. Experimental results. Behavior of the plant and the observer for $c_3 = 0$ (classical observer).

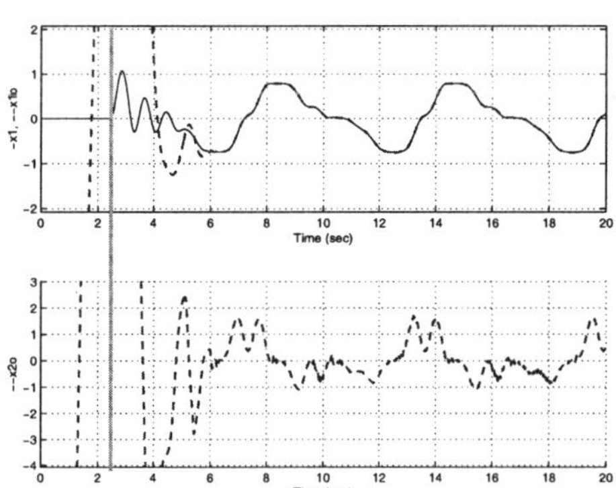

Fig. 2. Experimental results. Behavior of the plant and the observer for $c_3 = 10$ (proposed observer).

5. Conclusion

The main contribution of this work is the design of a discontinuous observer for a class of nonlinear systems. The observer displays good characteristics of robustness to bounded parametric variation and external perturbations. For the case of plants with perturbations this observer has better performance than the classic Luenberger observer and its design procedure is easier than some sliding mode observers.

Due to its discontinuous nature, the observer state vector displays chattering; however, in the experimental results chattering was not an important problem.

Also, this observer guarantees exponential rate of convergence to the state of the plant in spite of the existence of nonvanishing bounded perturbations. Therefore, this observer can be very useful in the implementation of feedback controllers when the plant has this kind of perturbations.

References

1. D. Luenberger, "An introduction to observers", *IEEE Transaction on Automatic Control*, Vol. AC-16, 1971, pp 596-602.
2. Da-Wei and Fu Wah Poon, "A robust state observer scheme", *IEEE Transaction on Automatic Control*, Vol 46, No. 12, December 2001, pp 1958-1963.
3. Sang Joo Kwon, Wan Kyun Chung, and Youngil Youm, "A combined observer for robust state estimation and Kalman filtering", *2003 American Control Conference (ACC)*, Denver, Colorado, 2003
4. Michel Fliess and Hebertt Sira-Ramírez, "State reconstructors, a possible alternative to asymptotic observers and Kalman filters", *Proceedings CESA 2003*, 2003.
5. Branicky Michael S., "Multiple Lyapunov functions and other analysis tools for switched and hybrid systems", *IEEE Transaction on Automatic Control*, Vol. 43, No. 4, 1998, pp. 979-982.
6. Liberzon Daniel, *Switching in Systems and Control*, Birkhäuser, Boston, 2003.
7. Joaquín Alvarez, Iouri Orlov and Roque Martinez, "A discontinuous control for robotic manipulators with Coulomb friction", *Proc. of the 15th Triennial World Congress*, Barcelona, Spain, 2002.
8. Barbot J-P, Djemai M., and Boukhobza T. *Sliding Mode Observers,* Chapter 4 of *Sliding Mode Control in Engineering*, Edited by Wilfrid Perruquetti and Jean Pierre Barbot. Marcel Bekker Inc. New York, 2002.
9. Utkin, V. I., *Sliding Modes in Control and Optimization*, Springer-Verlag, 1992.
10. Hassan K. Khalil, *Nonlinear Systems.* Prentice Hall, 2002.

Two Degree-of-Freedom of Self-Tuning Generalized Predictive Control Based on Polynomial Approach with Computational Savings

AKIRA YANOU

School of Eng., Kinki University,
1, Takayaumenobe, Higashi-Hiroshima 739-2116, JAPAN
E-mail: yanou@hiro.kindai.ac.jp

SHIRO MASUDA

Faculty of Eng., Tokyo Metropolitan Institute of Technology,
6-6, Asahigaoka, Hino 191-0065, JAPAN
E-mail: smasuda@cc.tmit.ac.jp

AKIRA INOUE

Faculty of Eng., Okayama University,
3-1-1, Tsushimanaka, Okayama 700-8530, JAPAN
E-mail: inoue@suri.sys.okayama-u.ac.jp

We propose a design scheme of two degree-of-freedom of self-tuning generalized predictive control based on polynomial approach with computational savings. When the identified plant parameters converge on true values, the proposed method reveals the effect of the integral compensation only if there exists modeling error or disturbance. Therefore, performance degradation due to an integral compensation can be avoided when the controlled system has no perturbation.

Keywords: Generalized predictive control, Two degree-of-freedom, Self-tuning control, Polynomial approach

1. Introduction

Generalized Predictive Control (GPC) is first proposed by Clarke and others in 1987.[1] The control law is derived by minimizing the performance index, which has the parameters of the prediction horizon and the control horizon and the weighting factor of the inputs. The prediction horizon is a period of calculating the prediction of the plant output and the control horizon is a period of finding the optimal inputs on the performance index. The control signals are derived by minimizing the performance index on the future control inputs. And they are re-calculated by receding from their horizons at each step. With these features, the control strategy follows the reference signal of the step robustly by including an integrator in the controller and it has been accepted by many practical engineers and applied widely in industry. Whereas, if the controlled plant is modeled accurately and there has no disturbance, the controlled output can track to the reference signal of the step without an integrator in it. And the effect of an integral compensation has the possibility of the change for the worse of the characteristics of the transient response or the increase of the control input. For safety therefore it is desirable that an effect of an integral compensation appears only if there exists modeling error or disturbance. In this paper, this feature is defined as two degree-of-freedom system because the characteristics of the output response and the disturbance response can be designed independently, that is, on one hand the characteristic of the output response is designed by minimizing the performance index which includes a control input and a tracking error between the reference signal and the output, on the other hand the characteristic of the disturbance response is designed by the gain of the integral compensator.

Although many papers have proposed two degree-of-freedom optimal servo systems[2,3] and the authors have proposed two degree-of-freedom of generalized predictive control based on state space approach and polynomial approach.[4–6] For two degree-of-freedom of GPC based on polynomial approach, it needs to solve Diophantine equation to derive the control law. In the case of designing self-tuning controller, the amount of computation to solve Diophantine equation increases according to the length of the prediction horizon. To apply actual systems, it is desirable to reduce the amount of computation to derive a control law. For this problem, although Saudagar and others have proposed the method to reduce the amount of computation to solve Diophantine equation in LRPC,[7] its controller has the possibility of the change for the worse of the transient response or the increase of the control input because it is derived by including an integral compensation in advance, as the conventional GPC.[1] Therefore this paper considers two degree-of-freedom of self-tuning GPC for single-input single-output systems with computational savings, based on the method by Saudagar and others.[7]

The proposed method is designed by the following steps. First, although GPC strategy proposed by Clarke and others has an integral compensation by including an integrator in the performance index, it makes the strategy proposed in this paper difficult, because in this paper the amount of the integral action must be calculated analytically on condition that there does not exist modeling error and disturbance. Therefore deriving the prediction of the output by solving Diophantine equation,[7] the controller is first designed with no integral compensation in the performance index, on condition that there is neither modeling error nor disturbance. Concretely, the performance index of two degree-of-freedom of GPC in this paper consists of the summation of the square of the following error of the output for the reference signal in the prediction horizon and the summation of the square of the variation of the control input from the steady state input in the control horizon, where the steady state input is calculated on the supposition that the output coincides with the reference signal. Second, a new controller is designed by adding an integral compensation to the controller in the first step. The effect of its integral compensation can be designed by a gain introduced in the controller at once. This controller always reveals the integral action and it may have an extra integral compensation because the integral action is merely added to the controller designed in the first step. Therefore calculating the extra integral compensation analytically and subtracting it from the controller designed in this step, performance degradation will be able to be avoided when the control system has no perturbation. In the third step, two degree-of-freedom GPC with computational savings is obtained by calculating the amount of the integral compensation for the controller designed in the second step on condition that there is neither modeling error nor disturbance on the nominal model, and subtracting its amount from the control input designed in the second step. When there is neither modeling error nor disturbance, its controller generates the control input designed in the first step, which has no integral compensation. And it reveals the effect of the integral compensation in the case that there exists modeling error or disturbance. This feature is obtained by subtracting the integral compensation calculated in the last step from the control designed by the second step in advance. Finally, by adding the parameter identification law to the controller derived in the third step, the self-tuning two degree-of-freedom GPC with computational savings proposed in this paper can be obtained.

The numerical example is also shown to verify the validity of the proposed method.

2. Problem Statement

Consider single-input single-output systems,

$$A[z^{-1}]y(t) = z^{-k_m}B[z^{-1}]u(t) \tag{1}$$

where $y(t)$ and $u(t)$ denote the output and the input. $A[z^{-1}]$ and $B[z^{-1}]$ are the n-order and m-order polynomials respectively. And the integrator $w(t)$ is given by the following equation,

$$w(t) = \frac{1}{\Delta}e(t) \quad (\Delta = 1 - z^{-1}) \tag{2}$$

where the tracking error $e(t)$ is defined by $e(t) = r(t) - y(t)$. The control objective is that the output $y(t)$ tracks to the reference signal $r(t)$.

3. Design Scheme of Two DOF of GPC with Computational Savings

Two degree-of-freedom of GPC with computational savings is designed by following steps. In the first step GPC with computational savings is designed without an integral action in the performance index. The controller given by this step can not achieve the control objective because it does not include an integral action, that is, if there exists modeling error or disturbance, $y(t)$ given by the controller in the first step can not track to $r(t)$. Therefore, in the second step an integral action is added to the controller given by the first step. Then because the controller includes an integral action, $y(t)$ given by its controller can track to $r(t)$ even if there exists modeling error or disturbance. But the integral action always acts, there may be an extra integral action, that is, the degradation of the transient response or the increase of the control input may occur. Therefore in the third step, by calculating the extra integral action and subtracting it from the controller in the second step, two degree-of-freedom of GPC controller with computational savings is given. The controller in the third step appears the effect of an integral compensation only when there exists modeling error or disturbance. By combining the controller in the third step with a parameter identification law, two degree-of-freedom of self-tuning GPC with computational savings, that is, the proposed controller can be obtained.

In this section it is assumed that the identified values of the plant parameters converged on the true values.

3.1. *GPC without an integral action*

The prediction $\hat{\tilde{y}}(t)$ is given for the deviation system of the plant Equation (1). The steady state values y_∞ of $y(t)$ and u_∞ of $u(t)$ are derived as follows.

$$A[z^{-1}]y_\infty = z^{-k_m}B[z^{-1}]u_\infty$$

From the previous equation of the steady state values, the deviation system of the plant Eq. (1) is obtained,

$$A[z^{-1}]\tilde{y}(t) = z^{-k_m}B[z^{-1}]\tilde{u}(t) \tag{3}$$

where the deviations $\tilde{y}(t)$ and $\tilde{u}(t)$ are defined as follows respectively.

$$\begin{aligned}\tilde{y}(t) &= y(t) - y_\infty\\ \tilde{u}(t) &= u(t) - u_\infty\end{aligned}$$

The prediction for Eq. (3) can be derived by the Diophantine equation,

$$1 = A[z^{-1}]E_N[z^{-1}] + z^{-N}F_N[z^{-1}] \qquad N \geq N_2 \tag{4}$$

$$E_N[z^{-1}]B[z^{-1}] = R_j[z^{-1}] + z^{-j}S_j[z^{-1}] \qquad j = N_1, \cdots, N_2 \tag{5}$$

where

$$R_j[z^{-1}] = r_0 + r_1 z^{-1} + \cdots + r_{j-1}z^{-(j-1)}$$

$[N_1, N_2]$ and $[1, N_u]$ denote the prediction horizon and the control horizon respectively. To derive the control law, although the conventional GPC[1] needs to solve the Diophantine equation Eq. (4) $N_2 - N_1 + 1$ times, the proposed method has only to solve it once.[7] Therefore in the case of designing self-tuning controller, it is found that the proposed method can save the computation to derive the controller.

The vector form of the prediction $\hat{\tilde{y}}(t)$ is given by the following equation.

$$\hat{\tilde{Y}} = R\tilde{U} + H \tag{6}$$

where

$$\begin{aligned}\hat{\tilde{Y}} &= [\hat{\tilde{y}}(t+N_1|t)\cdots\hat{\tilde{y}}(t+N_2|t)]^T\\ \tilde{U} &= [\tilde{u}(t)\cdots\tilde{u}(t+N_u-1)]^T\\ H &= [h_{N_1}(t)\cdots h_{N_2}(t)]^T\\ h_j(t) &= z^{-(N-j)}F_N[z^{-1}]\tilde{y}(t) + z^{-k_m}S_j[z^{-1}]\tilde{u}(t)\\ R &= \begin{bmatrix} r_{N_1-k_m} & \cdots & r_{N_1-k_m-N_u+1}\\ \vdots & & \vdots\\ r_{N_2-k_m} & \cdots & r_{N_2-k_m-N_u+1}\end{bmatrix}\end{aligned}$$

To simplify, it is assumed that $N_1 = k_m = 1$ and $N_2 = N_u = N$ in this paper.

The performance index for the deviation system Eq. (3) is considered under the condition of $\tilde{y}(t+j) = \hat{\tilde{y}}(t+j|t)$.

$$\begin{aligned} J &= \sum_{j=N_1}^{N_2} \tilde{y}^2(t+j) + \sum_{j=1}^{N_u} \lambda \tilde{u}^2(t+j-1) \\ &= (R\tilde{U} + H)^T (R\tilde{U} + H) + \lambda \tilde{U}^T \tilde{U} \end{aligned} \tag{7}$$

Minimizing the performance index J on $\tilde{U}$, the control law is derived.

$$\tilde{U} = -(R^T R + \lambda I)^{-1} R^T H$$

Because this equation is about the deviation system of the plant Eq. (1), the control law of the plant Eq. (1) is obtained as the following equation.

$$u(t) = H_0(z^{-1}) r(t) - F_0(z^{-1}) y(t) \tag{8}$$

where

$$\begin{aligned} H_0(z^{-1}) &= \frac{F_p[z^{-1}] + (1 + z^{-k_m} S_p[z^{-1}])K}{1 + z^{-k_m} S_p[z^{-1}]} \\ F_0(z^{-1}) &= \frac{F_p[z^{-1}]}{1 + z^{-k_m} S_p[z^{-1}]}, \quad K = \frac{A[1]}{B[1]} \\ [p_{N_1}, \cdots, p_{N_2}] &= [1, 0, \cdots, 0](R^T R + \lambda I)^{-1} R^T \\ F_p[z^{-1}] &= \sum_{j=N_1}^{N_2} z^{-(N-j)} p_j F_N[z^{-1}], \quad S_p[z^{-1}] = \sum_{j=N_1}^{N_2} p_j S_j[z^{-1}] \end{aligned}$$

3.2. *Two DOF of GPC with computational savings*

The controller including the integral compensation is expressed by the following equation.

$$u(t) = H_0(z^{-1}) r(t) - F_0(z^{-1}) y(t) + G z(t) \tag{9}$$

$z(t)$ is the term of the integral compensation and G is its gain. In this subsection $z(t)$ is derived, whose effect appears only when there exists modeling error or disturbance.

In the case that there exists no modeling error, the closed-loop system by the control law Eq. (8) is obtained,

$$y(t) = T(z^{-1}) r(t)$$

where $T(z^{-1})$ is the transfer function of the closed-loop system. If there exists no modeling error, then the tracking error $e(t)$ is given by the following equation.

$$e(t) = (1 - T(z^{-1}))r(t) \tag{10}$$

Now the integral compensation $z(t)$ is derived by Eqs. (2) and (10).

$$z(t) = w(t) - \frac{1}{\Delta}(1 - T(z^{-1}))r(t) \tag{11}$$

$z(t)$ in Eq. (11) has the feature that if there exists neither modeling error nor disturbance, $z(t)$ is always equal to zero. That is, the effect of the integral compensation does not appear.

4. Two DOF of Self-Tuning GPC with Computational Savings

This section considers the case that the plant parameters are unknown. In this case applying the following parameter identification law to Eq. (9), the self-tuning controller proposed in this paper can be obtained.

$$\hat{\theta}(t) = \hat{\theta}(t-1) + \frac{\Gamma(t-1)\psi(t-1)}{1 + \psi^T(t-1)\Gamma(t-1)\psi(t-1)}\epsilon(t) \tag{12}$$

$$\Gamma(t) = \Gamma(t-1) - \frac{\lambda\Gamma(t-1)\psi(t-1)\psi^T(t-1)\Gamma(t-1)}{1 + \lambda\psi^T(t-1)\Gamma(t-1)\psi(t-1)} \tag{13}$$

$$\Gamma(0) > \alpha I, \quad 0 < \alpha < \infty$$

$$\epsilon(t) = y(t) - \hat{\theta}^T(t-1)\psi(t-1) \tag{14}$$

where

$$\hat{\theta}(t) = \left[\hat{a}_1(t), \cdots, \hat{a}_n(t), \hat{b}_0(t), \cdots, \hat{b}_m(t)\right]$$

$$\psi(t-1) = [-y(t-1), \cdots, -y(t-n), u(t-k_m), \cdots, u(t-k_m-m)]$$

5. Example

In this section the numerical example is shown to verify the validity of the proposed method. Consider the controlled plant,

$$y(t) = \frac{z^{-1}(1.2 - 0.1z^{-1})}{1 - 1.0z^{-1} + 0.25z^{-2}}u(t)$$

Simulation steps are 300, the initial values of the output and the input are assumed to be zero, and the nominal values of the plant parameters are set to be 0.9 × the true values. The white gaussian noise with the variance

$\sigma^2 = 0.05$ is added to the controlled plant and each design parameter is given by follows.

$$N_1 = 1,\ N_2 = 10,\ N_u = 10,\ \lambda = 1$$

The integral gain G is set to be 0.5. The reference signal is a rectangular signal with amplitude 1 and the period of 60 steps. And it is assumed that the plant parameter changes after 170 steps as follows,

$$y(t) = \frac{z^{-1}(1.2 - 0.1z^{-1})}{1 - 1.1z^{-1} + 0.1z^{-2}}u(t)$$

where the parameter identification is cut after 170 steps, that is, the identified values of the plant parameters become constant after 170 steps.

In Figure 1 the solid line shows the output by the two DOF of self-tuning GPC with computational savings Eq. (9) proposed in this paper, and the dotted line shows the reference signal. Fig. 2 shows the conventional two DOF of self-tuning GPC[6] and each type of line is the same as Fig. 1.

In Fig. 1 the proposed method can track to the reference signal and the overshoot in the transient response can be controlled as Fig. 2.

6. Conclusion

In this paper, the design scheme of self-tuning two degree-of-freedom GPC based on polynomial approach with computational savings is proposed. If the identified plant parameters converge on true values, the proposed method has the characteristic that it reveals the effect of the integral compensation only if there exists modeling error or disturbance. Therefore, performance degradation due to an integral compensation can be avoided when the controlled system has no perturbation. In this paper, although the controlled plant is assumed to be single-input single-output systems, there is an extension to multi-input multi-output systems as the future work.

References

1. D. W. Clarke, C. Mohtadi and P. S. Tuffs: Generalized Predictive Control, Automatica, Vol. 23, No. 2, pp. 137–160 (1987)
2. Y. Fujisaki and M. Ikeda: Synthesis of Two-Degree-of-Freedom Optimal Servosystems, Transactions of SICE, Vol. 27, No. 8, pp. 907–914 (1991)
3. T. Hagiwara, M. Ichiki, M. Kanaboshi, K. Fukumitsu and M. Araki: Digital Two-Degree-of-Freedom LQI Servo Systems —Design Method and Its Application to Positioning Control of a Pneumatic Servo Cylinder, Transactions of ISCIE, Vol. 11, No. 2, pp. 51–60 (1998)

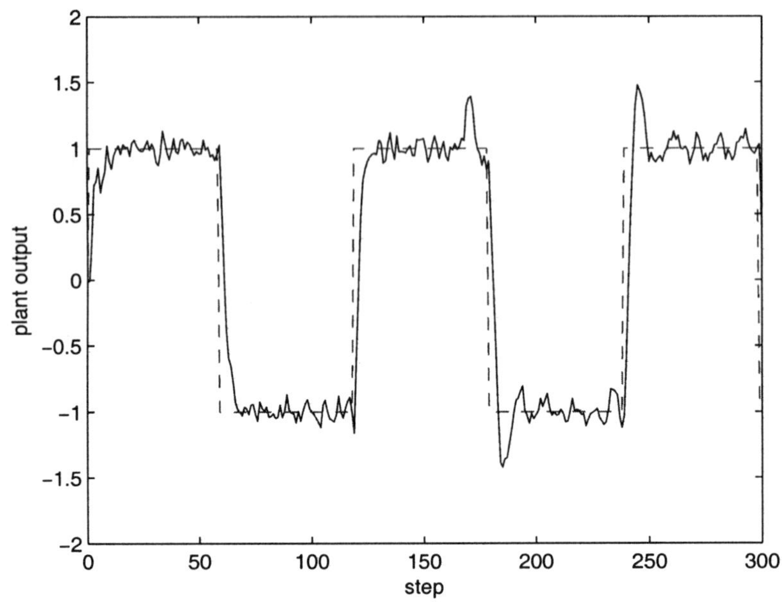

Fig. 1. Proposed method

4. A. Yanou, S. Masuda, A. Inoue and Y. Hirashima: Two Degree-of-Freedom of Generalized Predictive Control Based on State Space Approach, Transactions of ISCIE, Vol. 12, No. 2, pp. 106–114 (1999)
5. A. Yanou, S. Masuda and A. Inoue: A Design Scheme of Two-Degree-of-Freedom Generalized Predictive Control System in the Polynomial Approach, Preprints of 12th SICE Symposium in Chugoku Branch(In Japanese), pp. 98–99 (2003)
6. A. YANOU, S. MASUDA and A. INOUE: Two Degree-of-Freedom of Self-Tuning Generalized Predictive Control Based on Polynomial Approach, Proceedings of SICE Annual Conference 2005 in Okayama, pp. 1187–1190 (2005)
7. M. A. Saudagar, D. G. Fisher and S. L. Shar: Reduced Diophantine Predictors for Long Range Predictive Controllers, Proceedings of the American Control Conference, pp. 3688–3693 (1995)
8. E. F. Camacho and C. Bordons: Model Predictive Control in the Process Industry, Springer (1995)
9. S. Ohmatsu and T. Yamamoto(ed.): Self-Tuning Control(In Japanese), SICE (1996)

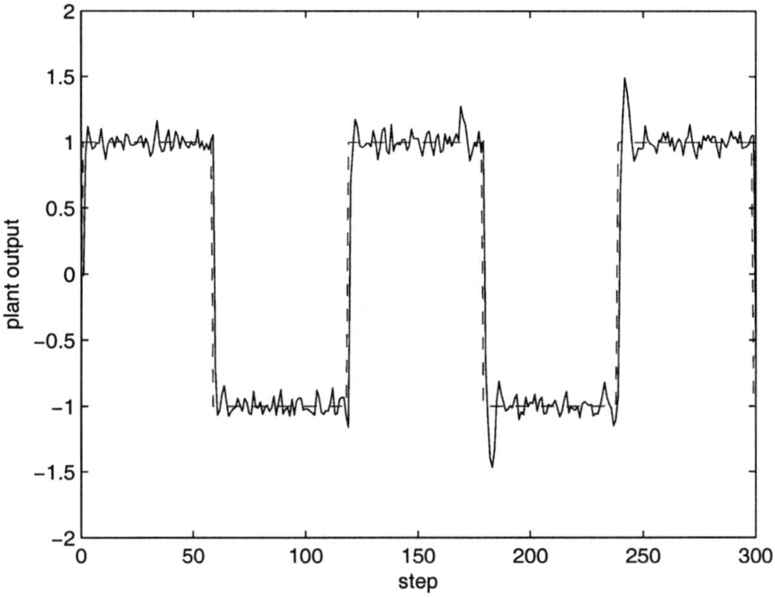

Fig. 2. Conventional method

A DESIGN METHOD OF GENERALIZED MINIMUM VARIANCE CONTROL CONSIDERING SAFETY OF SAMPLED-DATA SYSTEMS

TAKAO SATO

Division of Mechanical System, Department of Mechanical Engineering, Graduate School of Engineering, University of Hyogo, 2167 Shosha, Himeji, Hyogo 671-2201 JAPAN, E-mail: tsato@eng.u-hyogo.ac.jp

SHIRO MASUDA

Department of Management Systems Engineering, Faculty of System Design, Tokyo Metropolitan University, 6-6 Asahigaoka, Hino, Tokyo 191-0065 JAPAN, E-mail: smasuda@cc.tmit.ac.jp

AKIRA INOUE

Division of Industrial Innovation Sciences, Graduate School of Natural Science and Technology, Okayama University, 3-1-1, Tsushima-naka, Okayama, Okayama 700-8530 JAPAN, E-mail: inoue@suri.sys.okayama-u.ac.jp

This paper proposes a design method of Generalized Minimum Variance Control (GMVC) in sampled-data systems, where a continuous-time plant is controlled with a discrete-time controller. The proposed design method derives a discrete-time control law which minimizes a continuous-time performance function with a zero order hold. Because the proposed method evaluates not only sampling instants but intersample, the intersample performance of the proposed method is improved better than conventional design methods in discrete-time. Finally, we illustrate numerical simulation results in order to show the effectiveness of the proposed method.

Keywords: Sampled-data system; Generalized minimum variance control; Intersample output; Safety.

1. Introduction

This paper proposes a new design method for a sampled-data system, where a controlled plant in continuous-time is controlled with a controller updated in discrete-time. In the sampled-data system not only behavior on sampling instants but intersample have to be taken into account [1,2]. In this paper, to control the sampled-data system, we propose a design method of Generalized Minimum Variance Control (GMVC) [3] for the sampled-data system. The control law of GMVC is derived based on minimization of the variance of a generalized output including a dead-time predictive output. However, because discrete-time GMVC evaluates only behavior on sampling instants, intersample output may deteriorate. While GMVC in continuous-time was proposed [4], but because many control systems are controlled using digital computers operating in discrete-time, it is advisable to design GMVC as the sampled-data system. Hence, this paper proposes a design method of GMVC to control a continuous-time plant by using a sampler and a holder.

2. Problem Statements

Consider a plant given by the following single-input single-output continuous-time linear time-invariant state-space model including a dead-time.

$$\dot{x}(t) = Ax(t) + bu(t-L) \tag{1}$$

$$y(t) = c^T x(t) \tag{2}$$

where, $y(t)$, $u(t)$ and $x(t)$ are the plant output, the control input and the state variable respectively, and L is the dead-time.

The plant described by Eqs. (1) and (2) is controlled by using the following zero order holder.

$$u(\tau) = u[k] \quad (kT_s \le \tau < (k+1)T_s) \tag{3}$$

where, T_s is the sampling interval.

As a performance index, the following function including continuous-time variables is considered.

$$J[k] = \Phi[k]^2 \tag{4}$$

where generalized output $\Phi[k]$ is defined by

$$\Phi[k]=\int_{kT_s}^{(k+1)T_s} \{P(z^{-1})y(k_mT_s+\tau)+Q(z^{-1})u(\tau)-R(z^{-1})w(\tau)\}d\tau \tag{5}$$

where, k_m is the dead-time in discrete-time, $k_m \ge 0$, and $w(t)$ is a step-wise function defined by

$$w(t) = w[k] \quad (kT_s \le t < (k+1)T_s). \tag{6}$$

z^{-1} is backward shift operator, and $z^{-1}y[k] = y[k-1]$. $P(z^{-1})$, $Q(z^{-1})$ and $R(z^{-1})$ are the design polynomials of GMVC, and n_p, n_q and n_r are the order of $P(z^{-1})$, $Q(z^{-1})$ and $R(z^{-1})$, respectively.

The purpose of this study is to derive a control law updated at interval of T_s which minimizes the continuous-time performance function by using a sampler and a holder.

The following assumption is imposed on the dead-time.

Assumption 2.1. *A dead-time is integer multiple of a sampling interval and satisfies the next equation.*

$$L = k_m T_s \tag{7}$$

In order to simplify the representation, assuming the following.

Assumption 2.2.

$$n_p \leq k_m, \quad n_q \leq k_m \tag{8}$$

3. Derivation of a Control Law

3.1. *Prediction Form*

With the zeroth order hold the continuous-time model is transformed into the following discrete-time model.

$$x[k+1] = A_d x[k] + b_d u[k-k_m] \tag{9}$$

$$y[k] = c^T x[k] \tag{10}$$

$$A_d = e^{AT_s}, \quad b_d = \int_0^{T_s} e^{A\sigma} d\sigma \cdot b. \tag{11}$$

$k_m - j$ forward prediction can be calculated as

$$\begin{aligned} x[k+k_m-j] &= A_d^{k_m-j} x[k] \\ &+ \left[b_d \; A_d b_d \cdots A_d^{k_m-j-1} b_d \right] \left[u[k-1] \; u[k-2] \cdots u[k-k_m-j] \right]^T, \end{aligned} \tag{12}$$

and the following prediction form is given.

$$X[k] = \Phi x[k] + \Gamma U[k] \tag{13}$$

where,

$$X[k] = \left[x[k+k_m] \; x[k+k_m-1] \cdots x[k+k_m-n_p] \right]^T \tag{14}$$

$$\Phi = \left[A_d^{k_m} \; A_d^{k_m-1} \cdots A_d^{k_m-n_p} \right]^T \tag{15}$$

$$U[k] = \left[u[k] \; u[k-1] \cdots u[k-k_m] \right]^T \tag{16}$$

$$\Gamma = \begin{bmatrix} 0 \; b_d \; A_d b_d \; A_d^2 b_d \cdots A_d^{k_m-1} b_d \\ \vdots \; \ddots \; \ddots \qquad\qquad \vdots \\ 0 \cdots 0 \qquad b_d \quad \cdots A_d^{k_m-n_p-1} b_d \end{bmatrix}. \tag{17}$$

3.2. *Discrete Equivalent Performance Function*

The generalized output is rewritten as

$$\begin{aligned}\Phi[k] &= \sum_{j=0}^{n_p} p_j \int_{kT_s}^{(k+1)T_s} y(k_m T_s + \tau - jT_s)d\tau \\ &+ \sum_{j=0}^{n_q} q_j \int_{kT_s}^{(k+1)T_s} u(\tau - jT_s)d\tau \\ &+ \sum_{j=0}^{n_r} r_j \int_{kT_s}^{(k+1)T_s} w(\tau - jT_s)d\tau. \end{aligned} \tag{18}$$

The response between sampled outputs is calculated by the followings.

$$y(t + (k + k_m - j)T_s) = c^T A_d(t)x[k + k_m - j] + c^T b_d(t)u[k - j] \tag{19}$$

$$A_d(t) = e^{At}, \quad b_d(t) = \int_0^t e^{A\sigma} d\sigma \cdot b \tag{20}$$

Hence, the integral part in the first term of the right-hand side of Eq. (18) can be rewritten as follows.

$$\int_{kT_s}^{(k+1)T_s} y(k_m T_s + \tau - jT_s)d\tau = Mx[k + k_m - j] + Nu[k - j] \tag{21}$$

where M and N are defined as

$$M = \int_0^{T_s} c^T A_d(\tau)d\tau, \quad N = \int_0^{T_s} c^T b_d(\tau)d\tau. \tag{22}$$

Therefore, using Eq. (13) the first term of the right-hand side of Eq. (18) is given as

$$\begin{aligned}&\sum_{j=0}^{n_p} p_j \int_{kT_s}^{(k+1)T_s} y(k_m T_s + \tau - jT_s)d\tau \\ &= P^T \Lambda_M X[k] + P_u^T NU[k] \\ &= P^T \Lambda_M \Phi x[k] + P^T \Lambda_M \Gamma U[k] + P_u^T NU[k] \end{aligned} \tag{23}$$

where P, P_u and Λ_M are defined as

$$P = \left[p_0\ p_1 \cdots p_{n_p}\right]^T, \quad P_u = \begin{bmatrix} P \\ \mathbf{0}_{k_m - n_p, 1} \end{bmatrix}, \quad \Lambda_M = \begin{bmatrix} M & & 0 \\ & \ddots & \\ 0 & & M \end{bmatrix}. \tag{24}$$

And, the send term of the right-hand side of Eq. (18) becomes

$$\sum_{j=0}^{n_q} q_j \int_{kT_s}^{(k+1)T_s} u(\tau - jT_s)d\tau = T_s Q^T U[k] \tag{25}$$

$$Q = \begin{bmatrix} q_0 \; q_1 \; \cdots \; q_{n_q} \; \mathbf{0}_{k_m-n_q,1} \end{bmatrix}^T. \tag{26}$$

Further, the third term of the right-hand side of Eq. (18) becomes

$$\sum_{j=0}^{n_r} r_j \int_{kT_s}^{(k+1)T_s} w(\tau - jT_s)d\tau = T_s R^T W[k] \tag{27}$$

$$R = \begin{bmatrix} r_0 \; r_1 \; \cdots \; r_{n_r} \end{bmatrix}^T \tag{28}$$

$$W[k] = \begin{bmatrix} w[k] \; w[k-1] \; \cdots \; w[k-n_r] \end{bmatrix}^T. \tag{29}$$

Then, the vector form of the generalized output is obtained as

$$\begin{aligned}\Phi[k] = P^T \Lambda_M \Phi x[k] + P^T \Lambda_M \Gamma U[k] + P_u^T N U[k] \\ + T_s Q^T U[k] - T_s R^T W[k].\end{aligned} \tag{30}$$

3.3. *Control Law*

A control law minimizing the performance function is derived by

$$G^T U[k] = T_s R^T W[k] - P^T \Lambda_M \Phi x[k] \tag{31}$$

$$G^T = P^T \Lambda_M \Gamma + N P_u^T + T_s Q^T \tag{32}$$

Dividing the left side of Eq. (31) into past terms and a present term, a control input at k is calculated by

$$u[k] = \frac{1}{g_0}(T_s R^T W[k] - P^T \Lambda_M \Phi x[k] - G'^T U[k-1]) \tag{33}$$

$$g_0 = G^T \begin{bmatrix} 1 \\ \mathbf{0}_{k_m,1} \end{bmatrix}, \quad G'^T = G^T \begin{bmatrix} \mathbf{0}_{1,k_m+1} \\ I_{k_m} \; \mathbf{0}_{k_m,1} \end{bmatrix}. \tag{34}$$

Using the following full-order observer, the control law can be constructed using input and output signals.

$$\hat{x}[k+1] = A_d \hat{x}[k] + b_d u[k-k_m] + L(y[k] - c^T \hat{x}[k]) \tag{35}$$

where, $A_d - Lc^T$ has to be stable.

4. Simulation Result

A plant given by the following model is controlled by using both a discrete-time GMVC [5] and the proposed method.

$$\dot{x}(t) = \begin{bmatrix} -0.1 & -1 & 0 \\ 1 & 0 & 0 \\ 0 & 1 & 0 \end{bmatrix} x(t) + \begin{bmatrix} 1 \\ 0 \\ 0 \end{bmatrix} u(t-6) \tag{36}$$

$$y(t) = \begin{bmatrix} 0 & 0 & 1 \end{bmatrix} x(t) \tag{37}$$

The transfer function of Eqs. (36) and (37) is $P(s) = \frac{1}{s(s^2+0.1s+1)} e^{-6s}$.

In order to show the effectiveness of the proposed design method, the proposed GMVC is compared to the conventional discrete-time GMVC.

The sampling interval is set as $T_s = 3$, and the set-point to be followed by the plant output is set as an unit step-wise function. In designing both the proposed method and the conventional method, the design polynomials are set; $P(z^{-1}) = 1-0.3z^{-1}+0.05z^{-2}$, $Q(z^{-1}) = 8$, $R(z^{-1}) = P(z^{-1})$. The poles of an observer utilized in the proposed controller are set as $0, 0.1, 0.2$.

Control results are illustrated in Fig. 1 ~ Fig. 6. By using the proposed method the convergence of the plant output to the set-point is faster than that with the conventional method. The plant output with the conventional method vibrates between sampled outputs. On the other hand, the plant output with the proposed method converges to the set-point without vibration.

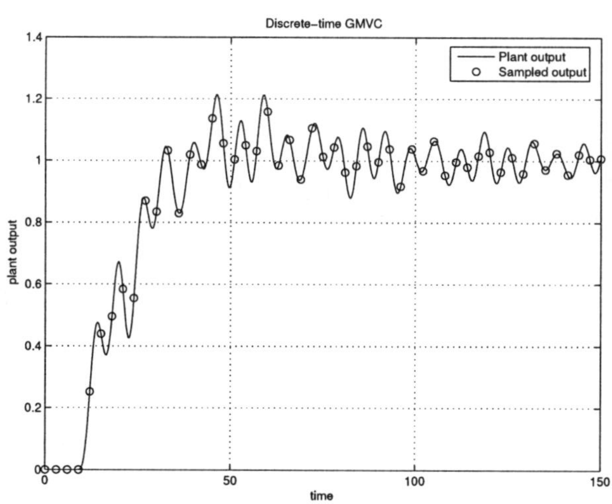

Fig. 1. Output result by using the conventional GMVC

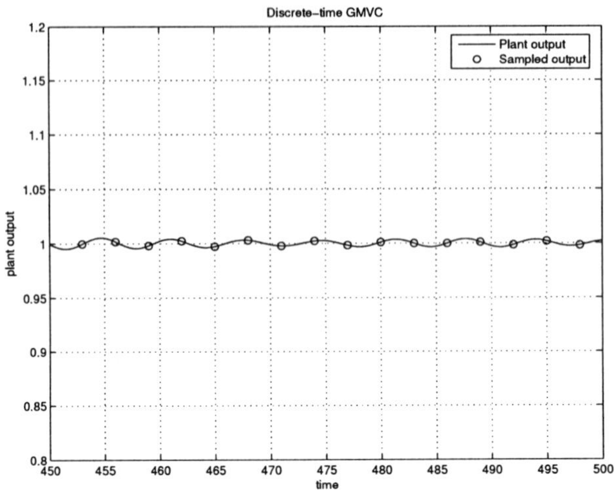

Fig. 2. Enlarged figure of Fig. 1

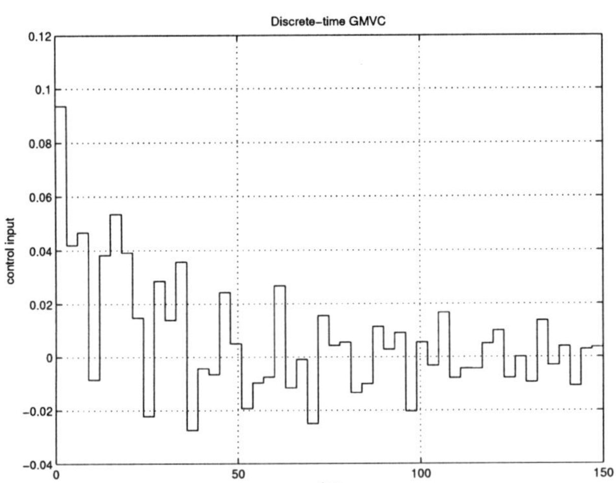

Fig. 3. Input result by using the conventional GMVC

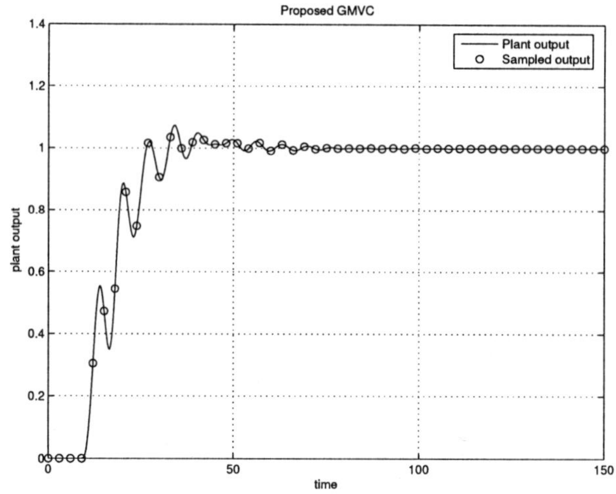

Fig. 4. Output result by using the proposed GMVC

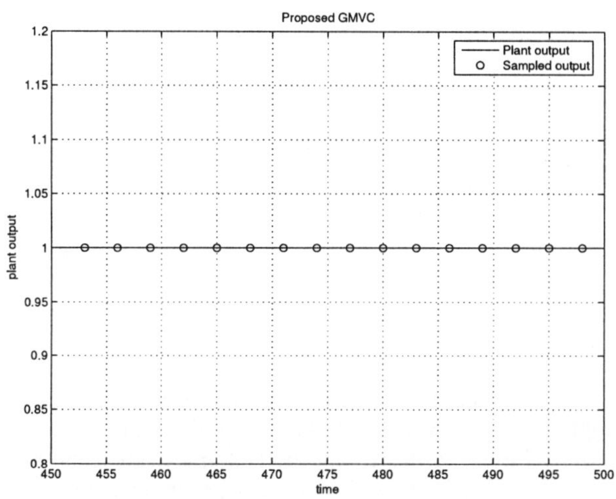

Fig. 5. Enlarged figure of Fig. 4

5. Conclusion

This paper have proposed a design method of Generalized Minimum Variance Control (GMVC) in sample-data systems. Because a dead-time is assumed as an integer multiple of a sampling interval, our future work is to extend the proposed design method to the case that a dead-time is a real

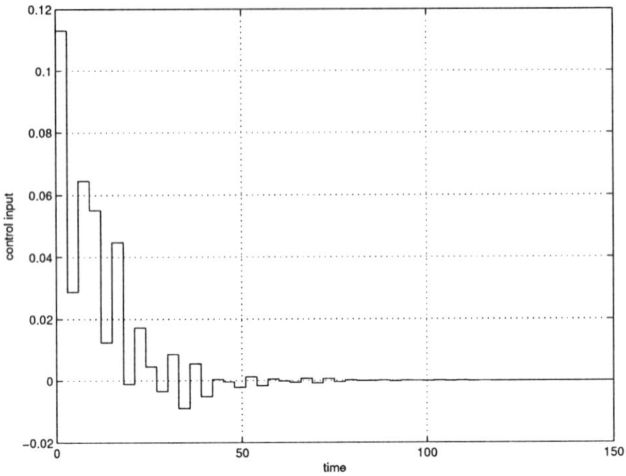

Fig. 6. Input result by using the proposed GMVC

number multiple of a sampling interval. Further, the proposed method is extended into a multirate system [6].

References

1. S. Masuda, A. Inoue, Y. Hirashima and R. M. Miller, Intersample performance improvement in generalized predictive control, in *IFAC International Symposium on Advanced Control of Chemical Processes*, 1997.
2. S. Kishino, T. Sato, S. Masuda and A. Inoue, Track-seeking control of a hard disk drive by using intersample performance improvement in multirate GPC, in *Proc. of SICE Annual Conference*, 2005.
3. D. Clarke, *Automatica* **20**, 501 (1984).
4. Y. Mori and F. Asami, *IEEJ Trans. EIS* **125**, 1743 (2005).
5. S. Omatu and T. Yamamoto (eds.), *Self-Tuning Control* (The Society of Instrument and Control Engineers, 1996). (in Japanese).
6. T. Sato and A. Inoue, A design method of multirate generalized minimum variance control, in *Proc. of SICE Annual Conference*, 2005.

TRACKING CONTROL SYSTEM FAULT DIAGNOSIS BY USING ROBUST RIGHT COPRIME FACTORIZATION AND ITS APPLICATION

MINGCONG DENG, AKIRA INOUE, TAKAHIRO KUWAMOTO AND NOBUYUKI UEKI *

Department of Systems Engineering, Okayama University
3-1-1 Tsushima-Naka
Okayama, 700-8530, Japan
E-mail: deng@suri.sys.okayama-u.ac.jp

Recently, for tracking control of nonlinear system, a technique based on the operator theoretic approach was studied. In details, a design method based on robust right coprime factorization is given, and this method has advantage of easiness to analyze robust stability of nonlinear control system. Based on this design scheme, design of feedback control system for a practical nonlinear plant can be realized. However, as applying the control scheme to real plants, system's failures are caused by various environmental factors. In this case, detecting fault signal or finding unusual state about safety of system is very important. This paper shows a method of designing nonlinear tracking control system fault diagnosis using robust right coprime factorization and applies the fault diagnosis method to a water level process system.

1. Introduction

Recently, nonlinear tracking control system design scheme [3] has been studied by using operator-based robust right coprime factorization [1], [2]. However, we need to check the consistency of the designed unit. A framework of fault diagnosis for the above tracking control system was given in [4]. The fault detecting method is an analytical method that uses processes information of the controlled process [5]. In this paper, we apply the fault detecting method to a process control experimental system. An experimental result to show effectiveness is also given.

*Work supported by grant in aid no. 16101005 of JSPS.

The outline of this paper is as follows. In Section 2, the process experimental system for water level control is shown, and the model of this system is given. Fault diagnosis system using operator theoretic approach is introduced in Section 3. In Section 4, experimental result confirm the effectiveness of the proposed fault diagnosis method.

2. Process Control Experimental System

2.1. *System Description*

In this section, a process control experimental system is introduced. The

Figure 1. The experimental system

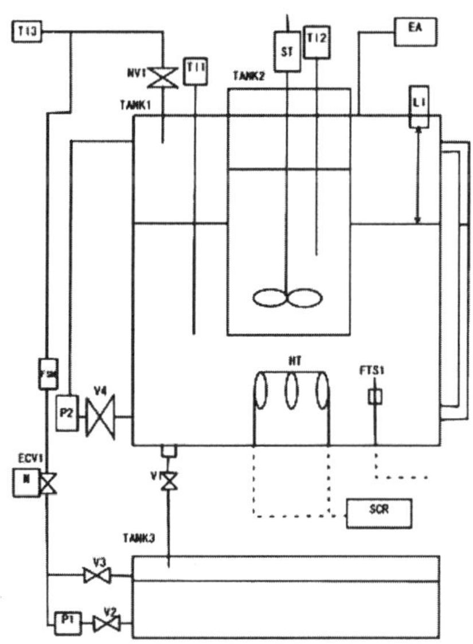

Figure 2. The equivalent diagram of Fig. 1

process control experimental system is shown in Fig. 1. Top part of the system has two tanks and bottom part of this system has a tank. The outside tank of top part is called TANK1, and the inside one is called TANK2. The bottom part tank is called TANK3 which stores water.

The diagram of the process control experimental system is shown in Fig. 2. TANK1 has a supersonic wave sensor (L1) and this sensor can measure water level. An inflow mouth on TANK1, and volume of water can be measured. Water is carried from TANK3 to TANK1. We can control the volume of water by using the valve (NV1). In addition, water is outflow from bottom of TANK1 to TANK3 through a drain pipe. Parameter of the system in Fig. 2 is given as follows.
Diameter of Tank1[m] : $D_1 = 0.3185$ [m]
Diameter of Tank2[m] : $D_2 = 0.1143$ [m]
Diameter of outlet[m] : d = 0.012 [m]
Level of Tank2 from Tank1[m] : $h_s = 0.2$ [m]
Gravity acceleration$[m/s^2]$: $g = 9.8$ $[m/s^2]$
The atmospheric pressure : $P_0 = 1013$[hPa]
Pressure in tank : $P(t) = 1013$[hPa]
Density of water : ρ
Water level in Tank1[m] : h
Inflow $[l/\text{min}]$: q_0
Outflow $[l/\text{min}]$: q

In this paper, we consider that water level change from $30[cm]$ to $72[cm]$ because a sensor can work by this range. Then, sectional square A of Tank1 and square a of drain pipe is given as follows.

$$A = \frac{{D_1}^2\pi}{4} - \frac{{D_2}^2\pi}{4} = 25.38 \times 10^{-3}[m^2] \tag{1}$$

$$a = \frac{a^2\pi}{4} = 0.1131 \times 10^{-3}[m^2] \tag{2}$$

2.2. *Modelling*

In the following, we will create a mathematical model of the water level process using parameter in previous section. Concerned with water level and outflow, the relationship is obtained by using Bernoulli theorem, and this theorem is shown in the following equation.

$$\frac{v^2}{2g} + z + \frac{p}{\rho g} = const. \tag{3}$$

If we consider that velocity of varying water level is V_1 and velocity of outflow is V_2, the following equation is obtained.

$$\frac{{V_1}^2}{2g} + \frac{P(t)}{\rho g} + h(t) = \frac{{V_2}^2}{2g} + \frac{P_0}{\rho g} \tag{4}$$

Volume of outflow equals that of lessen water in Tank1 as no input, and considering that $a \ll A$. We have

$$AV_1 = aV_2 \tag{5}$$

$$\frac{V_1}{V_2} = \frac{a}{A} \simeq 0 \tag{6}$$

In addition, if we consider that $V_1 = \dot{h}$, the following equation is obtained.

$$\dot{h} = \frac{a}{A}\sqrt{2gh} \tag{7}$$

Moreover, inflow q_0 is added in (7). Therefore, model of water level process is presented by the following equation.

$$\dot{h}(t) = \frac{q_0}{A} - \frac{a}{A}\sqrt{2gh(t)} \tag{8}$$

3. Fault Diagnosis System

Nonlinear tracking control using operator theoretic approach is proposed in [3] and a framework of fault diagnosis technique for the tracking operator is shown in [4]. In this paper, we design the system using the above techniques. Here, these methods are introduced as follows. Fig. 3 is the tracking control

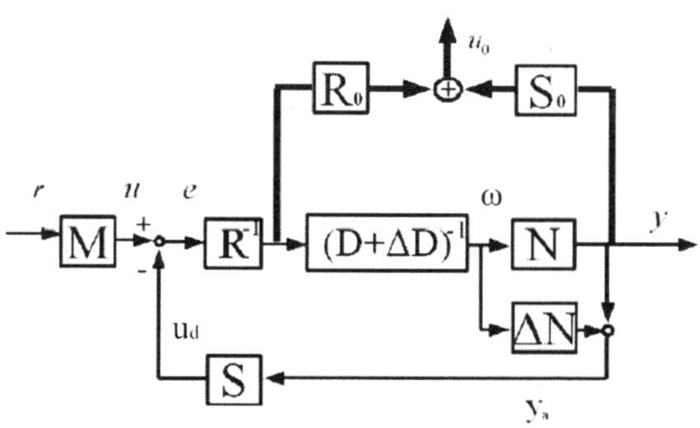

Figure 3. System of tracking control fault diagnosis, where we can observe u_0 to know the change of u [4]

and fault diagnosis system. If we consider a plant P, $P = ND^{-1}$ by right coprime factorization. Then N and D is obtained. In addition, N and D are satisfying the following Bezout identity.

$$SN + RD = I \tag{9}$$

In the real design, we need to design S and R such that Bezout identity (9) is satisfied. M is called tracking operator [4]. Tracking operator output u is input to the system. In this paper, we assume that S_0 and R_0 are as same as S and R, respectively.

In the following, we will introduce the fault diagnosis technique. If we input reference input r for tracking operator $M : r \to u$, tracking operator output is u. Because the feedback control system is satisfying Bezout identity, we have the following equation.

$$u(t) = RD(\omega)(t) + SN(\omega)(t) = e(t) + ud(t) \tag{10}$$

Under the condition $S = S_0$ and $R = R_0$,

$$u(t) = u_0(t) = R_0 D(\omega)(t) + S_0 N(\omega)(t) \tag{11}$$

When (10) and (11) are satisfied, we can check that if M is failure. That is, we check u_0 for knowing the tracking operator output u using the designed operator S and R.

The detailed design produce for the experimental system (8) is given as follows. According to the design method, we consider that q_0 is plant input u, h is plant output y and a function about u to y is plant $P(u)(t)$. That is, a function $P^{-1}(y)(t) : y \to u$ can be described as follows.

$$P^{-1}(y,t) = u(t) = A\dot{y}(t) + a\sqrt{2gy(t)} \tag{12}$$

From (12), $N(\omega)(t)$ and $D(\omega)(t)$ are selected as

$$N(\omega)(t) = y(t) = \frac{\omega(t)^2}{2ga^2} \tag{13}$$

$$D(\omega)(t) = u(t) = \frac{A}{2ga^2}\frac{d}{dt}\{\omega(t)^2\} + a\omega(t) \tag{14}$$

Design of controller S and R is presented as followews. S and R must be provided so that N and D satisfy Bezout Identify $SN + RD = I$. Based on the robust right coprime factorization approarch, two controllers S and R are given as

$$R = I(u) \tag{15}$$

$$S = a\sqrt{2gy(t)} - \{A\dot{y}(t) + a\sqrt{2gy(t)}\} \tag{16}$$

Consequently, we can obtain the control and fault diagnosis system shown in Fig. 3.

4. Experimental Result

In this section, experiment on fault detection is conducted for the process system with tracking filter. The initialized water level is about 33cm, the reference input of R is 35cm and the total time is 400 second. Let the operator M have a fault signal between 180 second and 240 second. Level control experimental result is shown in Fig.4.

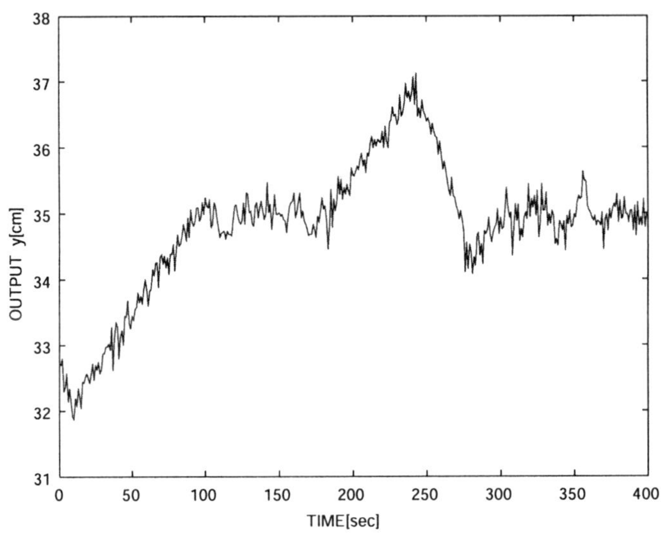

Figure 4. Response of water level control

By using the designed fault detecting system, operators R_0's output and S_0's output are shown in Figs. 5 and 6, where the output of M is shown in Fig. 7. Then, we obtain u_0, the equivalent signal to u, from the addition of output of operators R_0 and S_0. The result is given in Fig. 8. From Figs 7 and 8, the fault detecting system shows the desired performance. Moreover, normal tracking operator output $u^{'}$ is shown in Fig.9. Fig.10 presents the result of $u_0 - u^{'}$. The general checking rule is given as follows. If operator M is normal, $u = u_0 = u^{'}$. However if operator M is abnormal, $u = u_0 \neq u^{'}$. That is, we obtain $u^{'} - u_0 = 0$ when operator M is normal, and $u^{'} - u_0 \neq 0$ when operator M is failure. Fault detection is

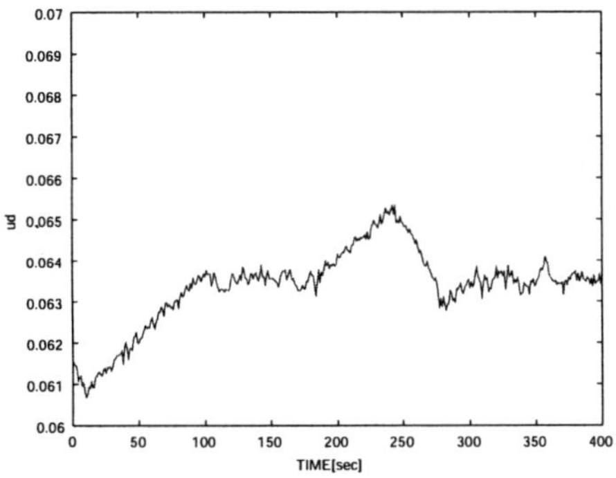

Figure 5. The output of S_0

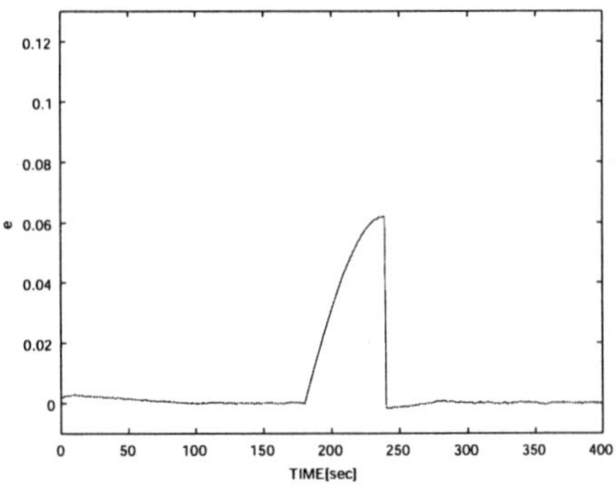

Figure 6. The output of R_0

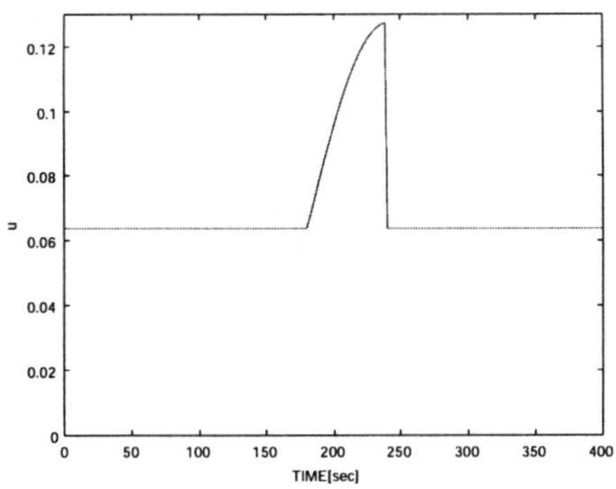

Figure 7. The output of M including failure signal

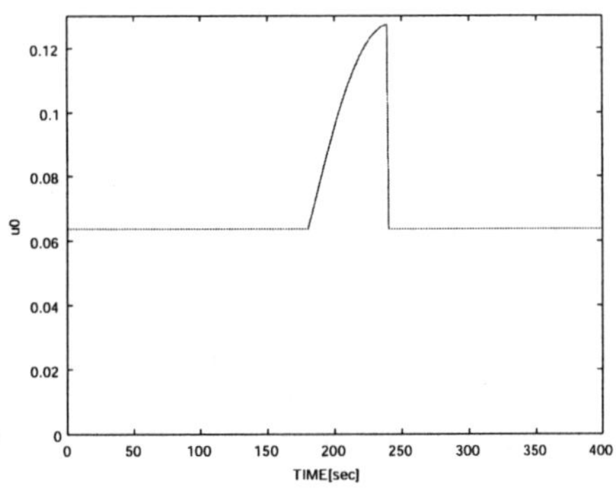

Figure 8. The signal u_0

successful from Fig.10, because the value is not zero while the operator M is abnormal.

5. Conclusion

A fault detecting system design problem of tracking operator in process control experimental system is considered by using robust right coprime factorization approach. The effectiveness of the proposed fault detecting method is confirmed by water level control experiment.

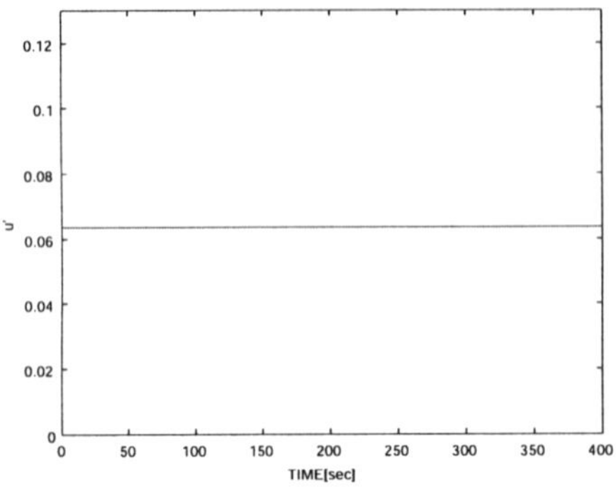

Figure 9. Normal tracking operator output u

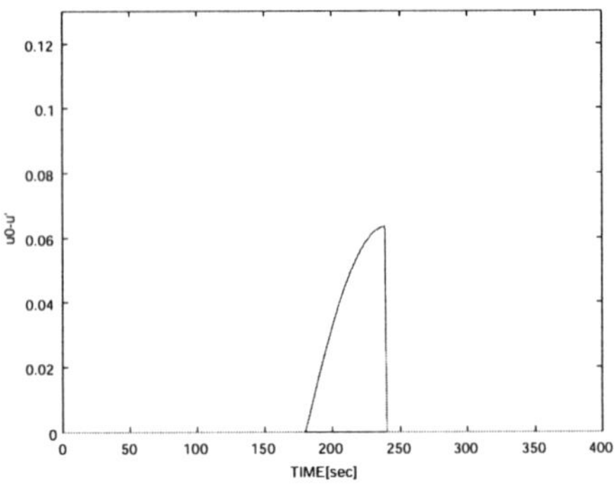

Figure 10. Difference between u_0 and $u^{'}$

References

1. G. Chen and Z. Han, Robust right coprime factorization and robust stabilization of nonlinear feedback control systems, *IEEE Trans. Automatic Control,* Vol. 43, pp. 1505-1510, 1998.
2. R. J. P. de Figueiredo and G. Chen, *Nonlinear Feedback Control System: An Operator Theory Approach.* New York: Academic Press, INC., 1993.
3. M. Deng, A. Inoue, K. Ishikawa and Y. Hirashima, Tracking of perturbed nonlinear plants using robust right coprime factorization approach, Proc. of the 2004 American Control Conference, pp. 3666-3670, Boston, 2004.
4. M. Deng and A. Inoue, Fault diagnosis in a nonlinear system using robust right coprime factorization approach, Proc. of the 2005 SICE Annual Conference in Okayama, pp. 2478-2482, 2004.
5. J. Korbicz, J. M. Koscielny, Z. Kowalczuk and W. Cholewa, *Fault Diagnosis, Models, Artificial Intelligence, Applications.* Springer, 2004.

FAULT ACCOMMODATION AND RECONFIGURATION IN VARIABLE SPEED DRIVES

D. U. CAMPOS-DELGADO AND E. PALACIOS
Facultad de Ciencias, UASLP, Av. Salvador Nava s/n, Zona Univ., C.P. 78290, S.L.P., Mexico, E-mail: {ducd,epalacios}@fciencias.uaslp.mx

D. R. ESPINOZA-TREJO*
Facultad de Ingeniería, CIEP, UASLP, E-mail: espinoza_trejo_dr@uaslp.mx

In this paper the problem of fault accommodation and hardware reconfiguration is discussed for variable speed drives (VSD) of electric motors (EM). First, the sensors and actuators used in VSD and their possible failure modes are reviewed. Next, the fault diagnosis strategies based on signal processing, model based techniques, and artificial intelligence are briefly discussed. Moreover, the diverse hardware reconfiguration schemes proposed in case of a semiconductor failure in the power actuators are detailed. A key piece in any VSD is the control algorithm, which can be severely affected in a fault conditions. Therefore, the fault accommodation algorithms suggested so far in the literature are illustrated.

1. Introduction

Electrical motors (EM) are fundamental pieces in industrial processes. In some conditions, the EM are needed to work either under a constant or a variable speed, and unknown load torque requirement. To achieve these goals, a control feedback system is implemented around the EM through a Variable Speed Drive (VSD), where electrical and mechanical sensors give information about the actual operating condition of the motor. Then, a control algorithm processes this information and compare it to the reference conditions, and finally the supplied voltage or current to the motor is modified through a power electronics actuator (see Figure 1) [26]. Hence the EM are the center elements to be controlled by the VSD. However, if the

*D.R. Espinoza-Trejo acknowledges the financial aid provided by CONACYT through a doctoral scholarship (# 166718).

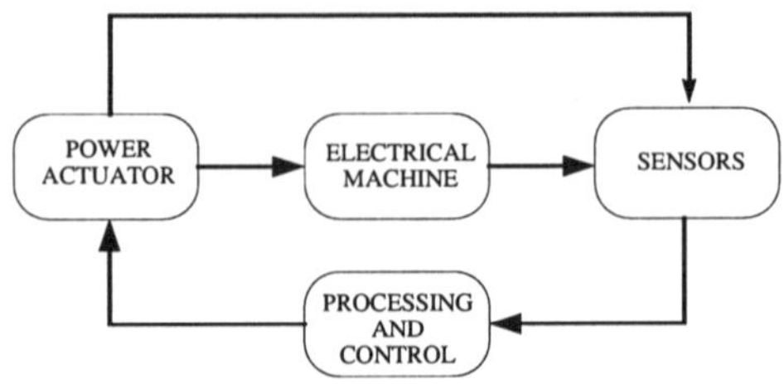

Figure 1. General Scheme of Variable Speed Drive.

automated process is running under harsh conditions or simply due to mechanical wearing and aging effects, faults can occur in the VSD system. In this case, to avoid accidents, it is important to detect as soon as possible the fault scenarios in order to judge their severity and take proper action. As a result, it is appealing the idea of running fault diagnosis algorithms in parallel to the control schemes. Moreover, for some faults, the control algorithm could compensate their appearance by replacing the erroneous information with virtual sensors, hardware reconfiguration, or switching among different control structures. This type of control strategy is called Fault Tolerant Control (FTC) [5,8]. As a result, in the signal processing and control community, the fault diagnosis and isolation (FDI) problem has attracted a lot of attention due to the industry applications and technical challenges [16,20]. Since the 80's, many strategies have been suggested looking to overcome the classical conflict of maximizing fault sensitivity against model uncertainty and unknown perturbations. Hence model-based techniques (unknown-input observers, Kalman filters, input-output parity equations, parameters estimation, etc.), statistical consistency, classification methods, principal components analysis, fuzzy reasoning, time-frequency analysis, etc. have been suggested in order to overcome the aforementioned design conflicts [5,10].

The rest of the paper is organized as follows. In Section 2, the fault scenarios in VSD's are detailed, but the discussion is limited to only actuator and sensor faults. Next, the fault diagnosis algorithms previously studied in the literature are illustrated in Section 3. Section 4 presents the accommodation schemes in terms of hardware reconfiguration and analytical compensation. Finally, the paper ends with some conclusions and final remarks in Section 5.

2. Fault Scenarios in VSD

The different fault conditions in the VSD can be classified as: actuator, sensor, mechanical and electrical faults. In Figure 2, a diagram is shown with the typical faults in VSD. In fact, the mechanical and electrical faults can be associated with new harmonics components in the stator currents, instantaneous input power or electrical torque [40]. At the same time, the faults can be classified according to their location, dynamical properties and capacity to be compensated from a control point of view (*recoverable* and *non-recoverable*). In this way, only a limited set of faults can be compensated in the control algorithm.

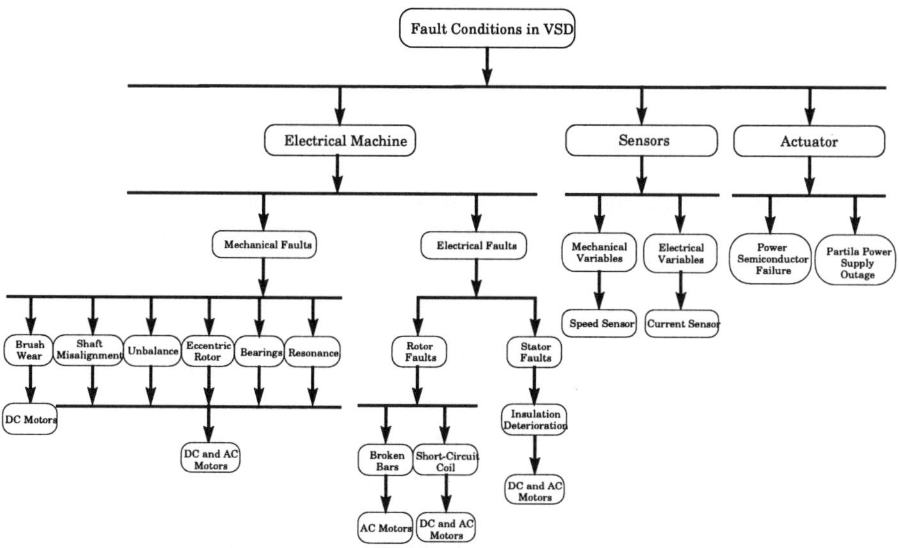

Figure 2. Fault Condition in Variable Speed Drive.

2.1. *Actuator Faults*

Nowadays, the actuators for EM use power semiconductor devices to achieve a variable DC or AC voltage to control the motors. These actuators are fed by three phase power supplies in order to provide a AC-DC, DC-AC or DC-DC power conversion. As a result, the typical faults are related either to the power supply or the converter [21]:

- Unbalanced three phase power supply (F_1^a),
- Input supply single line-to-ground (F_2^a),
- Open gate or base drive power semiconductor (F_3^a),
- Short-circuit in power semiconductor (F_4^a),
- DC link capacitor short-circuit (F_5^a).

For a standard power converter in DC and AC motors: DC-DC chopper, controlled power rectifier, cycloconverter and inverter, the faults F_4^a and F_5^a will blow the protection fuses, and the VSD will stop its operation. The occurrence of fault F_3^a will produce no control voltage (DC-DC chopper) or will reduce the efficiency of the converter (controller power rectifier), and an unbalanced control voltage delivered to the motor in AC applications [2,25,34]. Meanwhile, for faults F_1^a and F_2^a, the VSD still can operate but with a performance deterioration (*recoverable faults*).

2.2. *Sensor Faults*

In any VSD, there are electrical and mechanical sensors. Due to the mechanical coupling, the mechanical sensors are less reliable than the electrical ones. In general, the sensors can face the following fault conditions:

- Intermittent sensor connection (F_1^s),
- Complete sensor outage (F_2^s),
- DC Bias in sensor measurement (F_3^s),
- Sensor gain drop (F_4^a).

The more severe faults are F_1^s and F_2^s, since they imply a momentary or complete lack of information which is used for control purposes by the VSD. So, due to these faults, there is a risk of closed-loop instability if proper action is not taken. Nevertheless, if a sensor fault is detected, and there is still enough healthy sensors, a virtual sensor can replace the erroneous information by using an observer, or more elaborate accommodation schemes can be implemented.

3. Fault Diagnosis Approaches for VSD

The fault diagnosis algorithms in VSD are very diverse in terms of their theoretical grounds, advantages and drawbacks. Nevertheless, the main objective of the FDI block is to obtain a signal sensitive to faults (*residual*), but robust against model uncertainty, noise and unknown disturbances. In fact, in the literature there is still no global FDI algorithm that can overcome the load torque effects, measurement noise, parametric and model uncertainty, and intrinsical EM mechanical unbalance in VSD applications. For this reason different authors have suggested many FDI schemes, which can be roughly classified in three approaches:

- **Data-driven**: these methods rely on vibrational analysis, motor currents signatures (MCS), instantaneous partial and total power, and electrical torque harmonics analysis to determine, identify and quantify the fault scenarios [4,33,39,40]. In Section 2, it was mentioned that for mechanical and electrical faults, the frequency content of these measurements has been characterized in order to correlate frequency components with each specific fault condition. Thus, by using the electrical measurements of the EM, the frequency components are quantified in order to diagnose a faulty condition. The main technical problem of this approach is that all frequency components related to faults depends on the supply, rotor or slip frequencies, so in a variable speed and load torque application the frequency components are varying in time. Nevertheless, for a quasi-stationary condition, pattern recognition techniques, higher order statistics and amplitude modulation detectors have been suggested for the classification process [11,19,38]. The multi-resolution frequency decomposition with wavelets has received a lot of attention as a valuable tools to identify faults due to its applicability to non-stationary signals [12,41]. Moreover, using wavelet transform, the time signal can be analyzed in frequency bands in order to identify the faults according with their frequency components.
- **Knowledge-based**: these algorithms use input-output data in normal and faulty conditions to train universal approximators (neural networks, fuzzy systems, neuro-fuzzy systems) in order to recognize these patterns, or to reproduce the input-output dynamical behavior of the EM [1,15,17,23,24]. Specifically, neural networks (NN) have been applied to estimate the friction coefficient in the EM, in order to detect slow developing (incipient) faults [1]. In [23], NN were used to estimate the angular velocity and stator currents in induction motors, to obtain a residual by comparison with the measured data. Nevertheless, the accuracy in the fault detection and isolation process depends mostly on the training input-output data. On the other hand, the fuzzy logic and probabilistic networks have been also applied as classification tools [31].

- **Model-based**: complete or reduced mathematical models of the EM are used to synthesize observers to generate residuals in order to isolate faults [30,36]. Using these models, on-line parametric identification was suggested as a diagnosis tool [29,32], since the studied faults affected mainly the self and mutual inductances, and the friction coefficient. Moreover, using the dynamical relations described by the mathematical model, parity space residuals or high-gain observers can be derived to isolate sensor and actuator faults. However, the model-based approaches are highly sensitive to model uncertainty. Hence in order to compensate model uncertainty or disturbances, adaptive observers have been investigated for the induction motor at constant speed. Recently, the energy relations in the EM have been suggested using a bond graph representation as a natural way to determine abnormal conditions and isolate them [14].

Nevertheless, the best FDI algorithm depends strongly on the problem at hand, i.e. EM characteristics, type of measurements, hardware capabilities, implementation speed (on-line or off-line diagnosis), etc. However, the data-driven approach is still the most used diagnosis tool in industrial applications [39].

4. Fault Accommodation in VSD

For some mechanical, electrical, sensor and actuator faults might be possible to avoid performance degradation in a fault scenario or just simply to avoid the process stoppage, if the controller is designed with this goal in mind and the control hardware has some minimal redundancy. The fault accommodation strategies for VSD can be classified in two classes:

(1) Actuator hardware reconfiguration,
(2) Control accommodation or reconfiguration.

4.1. *Hardware Reconfiguration for Power Actuators*

In the VSD, the standard power actuators are sensitive to failures in the power electronic semiconductors and voltage supply [21]. So after a fault, the complete actuator must be replaced for repair, and consequently the automation process has to be stopped. To avoid this problem, new actuators architectures based on some hardware redundancy have been suggested to add reconfigurability to them [18]. As a result, if the analytical and hardware redundancy is combined, very effective FTC can be designed for critical applications. Thus, in the case that an inverter leg is lost and the machine neutral point is available for output connection, reconfiguration schemes have been proposed to operate the VSD with only two stator windings by connecting the neutral point to a fourth inverter leg, or to the middle point of a capacitor bank (see Figure 3.A and B) [6,37,28]. If the neutral point is not available, a fourth inverter leg is also suggested but bidirectional switches are employed to bypass the faulty inverter leg (see Figure 3.C) [3]. Another approach uses the cascade multilevel inverters in order to provide hardware redundancy to bypass the faulty cells (see Figure 3.D) [35].

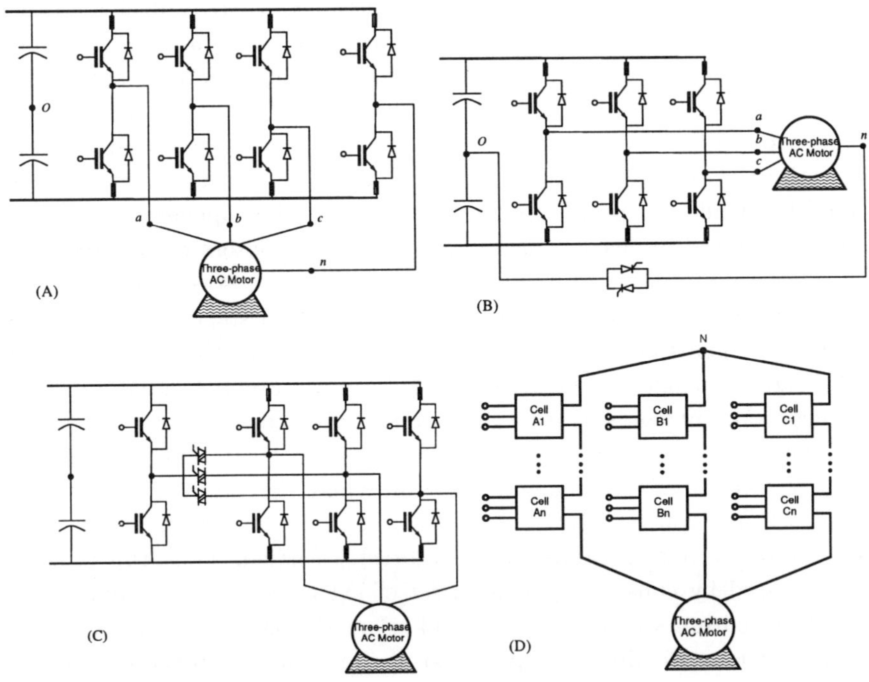

Figure 3. Power Actuators with Fault Reconfiguration.

4.2. *Fault Compensation Strategies*

For the recoverable faults, it is possible to compensate the nominal controller in order to prevent performance degradation or instability (Fault

Tolerant Control, FTC). There are two types of FTC: active and passive approach [5]. In the passive approach, an overall controller that can maintain acceptable controller against a set of faults is pursued. Meanwhile, the active FTC schemes look first to detect a general fault condition, and next isolate it (FDI stage), in order to provide the appropriate fault accommodation. These techniques rely mainly on analytical redundancy to achieve their objective. Comparing the research efforts in the FDI problem for VSD, the contributions in the FTC applications are sparse. Some of them are briefly described next.

One of the earliest approaches studies the operation of an inverter in single-phase mode for two faults: open base drive power semiconductor and device short-circuit. For this condition, a modified V/Hz control is applied looking to reduce the resulting pulsating torque by injecting odd harmonic voltages [22]. Another application is focused on sensor faults, in specific for intermittent disconnections, where a bilinear model of the induction motor is used to synthesize observers to generate the appropriate residuals [4]. However, the FTC scheme is reliable only in a limited operating condition of the EM. Meanwhile, the problem of fault detection and compensation of sensor faults in direct-torque controlled (DTC) induction motors was addressed in [27]. Thus an estimation for the velocity and current sensors is applied using gain-scheduled observers, and according to the FDI stage, the real or the estimated signals are applied in the DTC scheme. On the other hand, an FTC algorithm against velocity sensor failures was also presented, where an sliding-mode control is used for an encoder-based control and a fuzzy-logic V/Hz control is applied for a sensorless condition [42]. Recently, an implicit FTC was applied for the induction motor, where the failure modes are embedded in the control strategy [7]. This is the only work so far where rotor and stator mechanical faults has been directly studied. Thus, this problem is addressed by considering a finite set of extra harmonic components in the stator currents with a indirect field oriented controller to add a voltage compensation. Finally, an application of sliding-mode observers was considered for sensor faults but for a separately excited DC motor [13]. Hence the sliding mode observer is used to estimate the sensor fault vector which is then applied for compensation of a PI nominal controller. For the same application, an algorithm inspired on robust control uses the generalized internal model control (GIMC) to achieve an FTC architecture [9].

5. Conclusions and Final Remarks

In this paper an overview of fault accommodation and reconfiguration in VSD was presented. First, the fault scenarios in VSD were reviewed. Typically, they can be classified into EM faults, sensor and actuator faults. Due to the natural variability in the motor parameters, measurement noise, unknown perturbations (load torque) and unmodeled uncertainty, most of the FDI approaches rely on signal processing techniques (signatures matching) to identify certain frequency components associated with these faults. On the other hand, by adding hardware redundancy to the power actuators, these devices can be reconfigured in order to still maintain aceptable performance for some faults. The nominal control algorithm should also be updated in a fault event to avoid instability or severe performance deterioration. In this context, the induction motor, due to its popularity in industry applications, has received most of the attention. At this point, the results in FDI and FTC applications to VSD are growing, looking to reduce the dependance on less reliable sensors and to add extra electrical measurements that could be used in a fault event for diagnosis purposes.

References

1. S. Altug, M.Y. Chow, and H.J. Trussell, *IEEE Trans. Ind. Electronics*, **46**, 1069 (1999).
2. R.L. de Araujo Ribeiro, C. Brandao Jacobina, E.R. Cabral da Silva and A.M. Nogueira Lima, *IEEE Trans. Power Electr.*, **18**, 587 (2003).
3. R.L. de Araujo Ribeiro, C. Brandao Jacobina, E.R. Cabral da Silva and A.M. Nogueira Lima, *IEEE Trans. Ind. Electr.*, **51**, 439 (2004).
4. M.E.H. Benbouzid, *IEEE Trans. Ind. Electr.*, **47**, 984 (2000).
5. M. Blanke, M. Kinnaert, J. Lunze, and M. Staroswiecki, *Diagnosis and Fault-Tolerant Control*, Springer-Verlag, 2003.
6. S. Bolognani, M. Zordan, and M. Zigliotto, *IEEE Trans. Ind. Electr.*, **47**, 1134 (2000).
7. C. Bonivento, A. Isidori, L. Marconi and A. Paoli, *Automatica*, **40**, 355 (2004).
8. D.U. Campos-Delgado, and K. Zhou, *IEEE Trans. Aut. Cont.*, **48**, 832 (2003).
9. D.U. Campos-Delgado, S. Martinez-Martinez and K. Zhou, *IEEE/ASME Trans. Mechatronics*, **10**, 419 (2005).
10. J. Chen and R.J. Patton, *Robust Model-Based Fault Diagnosis for Dynamic Systems*, Kluwer Academic Publishers (1999).
11. T.W.S. Chow and H.Z. Tan, *IEEE Trans. Ind. Electr.*, **47**, 1051 (2000).
12. T.W.S. Chow and S. Hai, *IEEE Trans. Ind. Electr.*, **51**, 558 (2004).
13. C. Edwards and C.P. Tan, *Control Eng. Practice*, **14**, 897 (2006).
14. C. Fantuzzi, and C. Secchi, *Proc. 2004 American Control Conf.*, Boston, June 30 - July 2 (2004).

15. F. Fillippetti, G. Franceschini, C. Tassoni, and P. Vas, *IEEE Trans. Ind. Electr.*, **47**, 994 (2000).
16. P.M. Frank, *Automatica*, **26**, 459 (1990).
17. D. Fuessel, and R. Isermann, *IEEE Trans. Ind. Electr.*, **47**, 1070 (2000).
18. T. Gopalarathnam, H.A. Toliyat, and J.C. Moreira, *Proc. IEEE-IAS Annual Metting*, Rome, Oct. 8-12 (2000).
19. J. Ilonen, J.K. Kamarainen, T. Lindh, J. Ahola, H. Kalviainen, and J. Partanen, *IEEE Trans. Ind. Appl.*, **41**, 963 (2005).
20. R. Isermann and P. Ballé, *Control Eng. Practice*, **5**, 709 (1997).
21. D. Kastha, and B.K. Bose, *IEEE Trans. Ind. Appl.*, **30**, 1028 (1994).
22. D. Kastha, and B.K. Bose, *IEEE Trans. Ind. Electr.*, **41**, 426 (1994).
23. K. Kim and A.G. Parlos, *IEEE/ASME Trans. Mechatronics*, **7**, 201 (2002).
24. K. Kim, A.G. Parlos, and R.M. Bharadwaj, *IEEE Trans. Ind. Electr.*, **50**, 1038 (2003).
25. J. Klima, *IEE Proc.-Electr. Power Appl.*, **150**, 255 (2003)
26. R. Krishnan, *Electric Motor Drives: Modeling, Analysis, and Control*, Prentice Hall (2001).
27. K.S. Lee and J.S. Ryu, *IEE Proc. Control Theory Appl.*, **150**, 376 (2003).
28. T.H. Liu, J.R. Fu, and T.A. Lipo, *IEEE Trans. Ind. Appl.*, **29**, 910 (1993).
29. X.Q. Liu, H.Y. Zhang, J. Liu and J. Yang, *IEEE Trans. Ind. Electr.*, **47**, 1021 (2000).
30. K.A. Loparo, M.L. Adams, W. Lin, M.F. Abdel-Magied, and N. Afshari, *IEEE Trans. Ind. Electr.*, **47**, 1005 (2000).
31. L.J. de Miguel and L.F. Blazquez, *Eng. Appl. Artificial Intelligence*, **18**, 423 (2005).
32. O. Moseler and R. Isermann, *IEEE Trans. Ind. Electr.*, **47**, 1015 (2000).
33. S. Nandi, H.A. Toliyat, and X. li, *IEEE Trans. Energy Conv.*, **20**, 719 (2005).
34. R. Peuget, S. Courtine, and J.P. Rognon, *IEEE Trans. Ind. Appl.*, **34**, 1318 (1998).
35. J. Rodriguez, P.W. Hammond, J. Pontt, R. Musalem, P. Lezana and M.J. Escobar, *IEEE Trans. Ind. Electr.*, **55**, 1080 (2005).
36. H. Rodriguez-Cortes, C.N. Hadjicostis, and A.M. Stankovic, *Proc. 2004 American Control Conf.*, Boston, June 30 - July 2 (2004).
37. M.B. de Rossiter Correa, C.B. Jacobina, E.R. Cabral da Silva, and A.M. Nogueira Lima, *IEEE Trans. Ind. Appl.*, **37**, 873 (2001).
38. J.R. Stack, R.G. Harley, and T.G. Habertler, *IEEE Trans. Ind. Electr.*, **51**, 1097 (2004).
39. W.T. Thomson, and R.J. Gilmore, *Proceedings of the Thirty-Second Turbomachinery Symposium*, Houston, Texas (2003).
40. A.M. Trzynadlowski, and W. Ritchie, *IEEE Trans. Ind. Electr.*, **47**, 1092 (2000).
41. Z. Ye, B. Wiu, and A. Sadeghian, *IEEE Trans. Ind. Electr.*, **50**, 1217 (2003).
42. F. Zidani, M.E.H. Benbouzid, D. Diallo, and A. Benchaib, *IEEE Int. Electric Machines and Drives Conf.*, June 1-4 (2003).

DEVELOPMENT OF HARD LANDING DIAGNOSIS SYSTEM BASED ON ACCELERATION SENSING USING MEMS

A. FIROOZRAI, I. STIHARU and R. SEDAGHATI
Mechanical and Industrial Engineering Department, Concordia University, 1455 de Maisonneuve Boulevard, Montreal, Quebec, Canada
istih@vax2.concordia.ca

A new study on diagnosis of hard landing phenomena based on accelerations sensed by micro accelerometers attached to different locations of an aircraft is conducted. In order to consider all significant parameters involved in landing dynamics, a comprehensive model including the geometry of the aircraft and variations within the recommended loading conditions is developed using MSC.ADAMS software. Here, two different classes of aircraft are considered, a transport class and a military type. Moreover, a micro accelerometer is designed for this type of diagnosis system and environment. The accelerometer design is further optimized to provide the maximum sensor sensitivity. To verify the performance of the designed accelerometer, its dynamic model and the aircraft's model generated in ADAMS is integrated. The behavior of the model is then investigated during landing. Different simulations are conducted by changing important parameters affecting landing dynamics. This study illustrates the importance of considering lateral aircraft dynamics in the diagnosis of hard landing phenomena. An aircraft with a roll angle before touchdown will cause high lateral impact loads to the landing gears and the airframe.

1. Introduction

Hard landing happens whenever impact loads due to the landing phase of an aircraft, cause damage to whole or some parts of its structure. Although hard landing is documented versus the vertical speed of the aircraft at the touchdown, specific landing condition raise legitimate questions with regard to this formulation. One of the most crucial phases of flight, as seen in a number of accidents in the past decades, is the landing. According to some published statistics [1], close to 60% of the flight-accidents occur during the landing phase. Other investigations confirm that, for a typical fleet of aircraft, landing normally has the highest percentage of accidents and/or incidents and, therefore, is considered as the most crucial and risky phase of a mission.

Contributors to accidents could be categorized into two different sets. The first set is related to sensing errors, such as altitude estimate error, runway conditions, and orientations. The second is due to sudden changes in atmospheric conditions. Thus, gust and wind shear conditions are responsible for a high number of hard landings and accidents each year [1].

Such a situation was encountered at Hong Kong International Airport in 1999, in which the aircraft encountered a high cross and tailwind. This caused the aircraft to land hard enough that the right wing and landing gear completely detached from the fuselage [2].

The current process of deciding whether a hard landing has occurred or not, is based on subjective assessment by the pilot. The reactions perceived at the level of the flight deck often very badly convey the real load level applied to the aircraft as a whole. Considerable loads can be applied to the structure, without, however, giving rise to effects which can be felt at the level of the flight deck. Because of lack of reliable quantitative data, errors are made in the above assessment. Also, current manufacturer guidance for hard landing identification and operator maintenance readouts and analysis of flight data recorder following suspected hard landings may not be adequate to identify landings in which structural damage may have occurred. As a result, an airplane may be grounded unnecessarily, at a considerable cost of time and money, or conversely, a damaged airplane can continue in service. For obvious safety reasons, inspection of the structure of the aircraft, as well as repairing possible damages, has to be carried out whenever the aforementioned design loads are exceeded.

The technical literature regarding methods for determining hard landings can be divided into two classes. The first and most often cited method is to utilize kinetic measurements (acceleration, velocity or displacement indications).The second method is to utilize force measurements (pressure or stress/strain indications).

Attempts to utilize kinetic measurements (airplane vertical deceleration or sink rate sensed at aircraft's deck level) have failed because the flight deck kinetic conditions are only ancillary to parameters and not the cause of the structural damage especially in landing gears.

Force measurements are mostly preformed by using strain gauges. These sensors are fragile and are not performing well under impact situations like the one investigated in this study.

This work is focused on the replacement of the subjective assessment with different measurable parameters in which they not only indicate the occurrence of a hard landing, but also determine the extent of the damage by finding which parts are damaged. This will give the maintenance team an insight to the location of each inspection and the type of tests to be carried out, saving a great deal of money and maintenance time.

The landing dynamics is a highly complex problem. Forces are applied to the airplane structure from a multitude of causes.

Multibody dynamic aircraft analysis is a well known solution to these spatial nonlinear dynamic problems. Here MSC.ADAMS software is used to build virtual models to accurately model landing conditions.

Microelectromechanical accelerometers are widely used in automotive industry to predict impact loads for deployment of airbags. ADXL 50 was used in impact detection and it is easily manufactured using surface micromachining [3]. However, other technologies could also be employed to build micro-accelerometers to be used for hard landing detection.

These types of accelerometers could be rearranged to measure three axis accelerations and because they do not need to expose to the environment to sense the stimuli, they are especially suitable for this type of highly temperature and pressure changing environments encountered during flight.

After designing, the sensor is integrated within the multibody model to asses the performance of the sensor in virtual hard landing situations.

2. Multibody Modeling

In order to model dynamic properties, MSC.ADAMS software is used. Fig.1 demonstrates one of the models used in this work. As mentioned above, one of the obstacles in modeling landing dynamics is that, the initial conditions and aerodynamic properties depend on the class of aircraft. The hard landing phenomena, is investigated for two different aircrafts with two different landing characteristics.

The main focus of this work is on modeling of longitudinal and lateral dynamics. The effect of braking dynamics is not considered here.

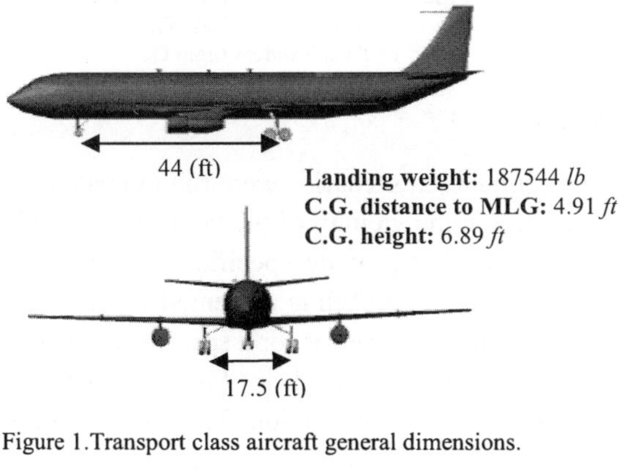

Figure 1.Transport class aircraft general dimensions.

2.1. *Landing gear types:*

Three types of landing gears were used. It is common to categorize landing gears based on their number of wheels. In this work, only those parts affecting landing dynamics are modeled. According to Fig.2 twin wheel landing gear is modeled without the steering degree of freedom and consequently shimmy vibrations caused by it.

All joints and connecting parts are assumed as rigid. Thus, compliances are related to tire and shock absorber deflections. Aircraft tires are subjected to a wide variety of high dynamic and thermal loads and their failure can have disastrous consequences. Therefore, development of a realistic mathematical model, describing vertical and lateral dynamic forces is necessary. Although the technical literature employs "the magic formula" model for tires [5], it is apparent that such model is more appropriate in the ground dynamics of the tire and it does not describe the impact with the ground accurately. A single-acting shock Absorber, which is the most commonly used design for commercial transports, is used. Its nonlinear behavior is shown in [4].

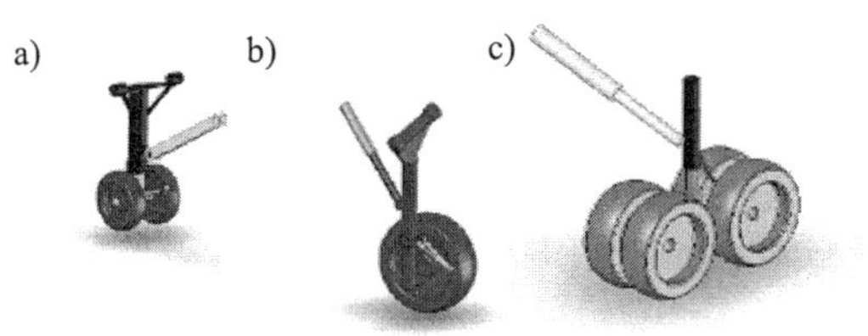

Figure 2.Types of shock absorbers: a) Twin wheel Nose Gear
b) Single wheel Main Gear
c) Twin Tandem Main Gear

3. Sensor Design

A number of simulation scenarios were run to evaluate the peak value of the recorded acceleration in the vicinity of the landing gears. The response time was also assessed to be able to provide specification for the micro-accelerometer. Below, a "classic" design approach is presented for illustrating the extent of the effort required for designing a sensor for a specific application.

This type of sensor illustrated in Fig.3, is using interdigitated finger capacitors to sense impact's acceleration. The most important advantage of this accelerometer is, its ability to be manufactured using surface micro machining [7].

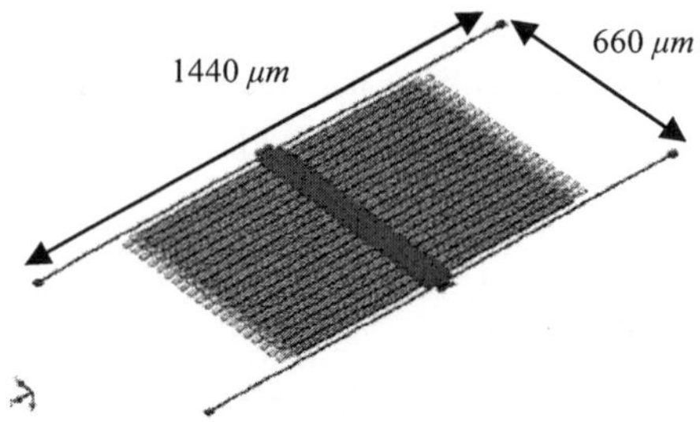

Figure 3.Interdigitated finger capacitive accelerometer.

First dimensions of the sensor are set by static design output. Then, the dynamic equations of motion are derived and optimization is performed to find the highest possible sensitivity of the sensor.

3.1. *Micro-structure model for dynamic analysis:*

Dynamic analysis of this micro-structure is performed using lump mass model. This model consists of a lump mass, a spring and a damper. There is an external vibration source which is the acceleration acting on the proof mass. Equation of motion is a second order differential equation:

$$m\ddot{x} + C_{av}\dot{x} + k_{eff}x = m\,a(t) \tag{8}$$

where *m* is the mass of the sensor, C_{av} is damping coefficient, k_{eff} is effective stiffness and *a(t)* is the input acceleration:

$$k_{eff} = k_m - k_{es} \tag{9}$$

$$k_m = 4\left(\frac{12EI}{l_t^3}\right) \tag{10}$$

k_m is mechanical spring constant calculated by lateral stiffness of four tethers. k_{es} is electrical stiffness which will soften the mechanical constant and is calculated as below.

$$F_{es} = \left|\frac{\partial U}{\partial x}\right| = \frac{1}{2}\left|\frac{\partial C}{\partial x}\right|V^2 \tag{11}$$

F_{es} is the relative force generated by the electrostatic field which is derivative of electrical energy with respect to mass displacement.

$$F_{es} = \frac{n\varepsilon l_0 t d V^2 x}{\left(d^2 - x^2\right)^2} \approx \frac{n\varepsilon l_0 t V^2 x}{d^3} \tag{12}$$

$$x^2 \approx 0\,, x^4 \approx 0 \tag{13}$$

For simplicity, the above assumptions are conducted to yield a linear equation relating force to the displacement [8]. This linearity will result in ordinary differential equation of motion. The linearization error is less than 5 percent. Further, as per our evaluations the electrical stiffness is expressed as:

$$k_{es} = \frac{n \varepsilon l_0 t V^2}{d^3} \tag{14}$$

It is further assumed that in this accelerometer damping is minor and it is modeled as shear damping as below:

$$C_{av} = \frac{\mu L B}{H} \tag{15}$$

here μ is the air viscosity and *H* is the sacrificial layer thickness. *L* and *B* are length and width of proof mass. The second order differential equation of motion is solved by MATLAB/SIMULNK software.

3.2. *Optimization:*

It is important to relate the design of the micro-sensor with the initial geometry selection that was made based on the static deflection of the supporting tethers under the peak acceleration assumption. However, the dimensions of the beam were selected more or less using the engineering common sense. The selection may yield good results for static output but may produce quite poor dynamic response for the structure. The objective of this section is to find within the range of selected dimension the best possible geometry that will enable the microstructure to perform both static and dynamic and yield a readable output for the selected situation.

An increase in proof mass width will decrease the capacitance and increase the mass. It is important to mention that increase of this parameter will decrease the tether length also, causing higher spring stiffness.

According to Fig.4, it can be seen that for 3.3 volts bias, upon an increase in center mass width there is a maximum sensitivity and then the curve will fall gradually. In this region mechanical and electrical stiffness are balanced in this maximum point which increasing further mass width will cause the sensitivity to fall because of reduction in static capacitance.

4. Interaction of the Sensor and Multibody Model

In this work, hard landing diagnosing system based on accelerations sensed in different parts of an aircraft is studied. The accelerometer designed in previous section can be easily rearranged to exhibit a three axis accelerometer by just using three of them positioned in three directions x, y and z.

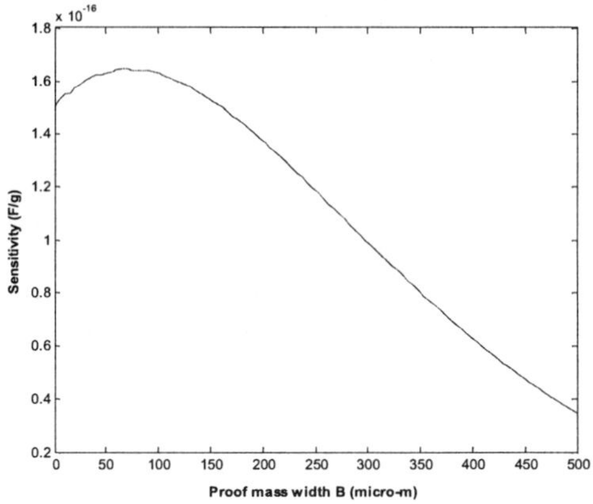

Figure 4.Sensitivity change with respect to proof mass width

The most important aspect of this study is to find how many sensors should be used and where should they be mounted. Landing gears experience high accelerations due to ground impact. So, accelerometers are attached to these three landing gears. They are mounted on the lower piston of the shock absorbers.

Multibody model simulated in MSC.ADAMS is connected to the sensors mathematical modeled in SIMULINK with the use of ADAMS control tool box.

Critical accelerations, in which beyond these values hard landing is said to be placed, is found by considering maximum allowable stresses. These allowable stresses and whether it is a tensile stress, or compressive or it is a maximum stress that a welded joint can bear is provided by the manufacturer. These stress amplitudes are reported and based on relation of forces with accelerations and stresses, one can find allowable accelerations in different directions and it will be the objective of study in the future work.

5. Simulation Results

After integrating the mathematical model for the sensor with multibody model, we are able to simulate the whole model in different landing situations. Main landing gear is affected mainly by parameters below:

1. *Sink speed:* Sink speed is one of the most important parameters affecting landing dynamics. Often, landing gears are designed with a maximum sink speed criteria. This value, according to [4] is 12 ft/sec. however, [6] mentioned that although sink speed is a very important parameter in landing dynamics it is not the only cause of hard landing.

2. *Roll angle and roll rate:* Possible cause of this roll angle could be a wind shear or a pilot mistake. Proper tire model should be used to see the real impact of the roll angle on landing dynamics. In this situation, landing impact energy is concentrated on single MLG and some of this energy is acted laterally from tires to shock absorber cylinder wall and to landing gear attached points to wing or fuselage. If these forces are large enough they will cause a catastrophic failure. Another parameter responsible of crashes and hard landings is the roll rate. After the first MLG touch down, the roll momentum will cause a more severe impact at the second MLG touch down.

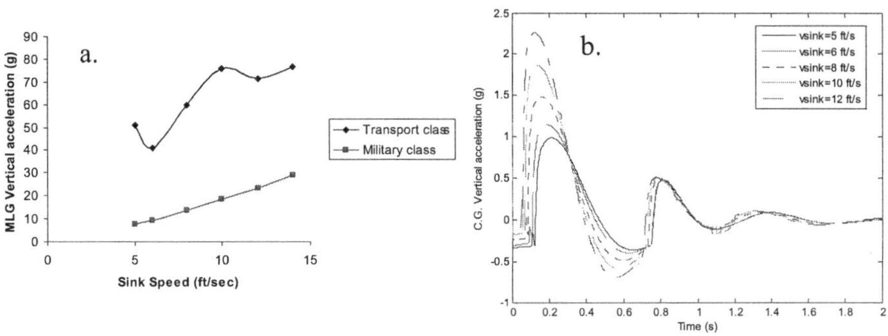

Figure 5.a) Effect of sink speed on main landing gear vertical acceleration
b) Effect of sink speed on transport-type aircraft's center of gravity acceleration

According to Fig.5a a raise in sink speed shows an increase in MLG vertical acceleration both for transport and military class, but it is interesting to see that the rate of this increase in transport class is more significant.

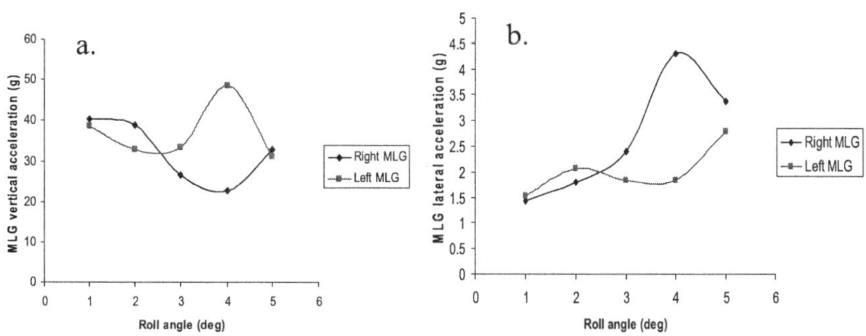

Figure 6.a) Effect of roll angle on vertical acceleration of main landing gears for transport class (right MLG touch down first), b)Effect of roll angle on lateral acceleration of main landing gears for transport class.

Different roll angles are examined using a 10 degree per second constant roll rate. According to fig.6a, vertical acceleration decreases for right land gear (first touchdown) due to lateral impact energy consumption. However, Left main landing gear encounters a more severe vertical impact due to roll momentum of the aircraft. This is shown in Fig.6b.

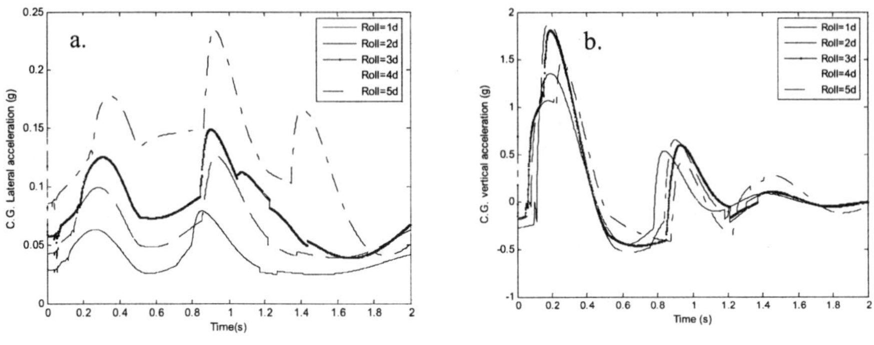

Figure 7.a) Effect of roll angle on lateral acceleration of center of gravity for the transport class, b) Effect of roll angle on vertical acceleration of center of gravity for the transport class.

6. Conclusion

This paper is a new study on diagnosis of hard landing phenomena based on accelerations sensed by micro accelerometers attached to different locations of an aircraft. Landing is a highly complex dynamic problem. In order to consider all important parameters involved in landing dynamics, comprehensive models containing all these parameters was developed using MSC.ADAMS software. In order to study in a broader view two different classes of aircraft, one a transport class and the other a military class was modeled.

Based on these models, a micro accelerometer suited for this type of diagnosis system and environment was designed. The design was optimized to give maximum sensor sensitivity with minimum noise generation.

Sensor's transfer function was integrated with mutlibody models using MATLAB/SIMULINK and MSC.ADAMS connection, to test the performance of designed accelerometer in action.

Different simulations were carried out by changing important parameters affecting landing dynamics. Results indicated that, although sink speed is a major affecting parameter in main landing gears, it wasn't the only one. Roll momentum of the aircraft was another important parameter which not only increased the vertical impact forces, but also generates a lateral impact to main landing gears. This study showed the importance of considering lateral aircraft dynamics in diagnosis of hard landing phenomena.

7. References

1. Malaek et. al. , Flight envelope expansion in landing phase using classic, intelligent, and adaptive controllers, Journal of Aircraft, Vol. **43**, No. **1**, Jan 2006.
2. Aircraft accident report: *Crush during landing federal express, Inc. Mcdonnell Douglas MD*-**11**, **N611FE**, Newark International airportNewark, New Jersey, July 31, 1997.
3. S. J. Sherman et. al. , *A Low Cost Monolithic Accelerometer*, Analog Devices Semiconductor, IEEE Symposium on VLSl Circuits Digest of Technical Papers, 1992
4. Currey, N. S., Aircraft Landing Gear Design: Principles and Practices, AIAA Education Series, Washington, 1988.
5. Anon., "*Aircraft Tires: Bias or Radials?*" Aerospace Engineering, Vol. **11**, No. **9**, September 1991, pp. **13 -14**.
6. Chester, D.H., Aircraft Landing Impact Parametric Study with Emphasis on nose Gear Landing Conditions, Journal of Aircraft 2002 ,0021-8669, Vol.**39**, No.**3**, pp.394-403.
7. W. Qu, C. Wenzel and K. Drescher, *Fabrication of low-cost capacitive accelerometers by 3D Microforming*, Institut f i r Hulbleiter- und Mikrosystemtechnik, 1997.
8. Innam Leea , *Development and analysis of the vertical capacitive accelerometer*, journal of Sensors and Actuators, 2005.

FDI IN THE INDUCTION MOTOR DRIVE UNDER VARYING LOAD TORQUE USING BOND GRAPHS

AGUILAR-JUSTO MARVING O.
GUERRERO-RAMÍREZ GERARDO V.
VELA-VALDÉS L. GERARDO

Centro Nacional de Investigación y Desarrollo Tecnológico, cenidet.
Interior Internado Palmira s/n, Col. Palmira. Cuernavaca, Morelos, 62490, México.

Squirrel cage induction motors (SCIM), supplied with power electronic converters, are widely used in industries, but they are prone to fail. A lot of work has been dedicated to fault diagnosis and isolation (FDI) of motors using signal-based techniques, like motor current signature analysis (MCSA). One disadvantage of the method is generation of harmonics without faults (causing false alarms), for example when there is a varying load or a power electronic converter supplying the motor. Another disadvantage is the long detection time occasioned because the frequency spectrum needs to be built with the signal in the stationary state. Moreover, the same technique (MCSA) cannot be used to diagnose faults in the power electronic converter, whose preferred fault diagnosis technique is the space vector loci. This work makes an FDI in the induction motor drive using the bond graph (BG) diagnosis method. As the method is based on the changes of signals' amplitudes, it can diagnose faults when a known varying load torque is present. Moreover, faults in the power electronic converter can be diagnosed too, because the method uses causal graph of the whole system, where the signals' amplitudes can be propagated.

1. Introduction

Presently most of the motors used in industry applications are squirrel cage induction motors (SCIM) due to their good behavior, low cost and robust characteristics [13]. Such motors, like other systems, are prone to fail because of corrosion, high temperatures, old age components, etc. [3]. A lot of work has been dedicated to the diagnosis of motors using motor current signature analysis (MCSA). As the motor works under a symmetrical ideal condition, the consequences of losing this condition has been analyzed, and normally some harmonic currents are induced in the motor's signals. The harmonics generated are commonly obtained by means of their frequency spectrum [15]. The advantage of this method is that the frequencies of the harmonics generated are known for different kinds of faults. One disadvantage of the method is the harmonics generated without faults (causing false alarms), for example when there is a varying load torque or a power electronic converter supplying the motor. Another disadvantage is the long detection time occasioned because the frequency spectrum needs to be built with the signal in the stationary state.

The fault diagnosis in the induction machine can also be executed with the space vector loci approach. Here, the *dq* stator currents are plotted to each other. In presence of a fault, an asymmetry of the three-phase signals is obtained and it impacts on the *dq* signals. The asymmetry provokes that the space vector loci displays a deformed figure [3].

Another approach for diagnosis is by means of adaptive neural networks (ANN). For example, in [14] an ANN is trained to yield negative sequence voltages when is supplied with negative sequence currents, positive sequence currents and positive sequence voltages.

In another context, some qualitative diagnosis tools have been implemented in different systems. References [7] and [10] have implemented a diagnosis tool that uses sign information in signals' amplitudes when there is a fault in their systems. From this information, a sign propagation procedure is executed through the elements of the system represented in a causal graph. This causal graph is derived from the Bond Graph (BG) model of the system, because this model has causal information in its elements. This paper is based on [1], where the BG diagnosis method was applied to the induction motor. This work considers the power electronic converter together with the induction motor as a global system. Although the system has two different subsystems (the inverter and the motor), it is expected to execute the diagnosis in just one scheme. This is because the BG modelling can combine subsystems of different nature in just one model.

2. BG Model of the Induction Motor Drive

There are a lot of ways to model the induction motor drive, but in the BG diagnosis it is important to have a BG model that includes all the parameters that are related with the faults under analysis.

2.1. *BG Model of the SCIM*

The three-phase SCIM model used is based on the model presented in [8]. This model has three windings in the stator and five bars in the squirrel cage rotor (Figure 1).

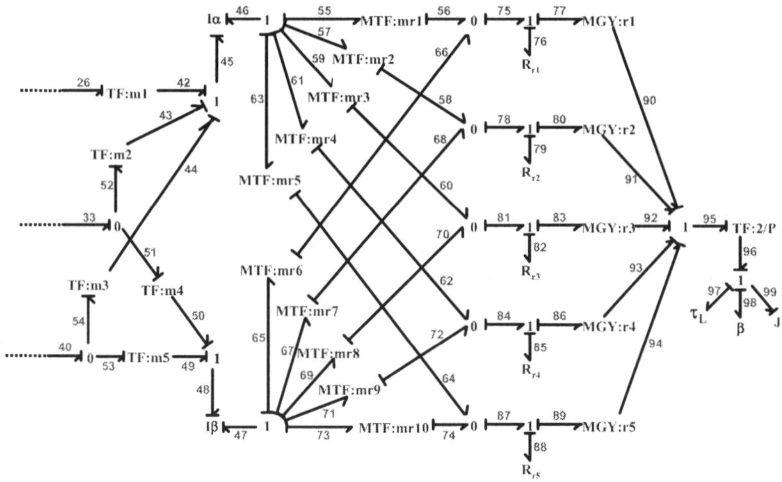

Figure 1. BG model of the three-phase squirrel cage induction motor.

In the left side there are five transformers that represent the αβ transformation (the stationary reference frame) of the stator currents. This transformation is used because the magnetic subsystem can treated as linear using two I-fields (Iα and Iβ). The points indicate the interconnection that exists with the power converter. The modulated transformers and the five 0-junctions to the right of these modulated transformers convert the αβ rotor currents to five rotor currents corresponding to the five rotor bars. The last transformer along with the modulated gyrators builds the electromagnetic torque expression and the final 1-junction builds the equation of torques in the mechanical subsystem.

A qualitative analysis has to be done in the BG diagnosis checking the increase or decrease of signals. All the variables of the induction motor BG model exhibit constant amplitudes in the steady state, except for the electrical angular position (it is a ramp). For this reason, the dependency on this variable was eliminated making $\theta_r = 0$.

2.2. *BG Model of the Inverter*

There are two ways to model switching elements. The first one consists of using switched bond graphs and it implies to have a bond graph element with variable causality (Figure 2a) [4]. This concept is inadequate for diagnosis with BG because involves 2^n causal graphs to analyze.

The second way of modeling a switching element is representing it like a modulated transformer along with a resistor (Figure 2b).

Figure 2. Bond graph model of a switching element.
a) Variable causality. b) Using a modulated transformer and a resistor.

Depending on the effort direction assigned to the switching element, its R element represents the ON or the OFF resistance [6]. The modulus of the transformer contains the activation signal value of the switching element.

The bond graph model of the power converter is based on a voltage source inverter (VSI), and it is shown in figure 3.

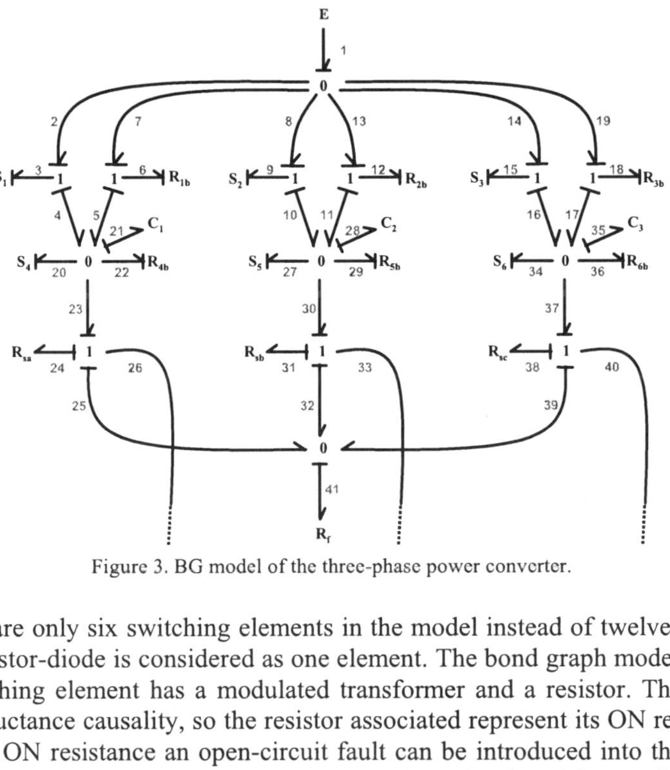

Figure 3. BG model of the three-phase power converter.

There are only six switching elements in the model instead of twelve, i.e., the pair thyristor-diode is considered as one element. The bond graph model used of the switching element has a modulated transformer and a resistor. This model has conductance causality, so the resistor associated represent its ON resistance. With the ON resistance an open-circuit fault can be introduced into the model. To simulate a short-circuit fault, an R_{OFF} resistor is placed in parallel to the switching element representing its OFF resistance. A fictitious resistor (with very high value) and three capacitors (with low values) have to be added in order to avoid causal conflicts. The points indicate the interconnection that exists with the induction motor.

3. Fault Diagnosis Method Applied to the Induction Motor

The fault diagnosis with bond graphs basically makes a qualitative analysis (state amplitude changes) through the causal graph obtained from the BG model of the system. The qualitative analysis gives information about sign changes of parameters (possibles responsible components) [7] [10].

From the BG model can be obtained the constitutive relations that form the dynamic model of the system. The causal graph is based on the constitutive relations, where the variable next to the arrow of a link is the consequent receiving qualitative information from antecedents and parameters. For example, the links 26, 42, 43, 44, 45 and 46 of the induction motor BG model can be used to build the corresponding part of the causal graph (Figure 4).

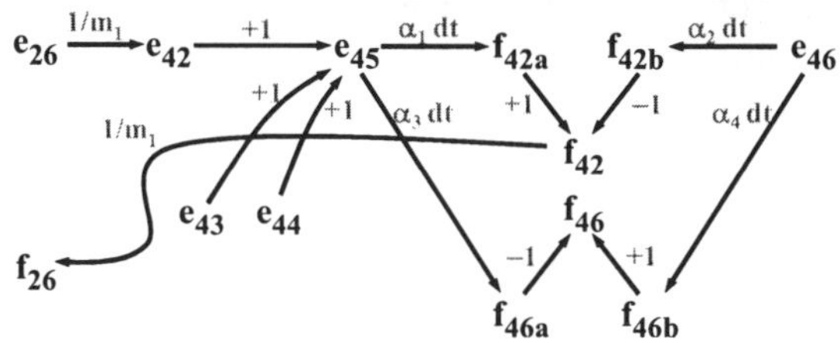

Figure 4. Part of the causal graph corresponding to links 26, 42, 43, 44, 45 and 46 of the BG model.

3.1. *Fault trees*

Fault trees are obtained making a backward propagation (a route opposite to the arrows direction) in the causal graph. Each measured or estimated variable has a fault tree. The beginning of the fault tree is the measured or estimated variable with the corresponding qualitative change: the plus (+) or minus (–) sign, depending on the increase or decrease of the variable with respect to the healthy state. In the propagation of this sign, possibles responsible parameters of the fault are recollected. The route should finish when a repeated variable be found [11]. In this work the route finish not only when a repeated variable is found, but when a measured or an estimated variable (with opposite sign to the true) is found too. On this way, the algorithm is on the alert of measured or estimated information and avoids recollecting parameters not responsible of the fault. The example of this explanation is shown in Figure 5, it is part of a fault tree obtained from an open-circuit fault in the stator winding of phase b (f_{30} is one measured variable, the stator current of phase b, and the qualitative sign – is recollected in the case of an R_{sb}^{+} fault, as the table 1 shows).

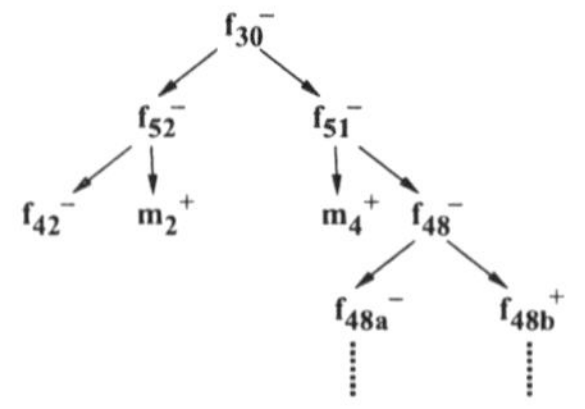

Figure 5. Part of the fault tree corresponding to the f_{30}^- variable for the R_{sb}^+ fault.

The information obtained from this fault tree is the following: if f_{30} decreases is because f_{52} decreases or f_{51} decreases; if f_{52} decreases is because f_{42} decreases or m_2 increases. As f_{42} is an estimated variable, it can be compared with the information of the fault tree under analysis. The estimation of f_{42} points out an increase on this variable (different to the predicted in the fault tree), so the analysis on this branch is stopped and the m_2^+ parameter is recollected as a possible responsible parameter of the fault. In the f_{51}^- branch it is obtained that f_{51} decreases because m_4 increases or f_{48} decreases. The estimation of f_{48} ensures that it decreases of qualitative value, so the analysis in this branch continues, but the m_4^+ parameter is not recollected to the set of possibles responsible parameters. Table 1 exhibits signatures with different kinds of faults, where in addition to the + and – signs, the 0 (no change) and x (don't care) signs are shown too.

Table 1. Simulation signatures of the induction motor drive.

	v_{S1} e_3	v_{S2} e_9	v_{S3} e_{15}	i_{sa} f_{23}	i_{sb} f_{30}	i_{sc} f_{37}	i_{r1} f_{75}	i_{r2} f_{78}	i_{r3} f_{81}	i_{r4} f_{84}	i_{r5} f_{87}	$i_{s\alpha}$ f_{42}	$i_{s\beta}$ f_{48}	$i_{r\alpha}$ f_{45}	$i_{r\beta}$ f_{47}	ω f_{96}
R_{sa}^-	0	0	0	+	–	+	–	+	–	+	–	+	–	–	+	+
R_{sb}^-	0	0	0	+	+	–	+	–	+	–	–	+	–	+	–	+
R_{sc}^-	0	0	0	–	+	+	–	+	+	–	+	–	+	–	+	+
R_{sa}^+	0	0	0	–	+	+	+	+	+	–	+	–	+	+	+	–
R_{sb}^+	0	0	0	+	–	+	+	+	–	+	+	+	–	+	+	–
R_{sc}^+	0	0	0	+	+	–	+	–	+	+	+	+	–	+	–	–
R_{r1}^+	0	0	0	–	+	+	–	+	+	+	+	–	+	–	+	–
R_{r2}^+	0	0	0	+	+	–	+	–	+	+	+	+	+	+	–	–
R_{r3}^+	0	0	0	+	–	+	+	+	–	+	+	+	+	–	+	–
R_{r4}^+	0	0	0	+	+	+	+	+	+	–	+	+	+	+	+	–
R_{r5}^+	0	0	0	+	–	+	+	+	+	+	–	+	–	+	–	–
β^+	0	0	0	+	+	+	+	+	+	+	+	+	+	+	+	–
R_{1a}^+	+	0	0	x	x	x	x	x	x	x	x	x	x	x	x	x
R_{2a}^+	0	+	0	x	x	x	x	x	x	x	x	x	x	x	x	x
R_{3a}^+	0	0	+	x	x	x	x	x	x	x	x	x	x	x	x	x
R_{4a}^+	–	0	0	x	x	x	x	x	x	x	x	x	x	x	x	x
R_{5a}^+	0	–	0	x	x	x	x	x	x	x	x	x	x	x	x	x
R_{6a}^+	0	0	–	x	x	x	x	x	x	x	x	x	x	x	x	x

The short-circuit fault signatures (not shown in table 1) are similar to the open-circuit ones corresponding to the opposite switching element of the same branch, that is, the signature of R_{1a}^{+} is the same that R_{4b}^{-}, the signature of R_{2a}^{+} is the same that R_{5b}^{-}, etc.

3.2. *Heuristic Knowledge*

If the cardinality of a set of possibles responsible parameters delivered by the fault tree stage is grater than one, an extra-stage has to be used to reduce it. In the classical methodology of diagnosis with BG the next stage is to reduce the sets of possibles responsible parameters using temporal graphs. In this work the temporal graphs stage was omitted because it yields erroneous information due to the time-varying signals and the short time constants of the system.

Instead of using the temporal graph stage, another methodology using the a priori knowledge of the system was investigated. This methodology seems to the expert systems approach but in this work the amount of data recollected is few. Indeed, it is only stored part of the signatures of table I (shaded regions). The knowledge used can be divided in four different cases.

Inverter faults: If the fault is in the inverter, one of the measured switches' voltages will change its value. With this condition all the parameters different to R_{1a}^{+}, R_{1b}^{-}, R_{2a}^{+}, R_{2b}^{-}, R_{3a}^{+}, R_{3b}^{-}, R_{4a}^{+}, R_{4b}^{-}, R_{5a}^{+}, R_{5b}^{-}, R_{6a}^{+}, R_{6b}^{-}. have to be discarded from the set of possibles responsible parameters. The next three cases take place when none of the measured switches' voltages changes its value.

Broken bar faults: In the fault free case all the rotor currents have the same amplitude, but if one of the bars gets broken much of the current distributes on the adjacent bars and a few on the other ones. In this case, the parameters different to R_{r1}^{+}, R_{r2}^{+}, R_{r3}^{+}, R_{r4}^{+} and R_{r5}^{+} have to be eliminated.

Open-circuit faults in the stator windings: When one of the windings gets opened, its current decreases to zero and the others increase their amplitudes. This makes to discard any parameter different to R_{sa}^{+}, R_{sb}^{+} and R_{sc}^{+}.

Viscous friction faults: When the friction in the mechanical subsystem increases, all the signals increase their amplitudes, except for the rotor speed. If this signature were presented in the detection stage, any parameter different to β^{+} will be discarded from the set of possibles responsible parameters.

4. Diagnosis Results

The results presented here were obtained in simulation. The parameters of the induction motor correspond to a 4-pole 3 hp machine [9] and they are shown in the appendix. The load torque changes from 5 to 15Nm in a time interval of 2 seconds. Signals supplied to the stator windings are sinusoidal PWM control voltages coming from the power inverter. The considered faults were stator winding turn-fault (R_{si}^-), open-circuit of a stator winding (R_{si}^+), broken bars (R_{rj}^+) and increase in the viscous friction coefficient (β^+) for the induction motor. For the power inverter was considered short-circuit (R_{yb}^-) and open-circuit (R_{xa}^+) faults of power electronic switches. With respect to the kinds of faults, these were single, multiplicative, sudden and non intermittent faults.

To detect the qualitative changes, rms currents and the speed were calculated for the motor variables to build enveloping signals. Average voltages across the electronic switches were calculated for the inverter variables. All these signals were filtered with a low-pass second-order butterfly filter. Figure 6 shows a stator current residue for a short-circuit fault (R_{sb}^- fault).

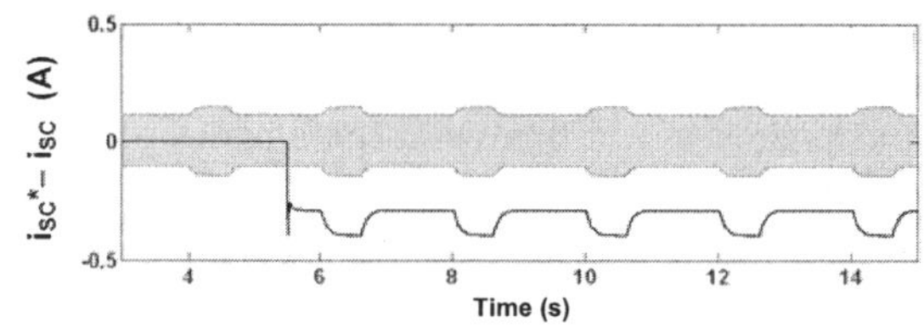

Figure 6. Residue of i_{sc}

To obtain the sets of possibles responsible parameters corresponding to the different measured and estimated values, a program was elaborated.

The diagnosis in the induction motor exactly determines the responsible of the fault for the considered cases. The exceptions are the short-circuit faults where the sets of possibles responsible parameters have two elements, but one of them is the correct one; however, both suspicious parameters are of the same kind of fault, so the diagnosis can mention what kind of fault is present. The diagnosis in the power inverter determines the leg responsible of the fault for the considered cases. It is impossible to reduce the set of possibles responsible parameters to just one element with the variables measured, because the signatures are equal for an open-circuit of the upper (or lower) switch and the short-circuit of the lower (or upper) switch of the same leg.

The information of heuristic knowledge has to be previously obtained to the operation of the induction motor and has to be stored in a memory device. The processor used has to be able to run the fault tree algorithm, and to estimate the two-phase variables and the rotor currents in the stationary reference frame. The three-phase stator currents and the rotor speed have to be measured.

5. Conclusions

In 75% of the faults of the induction motor, the sets of possibles responsible parameters were reduced to one element and it was the correct one. In the other 25% of the faults it was diagnosed the kind of fault present: short-circuit turns in one of the stator windings. In all the faults of the inverter the responsible parameter of the fault was isolated, but along with another parameter of the same leg (diagnosing the faulty leg). The BG diagnosis uses qualitative changes in the magnitude of signals, so it has the advantage of detecting and isolating a fault at the final of the transient provoked by the fault. The FDI with BG can be evaluated when the induction motor has a varying load, but it has to be known.

Appendix

Table 2. Parameters of the induction motor drive.

J	(Moment of inertia)	0.089 Kg·m^2
β	(Viscous friction coefficient)	0.01 N·m·s
R_{si}	(i = a, b, c)	0.435 Ω
R_{si}	(i = a, b, c) in short-circuit	0.305 Ω
R_{si}	(i = a, b, c) in open-circuit	435 Ω
R_{rj}	(j = 1, 2, 3, 4, 5)	0.816 Ω
R_{rj}	(j = 1, 2, 3, 4, 5) in open-circuit	816 Ω
Ls = Lr		0.0713 H
Lm		0.0693 H
f_{trian}	(Carrier frequency)	9.6 kHz
f_{sin}	(Sinusoidal frequency)	60 Hz
i_m	(Modulation index)	0.8
E	(Voltage supply in the inverter)	450 V
R_{xa}	(x = 1, 2, 3, 4, 5, 6)	0.1 Ω
R_{xa}	in open-circuit	500 kΩ
R_{yb}	(y = 1, 2, 3, 4, 5, 6)	2 kΩ
R_{yb}	in short-circuit	1 mΩ

References

1. M.O. Aguilar, G.V. Guerrero. and L.G. Vela, *Diagnóstico de fallas en el Motor de Inducción Trifásico usando Gráficos de Enlaces Energéticos*, Congreso Anual de la AMCA (2005).
3. M.E.H. Benbouzid, *A Review of Induction Motors Signature Analysis as a Medium for Faults Detection*. IEEE Transactions on Industrial Electronics, **47**, pp. 984 – 993, (2000)

4. C. Bidard, F.C. Favret, S. Goldstein and S.H. Larivière, *Bond Graph and Variable Causality*, International Conference on Systems Man and Cybernetics: Systems Engineering in the Service of Humans, **1**, pp. 270 – 275, (1993).
6. G. Dauphin-Tanguy and C. Rombaut, *Why a unique causality in the elementary commutation cell bond graph model of a power electronics converter*, International Conference on 'Systems Engineering in the Service of Humans, **1**, pp. 257 – 263, (1993).
7. P.J. Feenstra, E.J. Manders, P.J. Mosterman, G. Biswas and J. Barnett, *Modeling and Instrumentation for Fault Detection and Isolation of a Cooling System*, Proceedings of the IEEE Southeastcon 2000, pp. 365 – 372, (2000).
8. J. Kim and M.D. Bryant, *Bond Graph Model of a Squirrel Cage Induction Motor with Direct Physical Correspondence*, Journal of Dynamic Systems Measurement and Control, **122**, pp. 461 – 469, (2000).
9. P.C. Krause, *Analysis of Electric Machinery*, Ed. McGraw-Hill, (1987).
10. E.J. Manders and G. Biswas, *FDI of abrupt faults with combined statistical detection and estimation and qualitative fault isolation*, Proc. of the 4th Symposium on Fault Detection, Supervision and Safety for Technical Processes, pp. 1074 – 1079, (2003).
11. P.J. Mosterman, R. Kapadia and G. Biswas, *Using Bond Graphs for Diagnosis of Dynamic Physical Systems*, Sixth International Conf. on Principles of Diagnosis, pp. 81 – 85, (1995).
13. P.C. Sen, *Principles of Electric Machines and Power Electronics*, Ed Wiley, (1989).
14. R.M. Tallam, T.G. Habetler and R.G. Harley, *Stator Winding Turn-Fault Detection for Closed-Loop Induction Motor Drives*, IEEE Transactions on Industry Applications, **39**, pp. 720 – 724, (2002).
15. W.T. Thomson and M. Fenger, *Current Signature Analysis to Detect Induction Motor Faults*, IEEE Industry Applications Magazine, **7**, pp. 26 – 34, (2001).

AN INVESTIGATION INTO THE USE OF ACOUSTIC METHODS FOR LEAK DETECTION IN BLACK LIQUOR RECOVERY BOILERS

VILLE JÄRVINEN, JUHA MIETTINEN

Laboratory of Machine Dynamics, Tampere University of Technology, Box 589, FU-33101 Tampere, Finland

ROBERT HILDEBRAND, MATTHEW C. CARROLL

School of Engineering and Technology, Lake Superior State University, 650 West Easterday Avenue, Sault Ste. Marie, Michigan 49783, United States of America

Black liquor recovery boilers enhance the efficiency of paper pulp mills by burning the organic wastes recovered from the pulp making process and generating electricity. However, significant safety hazards exist when steam is used as the working fluid, in that the sludge from the waste incineration process contains sodium compounds. A strong need exists to detect steam leaks into the furnace area as they first begin to develop, and this paper investigates and evaluates the use of acoustic methods to detect these leaks, particularly in the bottom wall area of the furnace. The major conclusion of this investigation is that the sound attenuation at the frequencies usually monitored may be strong enough that a fault would be undetectable above the background noise level. Suggestions are then made for further investigation of this problem.

1. Background

1.1. *Principles of Kraft Recovery in Pulp Mills*

Vakkilainen in Reference [1] provides an extensive description of the Kraft recovery process. Basically, spent cooking chemicals and dissolved organic compounds are separated from pulp during the washing stage of the pulp making process. The resulting weak, black, alkaline liquor contains about 12 – 20% organic and inorganic solids. The black liquor is then concentrated by an evaporation process involving the direct or indirect heating and flashing of the liquor. Multiple effect, steam heated evaporators are commonly used.

The resulting concentrated black liquor can then be burned in a Kraft recovery boiler to produce energy for the generation of steam and electricity. In many cases sufficient energy is produced to provide for the steam and electricity needs of the entire pulp mill.

1.2. *Combustion of Black Liquor in the Kraft Recovery Boiler*

In a Kraft recovery boiler black liquor is a sprayed into a furnace through a number of liquor guns. In many combustion applications, the fuel is atomized to maximize combustion rates and temperature, but in a Kraft recovery boiler the liquor is sprayed as coarse droplets so that the unburned material can reach the char bed. Droplet diameters are typically between 2 and 3 mm.

Hupa and Solin [2] give a detailed description of the black liquor combustion process, which involves four stages. In the first, or drying, stage water is evaporated from the droplet. This takes from 0.1 – 0.2 seconds, and the limiting factor in the evaporation rate is the heat flux to the droplet. In the second, or devolatilization stage, gases such as methane, carbon dioxide, hydrogen, and hydrogen sulfide are released. The black liquor droplets swell considerably, and a visible flame appears. In the third stage, described in detail by Grace [3], the resulting "char," which primarily contains sodium compounds, continues to burn. All three of these stages occur in about a second or less, and the stages overlap significantly. For example, Frederick and Hupa [4] have determined that about 5% moisture remains at the onset of volatiles release. In the final stage the char combustion is finished and the inorganic residue remains. If this smelt is exposed to oxygen, small amounts of the sulfide in the smelt may be reoxidized to form sodium sulfate.

1.3. *Safety Hazards Associated with the Char Bed Sodium Content*

Significant to this study is the predominance of sodium, and sodium compounds, in the char bed composition. Heinävaara [5] has made measurements of char bed composition and determined that the char bed is slightly over 40% sodium by weight, and that over 90% by weight consists of the sodium compounds described above. In a Kraft recovery boiler, the sodium-rich smelt layer covers water tubes, and in the event of even a small water leak into the furnace, the mixing of water with the smelt compounds can cause a smelt-water explosion. Current practice is to monitor for water leaks in the bottom wall with the use of acoustic emission sensors, which are most sensitive at very high frequencies (100 kHz or more). Once a leak is detected, the automatic shutdown sequence described above is performed.

2. Problem Statement

It is the primary purpose of this investigation to answer the following question:

> **If a water leak in the bottom wall occurs, can this leak be detected by the acoustic emission sensors presently being used, in time to prevent an explosion in the recovery boiler?**

To be more specific, it is desired to evaluate the signal-to-noise ratio, that is, the ratio of leakage-induced vibrations to operation-induced vibrations ("background noise") reaching the transducers of an acoustic measurement system. The system would be expected to exist in the form of acoustic emission sensors placed at regular intervals along the bottom wall.

The situation is as pictured in the following figure. In this figure, all sources, except for a water leak and the tube cracking leading to such a leak, are regarded as background noise. The inset shows the structure of the bottom wall, in which tube and plate ("fin") elements alternate:

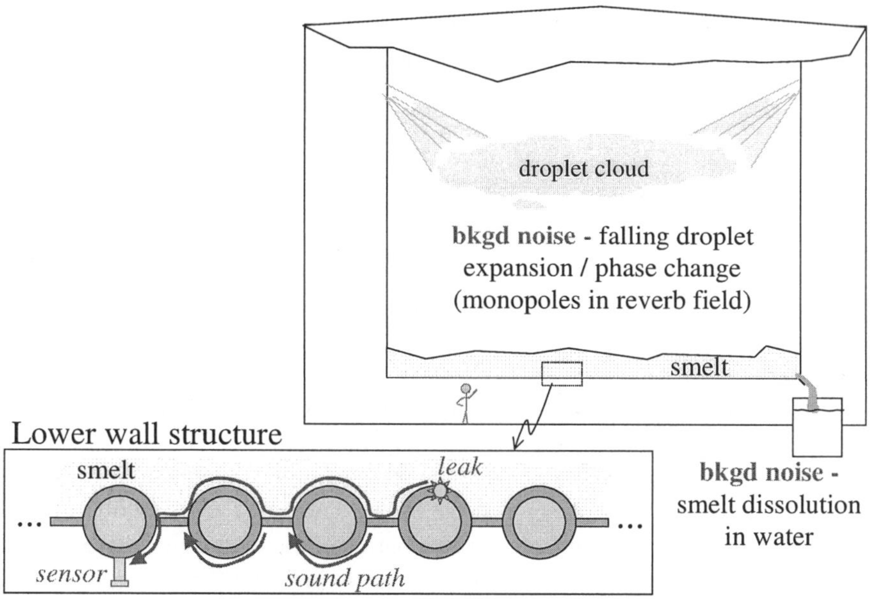

Figure 1. Vibroacoustic Sources and Paths in the Recovery Boiler.

As shown in the figure, vibrations will be induced at a specific point in the bottom wall, should metal cracking or a water leak occur. These vibrations then propagate along the bottom wall and eventually reach one of the transducers. On the other hand, background noise results from combustion of black liquor droplets, smelt dissolution, and other sources which are present in normal operation.

The relative magnitudes of the leak noise and background noise depend on three factors. These factors are the attenuation, or decay, of the leak vibrations with distance in the wall, the background noise vibration levels, and the vibratory power input of the leakage mechanism into the structure. This last factor will of course depend on the nature and magnitude of the leak.

This present work is primarily concerned with the attenuation, or decay, of the leak vibrations, with an additional consideration of the background noise levels in the experimental phase. As such, the study is not primarily concerned with the specific instrumentation, transducers or signal processing required of a monitoring system. Rather, it is concerned with the physical acoustics problem of how the sound and vibration fields generated by a leak propagate in the structure of the furnace. That propagation, and in particular the resulting strength of the leak-induced vibroacoustic field at the monitoring positions, in relation to the background noise level induced by operation, dictates whether or not detection is feasible at all, whatever the details of the monitoring system.

To determine this attenuation, two computational models are developed. One, an 18 degree-of-freedom wave model, is suitable for higher frequencies (30 kHz and upward). The other, for low frequencies ranging from 0 – 20 kHz, involves a finite difference formulation where the bottom wall is modeled as a series of lumped elements. Then, an experimental verification was conducted by two of the authors (Järvinen and Hildebrand) by performing measurements on an operating Kraft recovery boiler in Rauma, Finland. Acoustic excitation of the lower wall was effected by the use of an instrumented hammer with a force transducer, as well as such high frequency sources as the shattering of graphite. It was also possible to determine background noise levels in operating conditions.

Additional measurements of the transmission of sound in a boiler bottom wall were made at a boiler site not yet in operation at Skoghäll, Sweden. From these, it was also possible to estimate the attenuation of sound across the pipes, but without the operational noise in the background. While permitting a better signal-to-noise ratio in estimating the attenuation of the fault signal, those results could not take account of the smelt layer actually present in an operating boiler. Because the smelt is anticipated to have a damping effect on the structure-borne sound, however, the results obtained from this measurement may be regarded as conservative.

3. Theoretical Analysis

3.1. *18 Degree-of-Freedom Model for High Frequency Attenuation*

An 18 degree-of-freedom theoretical model suitable for determining the attenuation of high frequency sound is first developed. For this model, the structure of the bottom wall can be modelled as a periodic structure of the form ...fin - tube - fin - tube... Although the structure does, of course, have boundaries, it may be well approximated as infinite periodic, since its dimensions are very large with respect to the relevant wavelengths. The greater the decay and the higher the frequency, the better that approximation will be (for greater decay, reflections from the boundaries are much weaker than the direct field from a water leak; for higher frequencies, the wavelength becomes ever shorter with respect to the large dimensions of the gross structure).

Material damping in the structure can be determined by adding loss factors to the real-valued Young's moduli, as would be found in a handbook, to create complex Young's moduli for the fin and tube materials:

$$E_f = E_{f,nom}(1+\mathrm{i}\eta_f) \text{ and } E_t = E_{t,nom}(1+\mathrm{i}\eta_t) \qquad (1)$$

where $E_{f,nom}$ and $E_{t,nom}$ are the respective real-valued Young's moduli, and η_f and η_t are the loss factors, which would incorporate all forms of damping, including, for example, such things as friction at welds.

The starting point in this model is to model both the fin and tube as thin plates. Thus, rotational inertia and shear deformation are ignored in the treatment of bending, and the curvature of the tube is ignored. This provides only a rough qualitative estimate of the attenuation in the structure, but is useful in setting up the computational structure which can later be modified to accommodate three additional considerations:

- thick plate corrections for the fin,
- curvature (shell) and thickness corrections for the tube, and
- fluid interactions with the smelt and water.

Based on these initial assumptions, a computational model is developed which involves solving an 18-equation eigenvalue problem **Au** = **0**. The development of this model is discussed in detail in Ref. [6], and, for the sake of brevity in this article, will only briefly be summarized here.

First, bending stiffnesses, bending wave numbers, and quasi-longitudinal wave numbers are developed for the fin and tube, based on fin and tube thickness and Young's modulus, density, and Poisson's ratio for the material. Horizontal and vertical displacements are then derived.

The 18 equations mentioned above are then derived by considering the two junctions of the fin to the tube that occur in each of the repeating units. The displacement terms of the tube and fin derived in the previous step are now interpreted as waves and nearfields, with the fin-tube junctions representing local transmission-reflection problems for them. The equations stem from the fact that conditions of kinematical continuity within the tube (across a junction), and between the tube and the fin ("connectedness"), as well as equilibrium, must apply at each of the two junctions.

Solving the eigenvalue problem **Au=0** then gives the propagation constants g, and the attenuation constants are then a = Re(g). The decay may also be expressed in decibels per unit as

$$d = 20 \log e^{-a} \text{ dB} \qquad (2)$$

or in decibels per meter as

$$d = 20 \log e^{-a/(L+2R)} \text{ dB} \quad (3)$$

where L is the length of a fin and R radius of a tube (so that L + 2R will be the length of a repeating unit).

As stated earlier, corrections, discussed in detail in Ref. [6], are then made to account for fin and tube thickness, and for the interaction of the structure with the smelt and water, for which a classical interaction involving semi-infinite fluids is assumed. The smelt layer is not infinite, and is also not a fluid throughout the smelt layer. The water is also a finite volume (per unit length of a single pipe). This is precisely where the model falls short in propagation of sound at low frequencies, and is the major reason that the finite difference model discussed in the next section is needed. However, for frequencies above 30 kHz the wavelength is small compared to the dimensions of the fluid volumes, and the finiteness of the fluids becomes unimportant. Additionally, the solidified smelt layer is at a high temperature, and probably many orders of magnitude less stiff than the boiler wall material; pressure waves will thus be favorably radiated into it by the wall, while shear waves will not (because of the large impedance difference). Hence, the smelt will effectively act as a fluid. Because of these two considerations, it is believed that the classical treatment is an adequate approximation for the higher frequencies.

3.2. *Finite Difference Model for Low Frequency Attenuation*

At low frequencies, the 18degree-of-freedom wave model described in Section 3.1 will not be valid for reasons discussed in the previous paragraph. The bending wavelength, in that region, is of the order of, or larger than, the diameter of the enclosed water volume. Thus, another approach is needed. In this work, the frequency range 0-20 kHz has therefore been studied using a time-domain finite difference method, in which the fin and tube are simple lumped elements. The fin acts as a lumped stiffness, compliant for vertical and rotational relative motions of its opposite boundaries, but rigid against horizontal motions (thus, only bending, not longitudinal, motions are now possible in the structure). The tube acts as a lumped inertial element. In vertical translations (horizontal ones are not allowed), the mass is that of the tube and the water, combined. In rotations, only the tube's rotational inertia is included; shear stresses are not transmitted across the tube-water interface. All damping is ignored.

The bottom wall structure is thus modeled as a series of N repeating units as shown in Figure 2, where N is large. Note that the fin represents a lumped stiffness, and the tube a lumped inertial element in which the water only takes part in the translational motion.

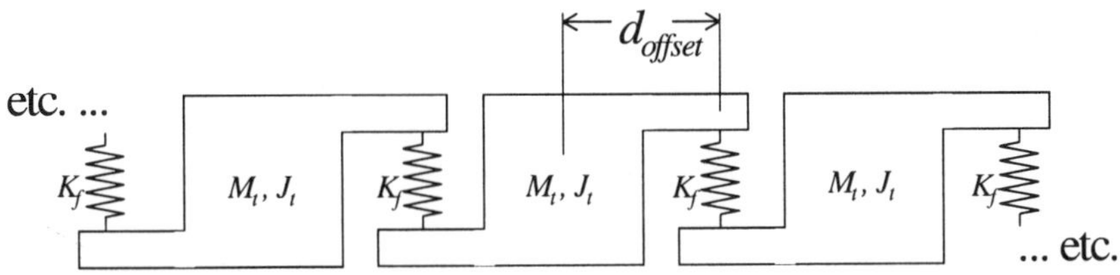

Figure 2. Schematic Model of the Bottom Wall as a Series of Lumped Elements.

For each repeating unit in the series, vertical and rotational dynamic equilibrium equations are written, and the time-dependent difference formulations are developed. The method then proceeds by specifying at rest initial conditions at all of the units. Next, the time is advanced by steps Δt, so that $t = m\Delta t$, as $m = 1, 2, 3 \ldots$, while forcing the first unit in the series to undergo a harmonic oscillation of unit rotational amplitude at a frequency f (or rotational frequency $\omega = 2\pi f$) of interest, and constraining all other degrees-of-freedom at the first and last (N-th) unit to be fixed. Using the equilibrium conditions expressed explicitly in terms of $v_{j,new}$ and $\theta_{j,new}$, for $j \in [2, N - 1]$, these variables can be solved in each step, knowing their values at all points in the preceding two time steps. This is possible for the entire time series, starting from the initial conditions, which supply two time instants.

3.3. *Computational Results*

Results using the two models discussed above are presented in Figure 3. Since the finite difference model is suitable at low frequencies, and the wave model at high frequencies, the two results together nearly cover the entire range of interest 0 – 200 kHz.

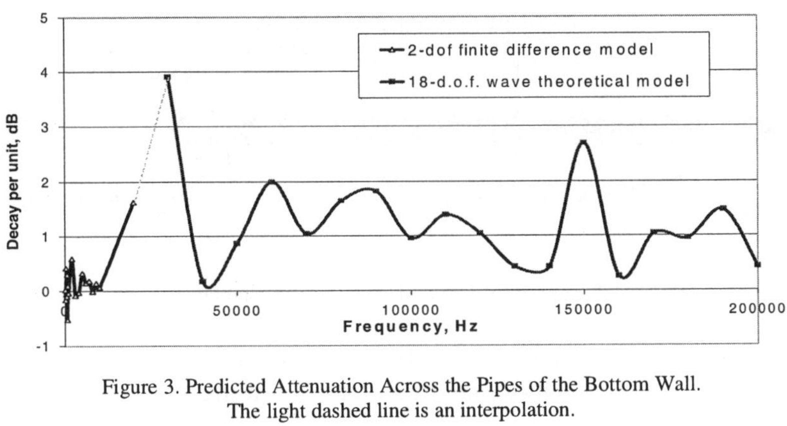

Figure 3. Predicted Attenuation Across the Pipes of the Bottom Wall.
The light dashed line is an interpolation.

Evidently, there is a low frequency pass-band, up to about 10 kHz, on which the attenuation is very close to zero. The scatter there is a consequence of resonant effects in the finite difference model, since that model is finite. Above about 4 kHz, there is a steep rise in the attenuation, indicating the beginning of a stop band.

The same stop band is also seen at 30 kHz in the wave model. The rest of the spectrum is dominantly a stop band, with decay in the range 1 – 2 dB per unit, but with narrow passbands occurring at 40 kHz, about 130 - 140 kHz, and about 160 kHz. This suggests that leakage monitoring could be concentrated on those frequency bands, since the leak-induced noise is least attenuated there.

The location of the minima are likely explainable by resonances within the repeating unit, and of the maxima by anti-resonances and by coincidence frequencies (at which there is a narrow peak in the radiation damping effect). The coincidence frequencies are at 57 kHz (smelt-fin), 130 kHz (smelt-tube), and 150 kHz (water-tube); there are, in fact, attenuation maxima at the first and third of these.

4. Experimental Verification

Measurements were performed at Skoghall, Sweden, in 2005 by Järvinen and Hildebrand. Measurements were focused to the bottom wall of the recovery boiler, and the AE-sensors were broadband ones, Fuji 1045S. The data acquisition was performed by a Wavebook 512 device at 500 kHz sampling rates. Measurement direction was chosen as perpendicular to the pipe direction, because the sound propagation in that direction clearly undergoes more attenuation. The measurement arrangement is as shown in Figure 4.

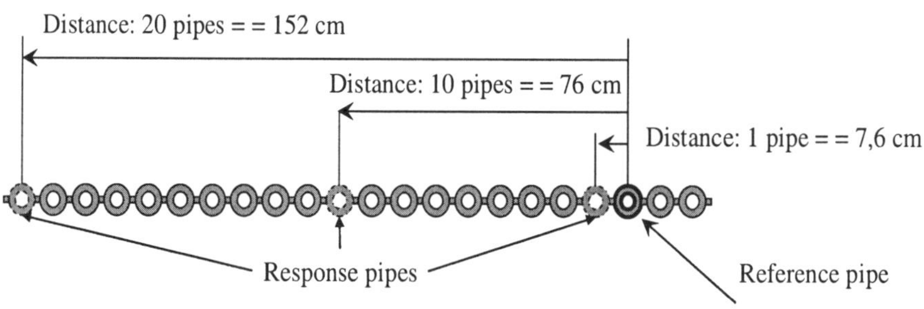

Figure 4. Measurement Arrangement for Leak Sound Propagation.

Results are as shown in Figure 5. Note that the solid line, indicating a moving average over the decay, is most significant for this investigation.

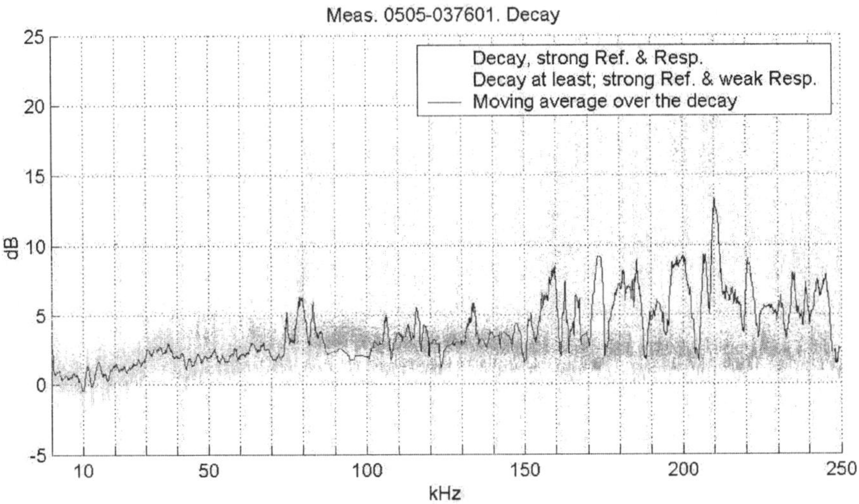

Figure 5. Experimental Results: Decay per Unit, As Function of Frequency.

One can conclude from the figure above that experimental results correlate quite well with those predicted by the theoretical model. Under 20 kHz there is a range in which little or no negative attenuation, and in some cases, amplification occurs. Significant attenuation begins at about 20 – 25 kHz, and at 70 – 90 kHz, a strong stop band is noticed. Beginning from about 100 kHz, with higher frequencies both the overall attenuation and the number of attenuation peaks increase. The narrow pass bands predicted by the theoretical model are also evident at these higher frequencies.

4. Conclusions

It is the basic conclusion of this investigation that a water leak in bottom wall area of a Kraft recovery boiler in many cases may *not* be detected by the acoustic emission sensors presently used in time to prevent significant damage to the boiler and hazardous conditions for its attendant personnel.

High attenuation occurs across the pipes in the bottom wall at the high frequencies typically monitored by acoustic emission transducers. This may result in a poor signal-to-noise ratio for water leaks that occur in that the nearest monitoring transducer would possibly separated from the leak by several pipes. Thus, the risk that such a leak would not be detected is quite high.

However, there is reason to believe that certain narrow passbands do exist; these would be promising bands for monitoring. The current work does predict a few such bands (40 kHz, 135 kHz, 160 kHz), but uncertainties of the smelt properties make these predictions tenuous.

5. Suggestions for Future Research

Additional research can be recommended for a deeper and more accurate understanding of the phenomena presented in this paper. An analytical study of the background noise from droplet expansion and combustion, and a study of the mechanism for generation of sound by the leak, as it already exists or is developing, are recommended. It is presently unknown whether the major source of this sound would be metal cracking, turbulent fluid flow through a fissure, localized chemical reaction of the water with the smelt, or localized rapid phase change of the water as it enters the smelt. A study of the relative power inputs of each of these vibrational sources would be very helpful.

Acknowledgements

The authors gratefully acknowledge the support of the Finnish Recovery Boiler Committee and the Laboratory of Machine Dynamics, Tampere University of Technology, in conducting the research described above, and the support of Lake Superior State University, Sault Sainte Marie, Michigan in the preparation of this article.

References

1. E. Vakkilainen, Kraft Recovery Boilers – Principles and Practice, Textbook: ISBN 951-764-64-6 (2003)
2. M. Hupa and P. Solin, Combustion Behavior of Black Liquor Droplets, TAPPI Proceedings of International Chemical Recovery Conference, New Orleans, Louisiana, Book 3 (1985)
3. T. Grace, J. Cameron, and D. Clay, Role of the Sulfate/Sulfide Cycle in Char Burning – Experimental Results and Implications, TAPPI Proceedings of International Chemical Recovery Conference, New Orleans, Louisiana, Book 3 (1985)
4. W. Frederick and M. Hupa, Combustion Properties of Kraft Black Liquors, Äbo Akademi, Department of Chemical Engineering Report 93-3 (1993)
5. A. Heinävaara, Recovery Boiler Analysis, Ahlstrom Machinery, Internal Report (1991)
6. J. Miettinen, V. Järvinen, and R. Hildebrand, Acoustic Leak Monitoring and Black Liquor Recovery Boilers, Tampere University of Technology (2005)

ANALYSIS OF A TWIN-GAS-CHAMBER HYDRO-PNEUMATIC VEHICLE SUSPENSION

DONGPU CAO, SUBHASH RAKHEJA, CHUN-YI SU, A.K.W. AHMED

CONCAVE Research Center, Concordia University, Montreal, Canada

A hydro-pneumatic suspension strut concept with integrated two gas chambers is proposed to realize nearly symmetric stiffness properties in compression and rebound, and progressively hardening roll stiffness characteristics. Fundamental stiffness properties of the proposed strut suspension are compared with the suspension involving one-gas-chamber struts with an anti-roll bar, in terms of suspension rate and roll stiffness. Dynamic responses are performed under a range of road inputs and vehicle velocities, and an excitation arising from a steady turning maneuver. The simulation results of stiffness properties indicate that the suspension rate of the twin-gas-chamber strut suspension can be designed to achieve soft vertical ride around static ride height and progressively hardening properties in both compression and rebound, which could help realize hardening effects in roll stiffness, compared with the softening effects in roll stiffness characteristics of suspensions with one-gas-chamber struts or commercial air springs. The dynamic responses demonstrate that the twin-gas-chamber strut suspension could considerably enhance the roll performance of heavy vehicles and slightly improve suspension travel responses, with negligible influence on vertical and roll ride.

1. Introduction

Suspension design of road vehicles involves a complicated compromise among different measures, related to ride comfort, handling, roll and directional performance [1]. Heavy vehicles, with their higher location of center of gravity (c.g.), and large weights and dimensions, exhibit relatively low roll stability limits [2]. The anti-roll characteristics of the suspension thus form one of the important design objectives. Moreover, the suspension must provide adequate attenuation of the road-induced vibration and shock for the protection of the cargos, and preservation of health, safety and comfort of the driver/passengers.

Heavy vehicle suspensions generally utilize leaf, rubber or air springs with nonlinear progressively hardening force-deflection properties and anti-roll bars, to realize a compromise among ride vibration, roll stability and directional responses [3-5]. The use of anti-roll bars, however, tends to add mass and be coupled with ride dynamics [6]. Alternatively, hydro-pneumatic struts, which could be interconnected in the roll plane, have also been investigated for enhancing the roll stiffness, while maintaining soft vertical ride [7-9]. These suspensions invariably exhibit asymmetric hardening and softening properties in compression and extension, respectively. Such properties that could help inhibit the motions in compression, however, tend to introduce a larger motion in rebound. Moreover, such nonlinear progressively hardening force-deflection properties could yield a softening effect of roll stiffness with increasing roll deflection [8, 9], which is much more undesirable considering that rollover generally occurs during relative large roll deflection.

A design concept of hydro-pneumatic strut involving two gas chambers is proposed in this study to realize nearly symmetric vertical stiffness properties in compression and rebound, and progressively hardening roll stiffness properties. The mathematical models of suspension forces are formulated. The properties and dynamic responses of hydro-pneumatic suspension involving such twin-gas-chamber struts are undertaken and compared with the suspensions with asymmetric stiffness properties and an anti-roll bar.

2. Roll plane model of a heavy vehicle

A roll plane model of a heavy vehicle has been developed to investigate the vertical and roll properties of different suspension configurations, and dynamic responses of heavy vehicles. The model (Fig. 1) considers a beam axle with vertical (z_u), lateral (y) and roll (θ_u) degrees-of-freedom (DOF), and a lumped sprung mass with vertical (z_s) and roll (θ_s) DOF. The sprung mass is assumed to roll about its roll center (RC_1), fixed to the unsprung mass. The vertical and lateral compliances of tires are represented by linear stiffness and damping elements, assuming point contact with the road. The left- and right- suspension strut forces are represented by f_L and f_R, respectively. The equations of motion for the roll plane model are formulated as follows:

$$
\begin{gathered}
m_s \ddot{z}_s = -f_L - f_R + m_s g \\
(I_{xxs} + m_s h_2^2)\ddot{\theta}_s + m_s h_2 \left[\ddot{y} + (h - h_2)\ddot{\theta}_u\right] = m_s g h_2 \theta_s - l_s f_L + l_s f_R + F_y h_2 \\
m_u \ddot{z}_u = f_L + f_R + k_t(z_{0l} + z_{0r} - 2z_u) + c_t(\dot{z}_{0l} + \dot{z}_{0r} - 2\dot{z}_u) + m_u g \\
\left[I_{xxu} + m_u h_1^2 + m_s (h - h_2)^2\right]\ddot{\theta}_u + m_s h_2 (h - h_2)\ddot{\theta}_s + \left[m_u h_1 + m_s (h - h_2)\right]\ddot{y} = \\
l_s (f_L - f_R) + F_y (h - h_2) + m_u g h_1 \theta_u + l_t \left[k_t (z_{0l} - z_{0r} - 2l_t \theta_u) + c_t (\dot{z}_{0l} - \dot{z}_{0r} - 2l_t \dot{\theta}_u)\right] \\
(m_u + m_s)\ddot{y} + m_s h_2 \ddot{\theta}_s + \left[m_u h_1 + m_s (h - h_2)\right]\ddot{\theta}_u = F_y - 2k_l (y - y_c) \\
k_l (y - y_c) - c_l \dot{y}_c = 0
\end{gathered}
\tag{1}
$$

where m_s and m_u represent the sprung and unsprung masses, respectively, and I_{xxs} and I_{xxu} are the corresponding roll mass moments of inertia. k_t and c_t are the vertical stiffness and damping coefficient of the tire, respectively. The horizontal spacing of the left and right struts with respect to the sprung mass c.g. is represented by l_s and the half tire track width is represented by l_t. z_{0l} and z_{0r} represent the road inputs at the left and right tires, respectively. h and h_1 represent the c.g. heights of the sprung and unsprung masses, respectively, and h_2 represents the vertical displacement between the c.g. and the roll center

(RC_1) of the sprung mass. F_y is the centrifugal force acting on the sprung mass due to a turning maneuver or crosswind. The series-coupled stiffness k_l and damping coefficient c_l represent the lateral compliance of the tire [6].

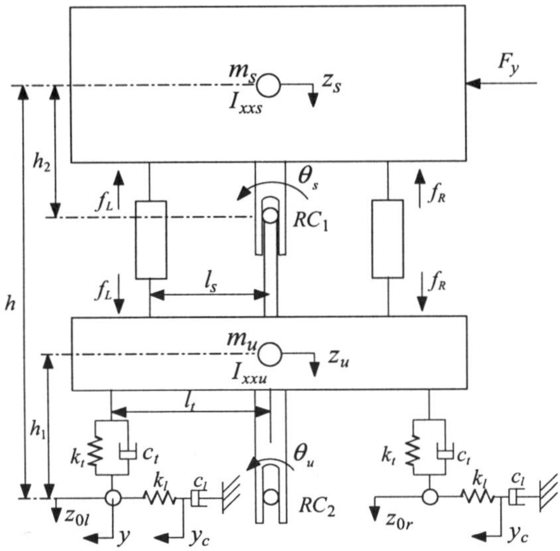

Fig. 1: Roll plane model of a heavy vehicle.

Figure 2 presents the schematic of a hydro-pneumatic strut comprising a gas chamber and the struts orientation in the roll plane (referred to as 'suspension *B*') [9]. Each strut consists of a number of damping orifices in the main piston separating the chamber 1 from chambers 2 and 3, while a floating piston isolates the hydraulic fluid in the chamber 2 from the nitrogen gas in chamber 4. The shim disc valves, consisting of shim packs, can be employed in conjunction with constant area bleed orifices to achieve variable flow resistance and thus the variable damping force. Such compact strut design eliminates the external gas chamber and damping valve, compared to those reported in [7, 8]. Moreover, the strut design could offer a relatively larger effective working area to significantly decrease the operating pressure corresponding to the load carrying capacity.

Figure 3 shows the schematic of the proposed twin-gas-chamber strut as well as the roll plane arrangement, referred to as 'suspension *A*'. In the strut, chambers 3 and 4 contain Nitrogen gas, and damping orifices or valves within the main piston to provide resistance to hydraulic flows between chambers 1 and 2. Under compression stroke, the gas in chamber 4 undergoes compression and tends to dominate the vertical suspension stiffness property. The spring rate in rebound may mostly be determined by the gas pressure in chamber 3, which undergoes compression during rebound. The proposed strut design concept, therefore, provide considerable potential to realize nearly symmetric spring rates in suspension compression and extension.

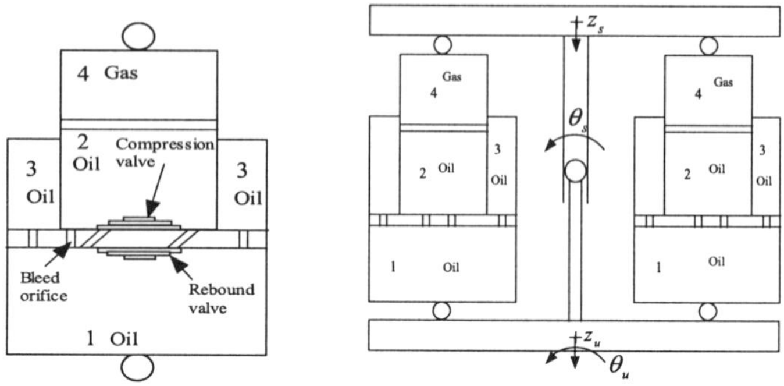

(a) (b)

Fig. 2: Schematics of (a) suspension strut with shim-disc valves; and (b) suspension *B*.

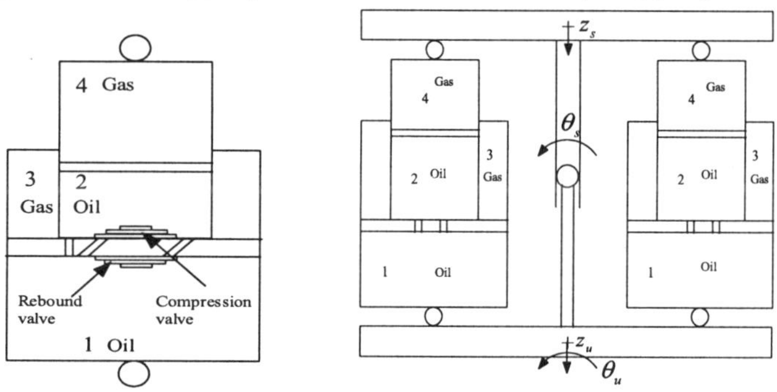

(a) (b)

Fig. 3: Schematics of (a) twin-gas-chamber strut with shim-disc valves; and (b) suspension *A*.

3. Suspension force formulations

The dynamic suspension forces developed by suspensions *A* and *B* can be developed by using similar flow and pressure equations, described in [9]. The major assumptions include incompressible hydraulic fluid, turbulent flows through the damping orifices, polytropic process of the gas in the gas chambers, negligible leakage across the pistons, negligible friction between the floating and main pistons and the cylinder, and negligible thermal expansion of the cylinder, piston and the oil. The dynamic suspension forces developed by suspension *A* can thus be derived, such that:

$$F_i = \left\{P_{40i}A_{1i}\left[\frac{V_{40i}}{V_{40i}+A_{1i}x_i}\right]^n - A_{1i}\right\} - P_{30i}\left\{A_{3i}\left[\frac{V_{30i}}{V_{30i}-A_{3i}x_i}\right]^n - A_{3i}\right\} - \frac{\rho A_{1i}^3}{2C_d^2 a_{12i}^2}|\dot{x}_i|\dot{x}_i \quad (i=l,r) \tag{2}$$

where the subscripts *l* and *r* refer to left- and right-struts, respectively, F_i is the dynamic suspension force developed by strut i ($i=l,r$), P_{40i} and P_{30i} are static gas pressures in chambers 4 and 3 of strut i, respectively, A_{ji} is the effective piston area reflected on the chamber j (j=1,2,3) side of the strut i, C_d is the discharge coefficient, a_{12i} is the total effective area of damping orifices between chambers 1 and 2 of strut i, and ρ is the mass density of hydraulic fluid, n is the polytropic exponent of the gas, V_{40i} and V_{30i} are the initial gas volumes in chambers 4 and 3 of strut i, respectively, $\dot{x}_l = \dot{z}_u + l_s\dot{\theta}_u - \dot{z}_s - l_s\dot{\theta}_s$ and $\dot{x}_r = \dot{z}_u - l_s\dot{\theta}_u - \dot{z}_s + l_s\dot{\theta}_s$, assuming small motions with positive direction being upward.

The dynamic suspension forces developed by suspension B can also be developed in a similar manner, such that:

$$F_i = P_{40i}A_{2i}\left\{\left[\frac{V_{40i}}{V_{40i}+A_{2i}x_i}\right]^n - 1\right\} - \frac{\rho A_{2i}^3}{2C_d^2 a_{12i}^2}|\dot{x}_i|\dot{x}_i - \frac{\rho A_{3i}^3}{2C_d^2 a_{13i}^2}|\dot{x}_i|\dot{x}_i \qquad (i = l, r) \tag{3}$$

where a_{13i} is the total effective area of damping orifices between chambers 1 and 3 of strut i.

4. Properties

The suspension rate and roll stiffness characteristics of different suspension configurations A, B and B with anti-roll bar (referred to as 'suspension B_{bar}') are investigated and compared, assuming symmetric load distribution on the left and right struts. All the suspension parameters are selected to achieve identical load carrying capacity, static suspension rate and static deflection. The chosen parameters reveal undamped bounce natural frequency of the sprung mass in the order of 1.5 Hz at the design ride height.

4.1. *Suspension rate*

The suspension rate of a hydro-pneumatic suspension is derived from the pressure-deflection relationships of the gas springs. The suspension rate k_{vi} due to each strut can be derived [8], such that:

$$k_{vi} = -dF_{si}/dx_i \qquad (i = l, r) \tag{4}$$

where F_{si} and x_i are the restoring force and relative deflection of strut i (i=l,r), respectively. The suspension rates are evaluated through solutions of the coupled differential equations of motion (Eq. 1) under a pure vertical displacement input (x_l=x_r). Figure 4(a) illustrates the suspension rates of different suspension configurations. Identical suspension rates for all the configurations are observed at the design ride height, while the configurations B and B_{bar} yield identical suspension rate over the entire deflection range. These configurations exhibit softening and hardening effects in extension and compression, respectively, similar to those reported for commercial air springs

[4, 5]. The configuration A comprising the proposed twin-gas-chamber struts, however, shows hardening effects in both suspension compression and extension.

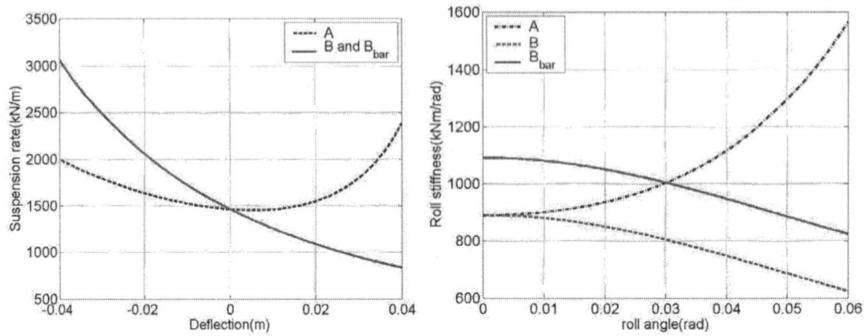

(a) (b)

Fig. 4: Properties of different suspensions: (a) suspension rate; and (b) roll stiffness.

4.2. *Roll stiffness*

The roll stiffness of a hydro-pneumatic suspension is derived from the restoring roll moment developed and the relative roll deflection. The effective roll stiffness k_R can be expressed as:

$$k_R = dM / d\theta \tag{5}$$

where M is the restoring roll moment developed due to restoring forces of both struts, and θ is the relative roll deflection of the sprung mass with respect to the unsprung mass. The roll stiffnesses of the three suspension configurations are derived using Eq. (5) and the static equilibrium equation in the vertical direction. The roll stiffnesses are evaluated through solutions of these equations under out-of-phase vertical displacement inputs (x_l=-x_r).

Figure 4(b) presents the roll stiffness comparisons of different suspensions. While the configurations A and B yield identical static roll stiffness, the suspension B exhibits softening tendency with increasing roll deflection, attributed to hardening and softening spring rates of the struts. The suspension A, however, exhibits progressively increasing roll stiffness with increasing roll angle, due to somewhat more symmetric spring rates of the struts in compression and rebound, as evident in Fig. 4(a). The use of an anti-roll bar could improve the static roll stiffness of the suspension. The roll stiffness of the configuration B_{bar}, however, decreases with roll deflection and could approach a value lower than that of suspension A. The results clearly show the benefit of the proposed twin-gas-chamber strut suspension on enhancing roll stiffness over the entire range of roll deflection, while the use of a relatively stiffer anti-roll bar would add considerable weight.

5. Dynamic responses

The relative dynamic responses of the suspension configurations (A, B and B_{bar}) are evaluated in terms of ride vibration, suspension travel and roll responses of the vehicle model described in Eq. (1). The equations of motion for the nonlinear roll plane vehicle model comprising the forces and moments due to the struts and the anti-roll bar are solved under excitations arising from tire-road interactions, and a steady turning centrifugal acceleration. While the stiffness properties of all the three suspensions have been shown in Fig. 4, the symmetric vertical mode damping properties are selected and tuned to be nearly identical for all the suspensions.

5.1. *Roll dynamics*

The relative roll dynamic responses of different suspensions can be assessed under a centrifugal force excitation arising from a steady turning maneuver or crosswind, which can be characterized by the lateral acceleration and represented by a rounded step input with steady amplitude of 3.5 m/s^2, as shown in Fig. 5(a). Due to the considerably higher c.g. height of heavy vehicles, the inhibition of sprung mass roll deflection could help reduce lateral load shift and therefore enhance roll stability.

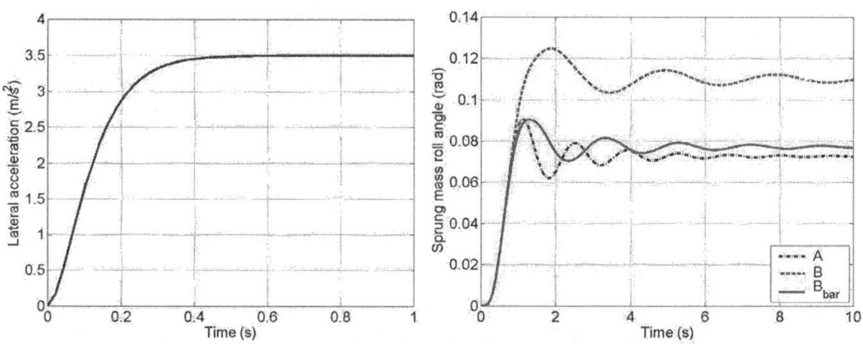

(a) (b)

Fig. 5: (a) rounded step lateral acceleration excitation; and (b) sprung mass roll angle responses.

Figure 5(b) presents the sprung mass roll angle responses of the heavy vehicle with different suspension configurations, subjected to the 3.5 m/s^2 rounded step lateral acceleration excitation. The suspension B yields significantly larger roll angle than the suspensions A and B_{bar}, due to its softening effects of roll stiffness with increasing roll deflection. Although the suspensions A and B exhibit identical static roll stiffness, the suspension A could considerably enhance the roll responses, contributed to the progressively hardening roll stiffness of the use of twin-gas-chamber struts. Furthermore, while the suspension A has similar peak roll angle response to that of the suspension B_{bar}, the steady response due to the suspension A is relative lower than that of B_{bar}, which also indicates the benefits of the use of the proposed twin-gas-chamber struts.

5.2. *Ride qualities and suspension travel under random road inputs*

The ride vibration and suspension travel responses are evaluated under a range of random road inputs and vehicle speeds, in terms of rms sprung mass bounce and roll accelerations, and rms left and right suspension travels. Three different road profiles are selected and termed as *smooth*, *medium-rough* and *rough* roads, based on their relative roughness values. Figure 6 illustrates the spatial power spectral density (PSD) of the roughness profiles of the three roads [10]. The nonlinear equations of motion for the roll plane model are solved under the selected road excitations and two different vehicle velocities (70 and 100 *km/h*).

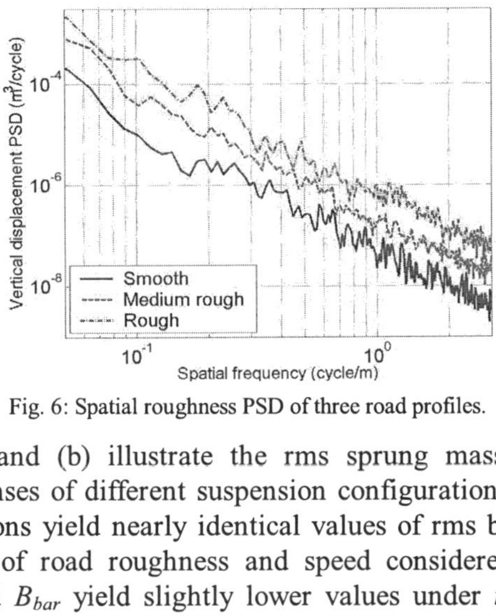

Fig. 6: Spatial roughness PSD of three road profiles.

Figures 7(a) and (b) illustrate the rms sprung mass bounce and roll acceleration responses of different suspension configurations, respectively. All the three suspensions yield nearly identical values of rms bounce acceleration under the ranges of road roughness and speed considered, except that the suspensions B and B_{bar} yield slightly lower values under *medium-rough* and *rough* road excitations. The suspensions A and B also exhibit very similar rms values of sprung mass roll acceleration due to their identical static roll stiffness. The addition of an anti-roll bar, however, yields about 3~14% higher roll acceleration, depending upon the vehicle speed and the road roughness, attributed to its higher roll stiffness for small roll deflections. Figures 8(a) and (b) present rms suspension travel responses of different suspensions. Relatively lower suspension travel could be obtained by using twin-gas-chamber struts, compared to the configurations B and B_{bar}. The results indicate that the use of twin-gas-chamber struts affects the vertical and roll ride only slightly, while it significantly enhances the roll dynamics performance and could also help improve suspension travel responses. The use of an anti-roll bar tends to deteriorate the roll ride quality, while its effect on the vertical ride and suspension travel is negligible.

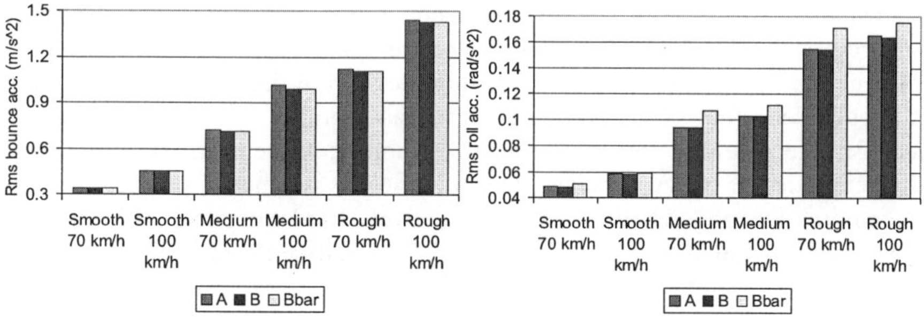

(a) (b)

Fig. 7: Rms sprung mass acceleration responses: (a) bounce; and (b) roll.

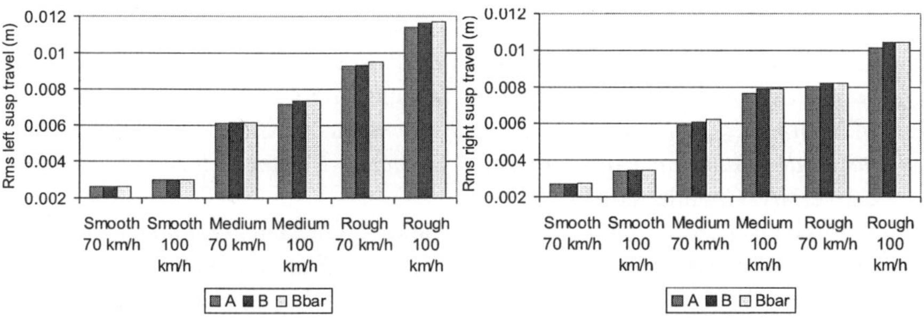

(a) (b)

Fig. 8: Rms suspension travel responses: (a) left; and (b) right.

6. Conclusions

A new concept of a twin-gas-chamber strut is proposed to realize nearly symmetric vertical suspension stiffness properties in compression and rebound, and progressively hardening roll stiffness characteristics. The results show that the proposed twin-gas-chamber strut suspension yield nearly symmetric but progressively hardening suspension rate in both compression and extension, as opposed to the conventional air suspensions that generally cause softening in rebound. Unlike the conventional suspension, the roll stiffness of the proposed struts also revealed progressively hardening effect with increasing roll deflection, attributed to the twin-gas-chamber design. The struts thus offer soft vertical ride in the vicinity of the static equilibrium and hardening roll stiffness to enhance the roll dynamics performance of vehicles, while eliminating the additional weight demands imposed by the use of anti-roll bar.

The dynamic responses of the road vehicle with suspensions involving the proposed and conventional struts with and without an anti-roll bar were further evaluated under different road roughness, vehicle speed and centrifugal acceleration excitations. The results showed that the suspension involving twin-gas-chamber struts could reduce suspension travel and considerably enhance roll response characteristics, and therefore roll stability, with negligible influence on the vertical and roll ride qualities. From the simulation results, it is concluded that the proposed twin-gas-chamber strut design offers considerable potential for enhancing roll stability, suspension travel and ride characteristics of heavy vehicles, due to its nearly symmetric vertical stiffness in compression and rebound, and progressively hardening roll stiffness. The proposed design also provides a light weight alternative to anti-roll bars.

References

1. D.J. Cole, *Fundamental issues in suspension design for heavy road vehicles*, Vehicle System Dynamics, 35, pp. 319-360, 2001.
2. R.D. Ervin, *The dependence of truck roll stability on size and weight variables*, Int. J. of Vehicle Design, Vehicle Safety Issue, 192-208, 1986.
3. M.N. Joarder, *Influence of nonlinear asymmetric suspension properties on the ride characteristics of road vehicle*, Master Thesis, Concordia University, Canada, 2003.
4. S. Rakheja, A.K.W. Ahmed, X. Yang and C. Guerette, *Optimal suspension damping for improved driver- and road-friendliness of urban buses*, SAE Paper No. 1999-01-3728, 1999.
5. H. Liu et. al., *A study on nonlinear stiffness characteristic of air spring for a bus*, SAE Paper No. 2002-01-3092, 2002.
6. D. Cebon, *Handbook of vehicle-road interaction*, Swets & Zeitlinger, the Netherlands, 1999.
7. J. Felez and C. Vera, *Bond graph assisted models for hydro-pneumatic suspensions in crane vehicles*, Vehicle Sys. Dynamics, 16, 313-332, 1987.
8. P.J. Liu, *An analytical study of ride and handling performance of an interconnected vehicle suspension*, Master Thesis, Concordia University, Canada, 1994.
9. D. Cao, S. Rakheja and C.-Y. Su, *Roll plane analysis of interconnected hydro-pneumatic suspension struts*. Proc. of AMSE Int. Mech. Engineering Congress, IMECE2005-80484, Orlando, 2005.
10. J. Carrier, *Releve du profil longitudinal de circuits d'autobus*, CRCAC report, Canada, 1999.

CONSTRUCTING OPERATIONAL RELIABILITY ANALYSIS MODEL OF UMT BASED ON PETRI NET*

ZHAO HUIXIANG†

Institute of Railway and Urban Mass Transit, Tongji University, Shanghai 200331, China

HU YONGSHENG

Institute of Railway and Urban Mass Transit, Tongji University, Shanghai 200331, China

LU ZHENGGUANG

Institute of Railway and Urban Mass Transit, Tongji University, Shanghai 200331, China

This paper presents an operational reliability model of urban mass transit (UMT). With the model, UMT can be decomposed into some object sets, and one place in the Petri net of the object sets can be replaced by low layer Petri net.

1. Guidelines

The methods of reliability analysis include mostly reliability block diagram (RBD) and fault tree analysis (FTA), but there are some defects in capability such as repetition or maintainability, and dynamic analysis [1]. Over the past 10 years, researchers both at home and abroad have tried to construct dynamic reliability analysis models based on dynamic FTA [2,3] and Petri net [4]. However, so far, there has been an absence of such research in the field of railway. In this paper, based on the operational characteristics of urban mass transit and the principles of hierarchical Petri net [5,6] and object-oriented Petri net [7] as well as the reliability modeling and analysis of complex correlative system [8] , we present a Hierarchical Object-oriented Colored stochastic Petri Net (HOPN) that can be used to operational reliability analysis of urban mass transit (UMT).

* This work is supported by the Science and technology key project of Shanghai Municipal Science and Technology Commission (No. 03DZ12046).

† E-mail: zhaohuixiang@263.net, mrzhhx@hotmail.com.

2. HOPN Model

Definition 1 A hierarchical object-oriented colored stochastic Petri net is a tuple

$$\text{HOPN} = (P, T, F, O, C, K, W, G, \Pi, M_0)$$

It satisfying the requirements below:

(1) (P, T, F) is a basic net;

(2) $P = P_D \cup P_B$, P_D is a finite set of places in the sub-net of objects, P_B is a finite set of place with the information of objects;

(3) $T = T_D \cup T_B$, T_D is a finite set of transitions in the sub-net of objects, T_B is a finite set of transitions with the information of object between objects, such that:

$$P \cup T \neq \varnothing, P \cap T = \varnothing$$

(4) $F = F_D \cup F_B$, F_D is a finite set of arcs in the sub-net of objects, F_B is a finite set of arcs with the information of object between objects, such that:

$$F_D \subseteq P_D \times T_D \cup T_D \times P_D, F_B \subseteq P_B \times T_B \cup T_B \times P_B$$

(5) $O = \{O_i, i = 1,2,\cdots I, I \in \mathrm{N}\}$, is a finite set of object types, and is in accord with Definition 1;

(6) $C : P \cup T \to \Gamma(D)$ Is color function. $\Gamma(D)$ is power set over color set D, so as to $\forall p \in P, C(p)$ is color set of the token with the p. $\forall t \in T, C(t)$ is arisen set with the t;

(7) $K : P_D \cup P_B \to D_{MS}$ is volume function, D_{MS} is multi-set over D, so as to $\forall p \in P$, $K(p) \in C(p)_{MS}$ is token quantitative set of p;

(8) $W : P \times T \to [D_{MS} \to D_{MS}]$ is arc weight function, $W = \{I_-, I_+\}$, thereinto: $I_-(p,t) \in [C(t)_{MS} \to C(p)_{MS}]$ is called input function, $I_-(p,t) = 0 \Leftrightarrow (p,t) \notin F$; $I_+(p,t) \in [C(t)_{MS} \to C(p)_{MS}]$ is called output function, $I_+(p,t) = 0 \Leftrightarrow (t,p) \notin F$;

(9) $G : T \times C(T) \to [\text{type}(G(t)) \to \{0,1\}]$ is guard function. It is defined from T into Boolean expressions, thereinto: $[\text{type}(G(t)) \to \{0,1\}]$ is a proposition function with the expressions;

(10) $\Pi : T \times C(T) \to [R^0 \to R]$ is a time-lapse distributing function with every T, thereinto: R^0 is a non-negative real set;

(11) $M_0 : P \to D_{MS}$ is called an initial marking.

Definition 2 object $O_K = (N_K, M_{0K})$ is a hierarchical net with K layer base on O_1, $K \in \mathrm{N}$. Thereinto: N_K is a hierarchical net with K layer base on N_1, and can be described as:

$$N_K = (((((((N_1, f_{\mathrm{P}1}, f_{\mathrm{T}1}), f_{\mathrm{P}2}, f_{\mathrm{T}2}), \cdots), f_{\mathrm{P}i}, f_{\mathrm{T}i}), \cdots), f_{\mathrm{P}K-1}, f_{\mathrm{T}K-1})$$

$f_{\mathrm{P}i}$ and $f_{\mathrm{T}i}$ are two displacements of place and transition with the N_i, and one of them will at least occur, $i = 1,2,\cdots,K$; such that:

$$M_{0K} = \begin{cases} M_{01}(p) & p \in P_1 \\ 0 & p \notin P_1 \end{cases}$$

Definition 3 $N = (P,T,F)$ is a basic net, having $f_{\mathrm{P}i}$ and $f_{\mathrm{T}i}$ is apart the operation below:

(1) $\exists p \in P$, the p and F_p are displaced by a trinity Petri net (P_1, T_1, F_1), viz. $f_{\mathrm{P}}(p, F_p) = (P_1, T_1, F_1)$, and make $(P_1, T_1 \bigcup {}^{\bullet}p^{\bullet}, F_1)$ turn into a Petri net, so f_{P} can be called as a displacement of place for the N ;

(2) $\exists t \in T$, the t and F_t are displaced by a trinity Petri net (P_1, T_1, F_1), viz. $f_{\mathrm{T}}(t, F_t) = (P_1, T_1, F_1)$, and make $(S_1 \bigcup {}^{\bullet}t^{\bullet}, T_1, F_1)$ turn into a Petri net, so f_{T} can be called as a displacement of transition for the;

(3) if one of f_{P} and f_{T} can occur at least, so the $(N, f_{\mathrm{P}}, f_{\mathrm{T}})$ can be called as a two layer Petri net;

thereinto: $F_p, F_t : F_x = \{(x, y) \in F\} \bigcup \{(y, x) \in F\}$ is a correlative set with x. ${}^{\bullet}p^{\bullet}, {}^{\bullet}t^{\bullet} : {}^{\bullet}x^{\bullet} = {}^{\bullet}x \bigcup x^{\bullet}$ is extension of x.

3. HOPN of UMT

3.1. *Reliability Model of Object*

According to the principle of categorizing object-oriented module, UMT can be categorized into six modules of object types: vehicle (VEH), station (STA), railway (RAI), power supply (POW), control center (CON), maintenance and repair department (RPM), among which the first five can be called task module. The tokens of individual objects in the task module are in different colors). The individual objects in the RPM module are the maintenance departments of various types of work, such as vehicle or track maintenance and repair department. Their tokens are also in different colors. The structure of the modules is a little bit different from the real one of UMT, but they are of the same function.

To reflect the reliability of the whole system, efforts are made to construct a model of failure state changes instead of a model of operation. Namely, the model of every object type embodies the module of failure section plane. Except RPM, all the other objects participate in the normal operation of the system. Their stability is reflected by the net model of figure 1(a). RPM is responsible for failure maintenance and the procedure of maintenance is illustrated by the net model of figure 1(b). The oval in figure 1 represents information place, the fillet rectangle the encapsulation of object modules.

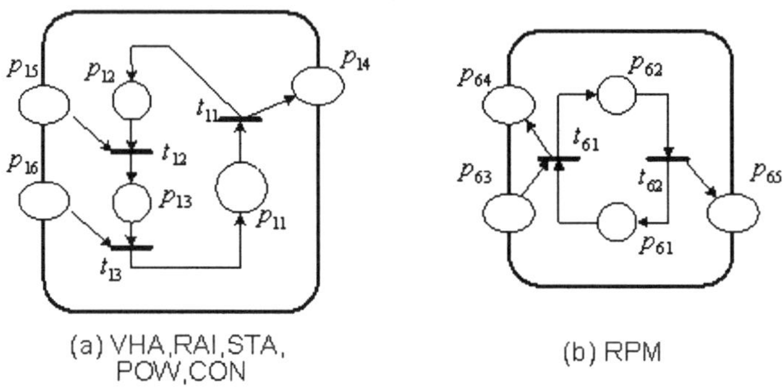

Figure 1. HOPN mold of object

In figure 1(a), p_{11}, p_{12}, p_{13} represent operation, failure and maintenance respectively. p_{14}, p_{15}, p_{16} respectively represent application for maintenance, response to the application, completion of maintenance, t_{11}, t_{12}, t_{13} respectively represent happening of failure, beginning of maintenance and changes of maintenance completion. When failure happens to an object, t_{11} startups, a token in the color of a corresponding object is put into p_{14}. When the response to maintenance reaches the p_{15}, t_{12} startups and the object begins to conduct maintenance. When the news of maintenance completion reaches the p_{16}, t_{13} startups to mark the completion of maintenance and the objects begins to be in operation. In figure 1(b), p_{63}, p_{64}, p_{65} respectively represent application for maintenance, response to the application and completion of maintenance, p_{61}, p_{62} respectively represent the maintenance and repair departments which are not in operation and those which are in operation (including the workers and equipment), t_{61}, t_{62} respectively represent the beginning and completion of maintenance. When p_{63} receives the application for maintenance from an object, t_{61} startups to make response to the maintenance application and begins to operate if the maintenance departments of related type of work are not in operation. Upon completion of maintenance, t_{62}

startups to mark completion of maintenance and stops operating. The Boolean expression in t_{61} is responsible for choosing the corresponding maintenance type of work. The time-lapse distributing function $\Pi(t_{62})$ of related type of work in t_{62} is responsible for timing of start-up. The Boolean expression in t_{62} is responsible for choosing related time-lapse distributing function $\Pi(t_{62})$ of related type of work. $\Pi(t_{62})$.

3.2. *Hierarchical Net in Object*

In Chapter2.1, the objects of every object type's module function as a whole, while the objects in a real system are composed of various units. The objects of the task module are composed of various equipment on the basis of reliability logic such as connection in series, connection in parallel, and redundancy etc. The equipment is composed of various parts, while the parts are composed of elements. The elements can be further decomposed into smaller units until they are decomposed into replaceable units. Thus, the objects can be decomposed into several layers, each of which is composed of a few units. On every layer, a Petri net model can be constructed and represented by identical figures. A sub-net model can be embedded into an upper net model according to the principle of displacement of place $f_{\mathrm{p}}(p_{11}, F_{p_{11}})$ presented in Definition3, as illustrated by figure 2. In figure 2(a) is a sub Petri net model derived from the equipment Petri net model (figure 2(b)) that displaced p_{11}.

In figure 2, p_{111}, p_{112} respectively represent the equipment in operation and that in failure state. The equipment is different colors. When failure happens to equipment, t_{111} startups, (If p_{111} is replaced by a sub net model, then t_{111} represents instant transition; if not, then t_{111} represents time-lapse transition and determines the time of start-up with the time-lapse distributing function $\Pi(t_{111})$ of the equipment in failure state. The Boolean expression in t_{111} is responsible for the choice of related time-lapse distributing function), and a token of corresponding color is placed in p_{113}, marking $M(p_{113}) \in C(p)_{\mathrm{MG}}$, namely, it is a multi-set over color set. In other words, failure may happen to several sub-objects at the same time or in sequence. With a Boolean expression of failure in t_{113}, t_{113} startups when the expression proposition is made true by marking $M(p_{113})$. The input function $I_{+}(p_{11}, t_{113})$ defined in arc (t_{113}, p_{11}) puts a token that indicates failure into p_{110} to show that the object in a failure state after t_{113} is made to startups, t_{11} startups when p_{11} in Figure 2(a) is in a failure state according to the marking defined in Definition 3, and the object gives signal of application for maintenance as in p_{14}, and waits for maintenance as in p_{12} until response is made to the application as in p_{15}, then

t_{12} startups and enters the state of maintenance as in p_{13}. Upon receiving the signal indicating completion of maintenance as in p_{16}, t_{13} startups and the object starts to be in operation again.

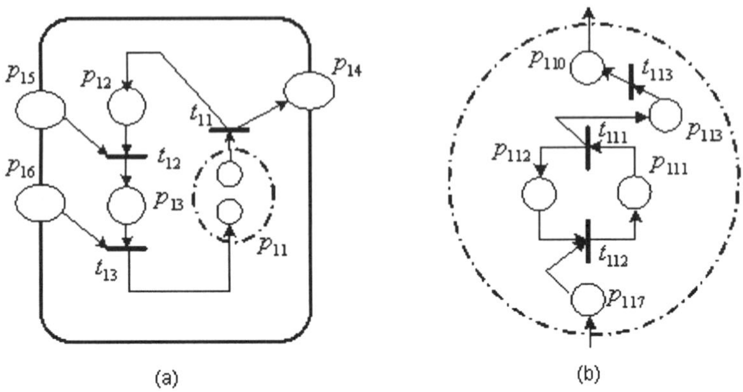

Figure 2. Mold of double-deck Petri net

The Boolean expression of failure in t_{113} determines the structure of reliability logic between the various equipment of the layer. The Petri net model in Figure 2(a) can represent such reliability logic structure as connection in series, connection in parallel, taking k out from N.

3.3. *Reliability Model of UMT*

With all the object types modules in Figure 1 assembled, a reliability HOPN model of UMT is constructed as in Figure 3. $t_{B1} \sim t_{B5}$ respectively represents the transition of application for maintenance, $t_{B6} \sim t_{B10}$ transition of response to application, and $t_{B11} \sim t_{B15}$ transition of completion of maintenance.

The maintenance of UMT can be decomposed into scheduled maintenance and unscheduled maintenance while unscheduled maintenance can be further decomposed into maintenance based on state and maintenance based on failure. Maintenance involved in this paper is maintenance based on failure, the time of which includes preparative time T_Z, maintenance time T_W (separating, disassembling, assembling, test etc) , and time of coming back to running plan T_H.

To fix the time for maintenance, the time lapse distributing function $\Pi(t_{B6}), \Pi(t_{B7}), \Pi(t_{B8}), \Pi(t_{B9}), \Pi(t_{B10})$ of maintenance preparative time is defined on $t_{B6} \sim t_{B10}$ to determine T_Z, the time lapse distributing function $\Pi(t_{B11}), \Pi(t_{B12}), \Pi(t_{B13}), \Pi(t_{B14}), \Pi(t_{B15})$ of the time for objects in failure state to resume operation is defined on $t_{B11} \sim t_{B15}$ to determine $T_{H.}$ The T_W of every object module is still determined by $\Pi(t_{62})$.

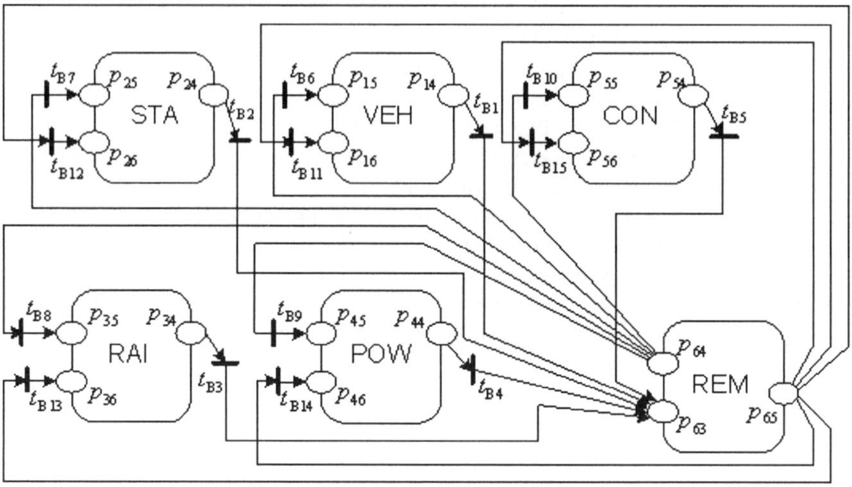

Figure 3. HOPN mold of reliability for UMT

4. Conclusion

(1) In figure 2, the embedded Petri net models at all levels is isomorphic, so the reliability model of every unit at all levels is represented by the standard figure of Figure 2(b). The figure can be a manifestation of operation/failure state of all units of the level and their reliability logic relationship.

(2) The operational reliability model of the whole UMT can be fully represented by Figure 1, Figure 2 and Figure 3. The number of facilities/equipment can be increased or reduced simply by adding or removing related colored tokens, facilities/equipment can be replaced or upgraded simply by changing the corresponding distributing function parameter without bringing about a change of the net model structure and swelling of a state space.

(3) HOPN is multi-functional. It can not only be used to construct operational reliability model of UMT, but also used to construct dynamic reliability model of other complex correlative system.

References

1. O'Connor P D T. *practical reliability engineering*. 4th ed. Chichester: John Wiley & sons Inc., (2002).

2. Siu N. Risk assessment for dynamic systems: An overview. *Reliability Engineering and System Safety*. **43**: 63-73 (1994).
3. Joanne Bechta Dugan. Modular Techniques for Dynamic Fault-tree Analysis. *Proceedings of the 1992 Annual Reliability and Maintainability Symposium*. Patterson-Hine, F A, 105-111.
4. Angela A, David H. Failure and safety assessment of systems using Petri nets. *IEEE International Conference on Robotics and Automation*. Washington DC: Institute of Electrical and Electronics Engineers Inc., (2002).
5. Zhang J J, WU zhehui. Hierarchical recursive model of Petri net. *Journal of System Simulation*. **15**(8): 89-92 (2003). (in Chinese)
6. Jensen K. *Colored Petri Nets*. 2nd ed. Berlin: Springer, (1997).
7. Wang L C. Objected oriented Petri nets for modeling of automated manufacturing systems. *Computer Manufacture Systems*. **26**(2): 111-125 (1996).
8. Wu X Y. *Reliability Modeling and Analysis of Complex Correlative System*. Changsha: National University of Defense Technology, (2000). (in Chinese)

TRAILER SWING WITH FLEXIBLY LASHED CARGO

ROBERT HILDEBRAND *, JOSÉ ANTONIO ROMERO NAVARRETE, MIGUEL MARTINEZ MADRID

Mexican Institute of Transportation (IMT), Apartado Postal 1098 Querétaro, Qro. C.P. 76000, Mexico

* *Currently at: Lake Superior State University, Sault Ste Marie, MI 49783, USA*

Yaw instability, resulting in cargo truck trailer swing, can have disastrous consequences for vehicles in other lanes. If the cab maintains its yaw stability (so that a jackknife does not occur) and the cab and trailers maintain their roll stability (so that no roll-over occurs), there still remains a critical braking deceleration at which trailer swing will ultimately occur. That critical deceleration is reduced when the cargo is flexibly lashed, as is shown in the present work. More generally, the trailer stability is found to depend on the friction between the tire and the road surface, the cargo placement and the compliance with which it is secured, the distances between axles, the braking distribution, tire compliance, and vibration. The present work provides a predictive method to estimate the critical deceleration, given sufficient information on the factors mentioned.

1. Introduction

Yaw instability of a tractor-semitrailer is a familiar mode for accidents and tends to occur before roll instability when the center of gravity of the cargo is low. As [2] notes, for trailers "...loaded with high density materials (e.g., steel plate) the total vehicle CG height is much lower and the rollover threshold is increased accordingly." [1] estimates rollover thresholds in the vicinity of .75 g lateral acceleration for such trucks. It is precisely for these kinds of trailers that yaw instability is likely to be the relevant accident mode.

The mode of yaw instability may be either jackknifing or trailer swing. [6] suggests that the decisive factor is the distribution of braking moment between the various axles: usually, locking of the rear tractor wheels will lead to jackknife, whereas locking of the semi-trailer wheels will lead to trailer swing.

The latter (trailer swing) is the topic of this paper. The critical deceleration at which it occurs is likely to be reduced by any of several possible "aggravating factors": lateral cargo tie-down flexibility (which permits the CG to swing out more easily), longitudinal cargo tie-down flexibility (causing an unloading of the rear tires), low lateral tire stiffness (due to under-inflation), low friction between the tire and the road surface (due to ice, rain, loose soils), trailer tire braking (which reduces the effective friction available to oppose lateral motion), and trailer vibration (also reducing the effective friction).

The discussion of section 2, below, is confined to the effects of tie-down flexibility, for which a simple formula is provided. Section 3 then generalizes to account for the other aggravating factors, using a time-integration approach.

2. Effect of Cargo Tie-Down Flexibility

Firstly, consideration is made of the yaw stability of a trailer in which the cargo is lashed by flexible belts, or other elastic restraints.

Notwithstanding the non-linearity and friction inherent in a belt tie-down (see [4,5]), the restraint elasticity is approximated here as linear, and friction between the cargo and the trailer bed is ignored. Often, the frictional restraint is unreliable, because of oil and dust contaminants [2], as well as vibration [4]. Moreover, the issue of stability may often be a small-angle phenomenon.

Figure 1(a) illustrates an idealization of the semitrailer, or trailer, as a column in which the cargo mass is concentrated at a CG position, which is

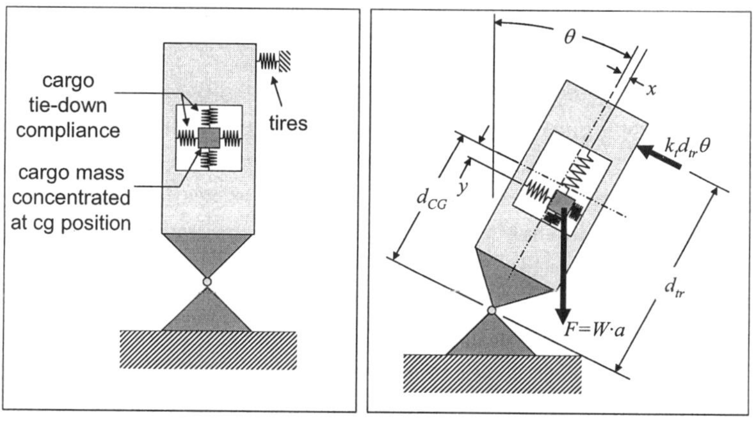

Figure 1. (a) Idealization of the trailer and cargo. (b) Forces that produce moments about the fifth wheel in the trailer bed plane, when swung out to an angle θ. Geometric dimensions are defined.

permitted to shift with respect to the trailer, impeded only by linear springs of stiffness k_x and k_y in the lateral and longitudinal directions, respectively. Moreover, the tires provide a lateral stiffness k_t opposing swinging of the trailer about the fifth wheel, which is a center of rotation in the plane of the trailer bed. The trailer itself is regarded as a rigid structure, and the tare mass ignored.

Suppose braking is described by a longitudinal deceleration a of the tractor, without lateral or yaw motion. Describing the semitrailer in the tractor reference frame, this is equivalent to a fixed fifth wheel and an effective force $F = Wa$ on the cargo, where W is cargo weight, and a is expressed in multiples of the gravitational acceleration $g = 9.81$ m/s^2. If the trailer centerline makes an angle θ with the tractor centerline, in the plane of the trailer bed, the forces acting on the cargo and the trailer, and which produce moments about the center of rotation, are as shown in fig. 1(b). The figure also defines the distances from the fifth wheel to the cargo CG position (d_{CG}), and to the semi-trailer axle (d_{tr}).

Replacing the tires by an equivalent rotational spring k_θ at the fifth wheel,

$$k_\theta = k_t d_{tr}^2 \,. \tag{1}$$

The lateral and longitudinal cargo shifts are,

$$x = \frac{F\sin\theta}{k_x} \text{ and } y = \frac{F\cos\theta}{k_y} \,. \tag{2,3}$$

Balancing moments about the center of rotation, and using (2) - (3)

$$F([d_{CG} - y]\sin\theta + x\cos\theta) = Fd_{CG}\sin\theta + F^2\left(\frac{k_y - k_x}{k_x k_y}\right)\sin\theta\cos\theta = k_\theta\theta \,. \tag{4}$$

Differentiating, introducing $k_{eff} = k_x k_y / (k_y - k_x)$, and solving for $d\theta/dF$,

$$\frac{d\theta}{dF} = -\frac{d_{CG}k_{eff}\sin\theta + 2F\sin\theta\cos\theta}{Fd_{CG}k_{eff}\cos\theta + F^2\cos 2\theta - k_\theta k_{eff}} \equiv \frac{N}{D} \,. \tag{5}$$

The swing angle grows without bounds, $\frac{d\theta}{dF} \to \infty$, when $D \to 0$; i.e., when

$$F^2\left(\frac{1}{k_{eff}}\cos 2\theta\right) + F(d_{CG}\cos\theta) - k_\theta = 0 \,. \tag{6}$$

Thus, as $\theta \to 0$, the critical braking force F is

$$F = \frac{1}{2}\left(\sqrt{d_{CG}^2 k_{eff}^2 + 4k_\theta k_{eff}} - d_{CG}k_{eff}\right) \tag{7}$$

For $k_y \to \infty$, i.e., rigid longitudinal attachment, $k_{eff} \to k_x$. The critical braking force F, normalized to that for the case of rigidly fixed cargo F_{rigid}, is plotted in fig. 2 as a function of the normalized lateral stiffness.

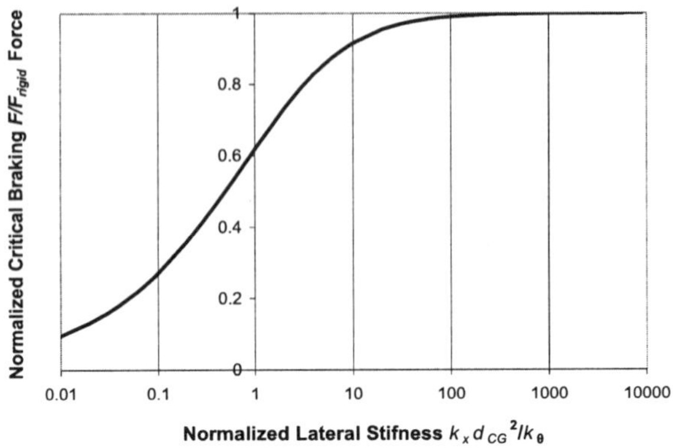

Figure 2. Yaw stability with flexibly attached cargo (laterally).

3. Trailer Swing with Tire Slide

3.1. *Approach*

The discussion is now generalized to add the effects of limited road-tire friction, braking, and rear wheel unloading due to the longitudinal shifting. See fig. 3.

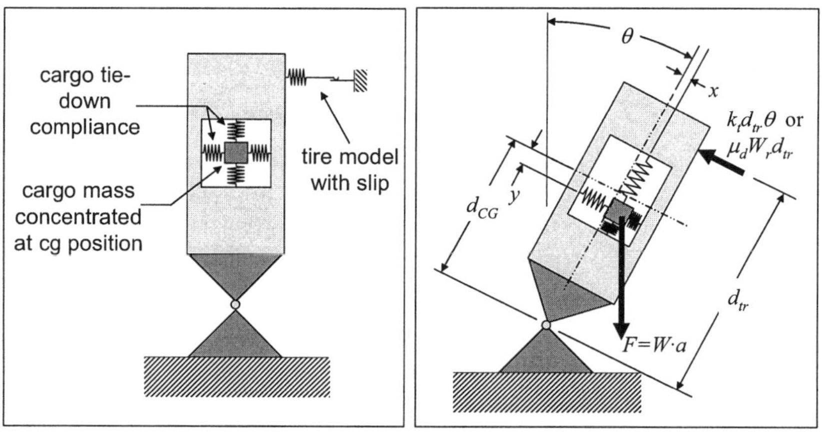

Figure 3. (a) Symbolic depiction of trailer, with tire slip allowed. (b) Equilibrium of the trailer.

The axle load is

$$W_r(y) = W(d_{CG} - y)/d_{tr}, \tag{8}$$

where y, as before, indicates the longitudinal position of the cargo CG.

As [3], p. 364, illustrates, partial skidding in the longitudinal direction, a consequence of braking, reduces the friction available to oppose lateral skidding. Consider a distribution of braking force amongst the axles of the entire vehicle, such that the braking force on the semitrailer tires is

$$F_{tr} = b_{tr} F \cdot \tag{9}$$

For a dynamic coefficient of dry friction μ_d, the dynamic friction load at the tire-roadway contact is $\mu_d W_r$. The percentage of skid at is

$$S_r = \frac{b_{tr} W a}{\mu_d W_r} \cdot \tag{10}$$

It is useful to define an "apparent friction" available to oppose lateral skid,

$$\mu_{dr,x} = \mu_d \sqrt{1 - S_r^2} \cdot \tag{11}$$

Considering figure 3(b), and recalling eq. (4), the moment M_{eq} needed to maintain equilibrium, when the trailer has swung out to an angle θ, is

$$M_{eq}(\theta) = F d_{CG} \sin\theta + F^2 \left(\frac{k_y - k_x}{k_x k_y} \right) \sin\theta \cos\theta \tag{12}$$

The resisting moment M_{res}, available from the tires to oppose a swing, is

$$M_{res}(\theta) = \begin{cases} k_\theta (\theta - \theta_{eq}), & \text{for } k_\theta \theta \le M_s(y) \\ M_d(y), & \text{for } k_\theta \theta > M_s(y) \end{cases}, \tag{13}$$

in which

$$M_s(y) = \begin{cases} \mu_s W_r(y) d_{tr}, & S_r = 0 \\ M_d(y), & S_r > 0 \end{cases}, \tag{14}$$

and

$$M_d(y) = \mu_{dr,x} W_r(y) d_{tr} \tag{15}$$

is that due to dynamic friction once sliding has begun. θ_{eq} is the initial angle of the trailer, or any angular position it assumes whenever sliding ceases.

The rotational moment I_θ of the trailer about the fifth wheel is approximately

$$I_\theta = W d_{CG}^2 / g \cdot \tag{16}$$

3.2. *Time history of trailer angle*

Given initial conditions $\theta(0) = \theta_0$, $\theta_{eq}(0) = 0$, and $\omega(0) = \omega_0$, for the trailer angle, trailer equilibrium angle, and trailer angular velocity, respectively, and a known braking deceleration a (or force $F=W \cdot a$), perhaps as a function of time, then the time history of the subsequent development of the trailer angle θ can be found from a consideration of the simultaneous integral equations

$$\theta(t) = \theta_0 + \int_0^t \omega(\tau)d\tau \quad \text{and} \quad \omega(t) = \omega_0 + \int_0^t \frac{M_{eq}(\theta[\tau]) - M_{res}(\theta[\tau])}{I_\theta} d\tau, \qquad (17,18)$$

and

$$\theta_{eq}(t) = \begin{cases} \lim_{\delta \to 0} \theta_{eq}(t-\delta), & \text{for } |\omega(t)| > 0, \text{or} \lim_{\delta \to 0} k_\theta \left[\theta(t) - \theta_{eq}(t-\delta)\right] < M_s \\ \theta(t), & \text{for } \omega(t) = 0, \text{and} \lim_{\delta \to 0} k_\theta \left[\theta(t) - \theta_{eq}(t-\delta)\right] > M_s \end{cases}. \qquad (19)$$

An expedient approach is to re-express these in the form of difference equations, and, choosing a suitably small time step, track the development of θ, θ_{eq} and ω using a digital computer. The equilibrium angle θ_{eq} is updated to equal θ whenever the tire has been sliding up to that point (fulfilled if $\lim_{\delta \to 0} k_\theta \left|\theta(t) - \theta_{eq}(t-\delta)\right| > M_s$) and the angular velocity is now zero (the slide has just stopped); then static friction is assumed to have been restored. Otherwise, at any other instant, θ_{eq} remains unchanged.

By such a time integration procedure, it is possible to identify critical decelerations a_{crit}, above which the semi-trailer will swing for any steering angle θ. Moreover, for subcritical decelerations, it is possible to identify critical steering angles θ at which the trailer will swing.

3.3. *Soil surfaces*

A soil surface is another aggravating factor that has the potential to reduce the stability of the trailer. These surfaces can be treated by finding an equivalent coefficient of friction. A particularly straightforward shear failure model that describes a wide range of soils is the Mohr-Coulomb-Terzaghi brittle failure model. According to that model, the failure shear stress of the soil is

$$\tau_f = c + \sigma \tan \varphi, \qquad (20)$$

where σ is the normal stress, and c and φ are soil properties called the cohesion and the friction angle, respectively. Defining an equivalent static coefficient of friction as $\mu_{s,eq} = \bar{\tau}_f / \bar{\sigma}$, in which the bars indicate averaging over the tire-soil contact zone, then

$$\mu_{s,eq} = \frac{c}{\bar{\sigma}} + \tan \varphi = \frac{cA_r}{W_r} + \tan \varphi \qquad (21)$$

where A_r is the area of the contact zone; that may have the load dependence

$$A_r = K_{TS} W_r^{1/2}, \qquad (22)$$

from which

$$\mu_{s,eq} = K_{TS}\frac{c}{\sqrt{W_r}} + \tan\varphi\,. \tag{23}$$

For a friction soil ($c \equiv 0$, $\varphi > 0$), $\mu_{s,eq}$ is independent of the tire loading, while for cohesive soils ($c > 0$) the apparent friction decreases with tire load.

For most soils, the dynamic friction will be enhanced "plowing" soil ahead of the tire; thus, the dynamic friction is seldom much less than the static friction,

$$\mu_{d,eq} \approx \mu_{s,eq}\,. \tag{24}$$

3.4. *Trailer vibration*

Vibration causes the tire loads to vary with time. Obtaining the dynamic contact force requires a model for the dynamic stiffness of the vehicle at the tire. For simplicity, consider a single degree-of-freedom model, with a specified natural frequency f_n and damping ratio ζ, which, given that the mass is already an input parameter, uniquely define the suspension properties.

The truck traverses the surface at some instantaneous speed v. Suppose that the surface has a roughness pattern with a Power Spectral Density $S_{zz}(\omega)$. Because the trailer response is expected to be fairly narrow-band, it is a good approximation to regard the roughness as white noise *with a constant spectral density* $S_{zz0} = S_{zz}(2\pi f_n)$. Then:

(1) A random number generator, with output scaled so that the rms-value is the square-root of S_{zz0}, provides a sample time history $z(t)$ of the roughness.

(2) The random force $W_{rand}(t) = (4\pi f_n \varsigma m)\dot{z}(t) + (2\pi f_n^2 m)z(t)$ acting on the equivalent fixed s.d.o.f. system is calculated, using a difference method.

(3) The trailer response $z_t(t) = \int_{-\infty}^{t} W_{rand}(\tau)h(t-\tau)\mathrm{d}\tau$ is found, where $h(t) = \mathrm{e}^{-\varsigma\omega_n t}\sin\omega_d t / m\omega_d$ is the impulse response function.

(4) The time history of the narrowband part of the contact force, $W_{nb}(t) = -(4\pi f_n \varsigma m)\dot{z}_t(t) - (2\pi f_n^2 m)z_t(t)$, is calculated.

(5) The dynamic contact force is $W_{rd}(t) = W_{rand}(t) + W_{nb}(t)$.

(6) The total tire load is $W_r(t) = W_{r,\,static} + W_{rd}(t)$, where $W_{r,\,static}$ is from (8). $W_r(t)$ may then be used in the method of section 3.2.

3.5. *Example results*

Consider a "reference case" (12 m semi-trailer; axle 2 m from rear of trailer; fifth wheel 1m behind front of trailer; load centered in semi-trailer, i.e., 6 m behind front of trailer), defined by the following input properties:

$d_{tr} = 9$ m, $d_{CG} = 5$ m, $\mu_s = .7$, $\mu_d = .55$,
$W = (30000 \text{ kg}) \times (9.81 \text{ N/kg}) = 2.94\times10^5$ N (represents 30000 kg cargo),
$k_t = 120$ kN/m, $b_{tr} = .33$, $k_x \rightarrow \infty$, $k_y \rightarrow \infty$ (cargo rigidly fastened),
Smooth road surface.

Example results, using the method of section 3.2, are presented in table 1. These represent a wide range of the variables of interest. In each case, the input parameters are as for the "reference case", except for indicated deviations.

Table 1. Examples of stability determined by general model, method of section 3.3.

Example Case Description	Parametric Deviations from Reference Case	Critical Decel'n a_{crit} (in g)	Critical angle at $a = a_{crit}$
Reference case	----	1.17	0°
Pure Tire Roll	$b_{tr} = 0$	6.61	6.2°
Icy road	$\mu_s = .1$, $\mu_d = .07$	.17	0°
Lateral Cargo Compliance	$k_x = 5$ kN/m	.70	3.9°
Longitudinal Cargo Compliance	$k_y = 100$ kN/m	.69	0°
Load towards back of trailer	$d_{CG} = 7$ m	1.65	0°
Soft tires	$k_t = 100$ kN/m	.68	41°
Cohesive soil surface (clay)	$\mu_s = \mu_d = K_{TS}c / W_r^{1/2}$	.26	0°

The reference case, believed to be close to "typical" driving conditions, is stable up to 1.17 g, well beyond what the brakes are capable of, so that there is no risk of braking too hard. On the other hand, since the critical angle reduces to 0° by the time the critical deceleration is reached, the steering accuracy becomes more and more important, the harder the driver brakes.

Evidently, braking of the trailer tires is a strong factor. That is clear from the much higher critical deceleration for the pure tire roll case, as compared to the reference case in which the trailer tires are in a partial skid.

The icy road is particularly demanding, the critical deceleration having been substantially reduced with respect to the reference case.

Reasonable parameters for the tie-down compliance cases come from [4] and [5]. The laterally-compliant case represents a relatively loose webbing. The longitudinally-compliant case represents a relatively tight chain. The trailer becomes unstable in these cases at decelerations considerably lower than for a rigidly fixed cargo. For the longitudinally-compliant case, the loss of tire loading outweighs the benefit gained from reducing the distance d_{CG}.

Similarly, load moved towards the back of the trailer improves the stability, i.e., the benefit of loading of the rear tires outweighs the CG location shift.

Very soft tires, perhaps representing considerable deflation, result in a reduction of the critical deceleration as well. Interestingly, the critical angle is still very high as the deceleration approaches the critical. Thus, soft tires seem to limit braking more than they demand steering accuracy.

The last case represents an overconsolidated clay with c = 50 kN/m; assuming K_{TS} such that A_r = 1000 cm^2 at a load of W_r = 6.8 kN ([6], p.82, fig.2.7), then $\mu_{s,eq} = \mu_{d,eq} = .15$. The stability behavior is considerably worse than the reference case, but somewhat better than for an icy road.

A time history for the reference case, but with road roughness induced vibration, is plotted in figure 4. In this case, the braking deceleration increases linearly to .75 g over 4 seconds, and the trailer is assumed to have a suspension with a natural frequency f_n = 2 Hz and damping ratio ζ = .2. The truck traverses a gravel road at 30 m/s with a white-noise roughness pattern having a constant Power Spectral Density $S_{yy}(\omega)$ = S_{yy0} =1.28·10^{-3} m^2/[cyc/m], a value representative of a wavelength of .067 m on such a surface [6]; that wavelength coincides with the trailer's natural frequency at a speed of 30 m/s. The trailer proves to be unstable, which it is not the case when it traverses a smooth road.

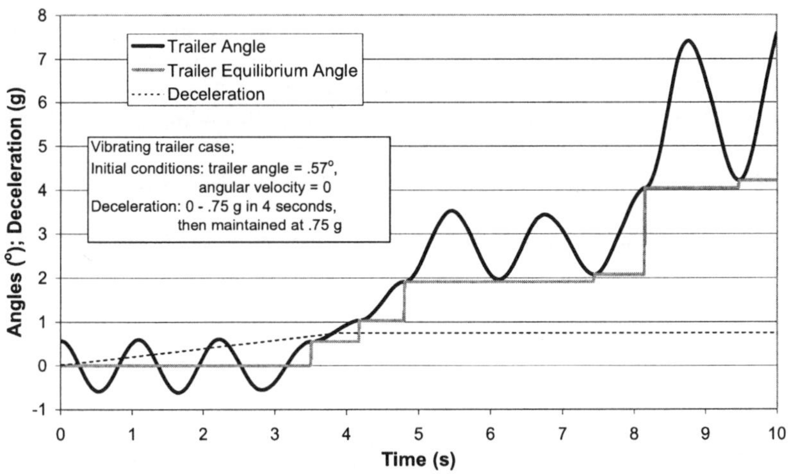

Figure 4. Trailer instability due to roughness-induced vibration on a typical gravel road. Same parameters as for the reference case, except for the white-noise road roughness pattern with a frequency-independent PSD of $S_{yy}(\omega) = S_{yy0}$ =1.28·10^{-3} m^2/[cyc/m].

4. Conclusions

- Braking distribution has a dramatic effect on trailer stability.
- A low friction road surface (e.g., rain, ice) substantially reduces yaw stability.
- Tire underinflation reduces the critical deceleration (i.e., the driver cannot brake as hard), but does not affect the critical angle (so that steering is not made any more demanding than otherwise, at deceleration levels below the critical).

· Flexibly lashed cargo reduces the trailer stability as well; the effect appears to be significant for values of the tie-down stiffness that might be seen in practice.
· Longitudinal cargo tie-down flexibility reduces the trailer tire loads, which is more harmful than the forward shift of the CG position is helpful.
· Vibration, induced by road roughness for example, reduces the effective tire-road friction, and thereby degrades the trailer stability as well.

References

1. T. Gillespie and R. D. Ervin, *Comparative Study of Vehicle Roll Stability* University of Michigan Transportation Research Institute, Technical Report UMTRI-83-25 (1983).
2. T. Gillespie, *Engineering of Cargo Restraint on Commercial Highway Trucks.* University of Michigan Transportation Research Institute, Technical Report UMTRI-87-28 (1987).
3. T. Gillespie, *Fundamentals of Vehicle Dynamics.* SAE International ISBN-1560911999 (1992).
4. J.A. Romero, S. Rakheja, A.K.W. Ahmed and A. Lozano, "Restrained Cargo Dynamics in Road Transportation: Indirect Tie-Downs", *Heavy Vehicle Systems*, a series of the *International J. Vehicle Design*, Vol. **9**, No. 2, p. 93-114 (2002).
5. J.A. Romero, A. Lozano, S. Rakheja, A.K.W. Ahmed and H. Hong, "Restrained Cargo Dynamics in Road Transportation: Direct Tie-Downs", *Heavy Vehicle Systems*, a series of the *International J. Vehicle Design*, Vol. 11, No. 2 (2004).
6. J.Y. Wong, *of Ground Vehicles.* John Wiley & Sons, ISBN 0-471-03470-3 (1978).

NON LINEAR CONTROL BASED IN AN OBSERVER; APPLICATION TO SUGAR EVAPORATION

ANSELMO OSORIO M.[1], ENRIQUE ARCE M.[2] AND JOSE CARRILLO A.[3]

[1]*Facultad de Ciencias Químicas, Universidad Veracruzana. Prolongación Oriente 6, No. 1009, Col R. Alvarado, Orizaba, Ver. 94360, México, anselmo_osorio@yahoo.com*

[2]*Escuela Superior de Ingeniería Química e Industrias Extractivas. Instituto Politécnico Nacional. Edif 7. UPALM, Av. IPN, Col. Zacatenco, Méx., DF, 07738, México. earcem@yahoo.com.mx*

[3]*División de Est. de Posgrado e Inv., Instituto Tecnológico de Orizaba. Oriente 9, No. 852, Col E. Zapata, Orizaba, Ver. 94320, México.*

In many technological processes the knowledge of all state variables is very important in order to design robust and efficient controllers. However, in practical applications, only some of the state variables are available for on-line measurement. The unknown variables can be estimated by means of state observers using a dynamic model in connection with measured variables. The purpose in this paper is to introduce a controller based on a nonlinear observer to a single sugar evaporator. The controller-observer scheme performance for the output concentration of the evaporator is verified by means of numerical simulations.

1. Introduction

The presence of unknown states in industrial processes becomes a difficulty that can be solved by means of a suitable state estimator. Highly nonlinear processes, like chemical processes, has given rise to the development of nonlinear observers to face many typical obstacles like restrictive conditions, uncertain performance, robustness, and poor estimation, as shown by [1, 4, 5]. A state observer is a replica of the process model, driven by the error in the estimate of the available measurements [2].

In a sugar cane factory, the evaporation process with multiple stages is used to reduce the water content in the juice or syrup, which has a sucrose concentration approximately of 15% (known as Brix). The evaporation product is a concentrated solution that has a concentration higher than 70%. Since sugar production cost depends on steam consumption, it is required to optimize the energy consumption by means of a controlled operation to obtain a quality final product at low cost.

The objective of this work is to present an efficient controller-observer scheme applied to the sugar evaporation process. The design procedure is based in a reduced dynamic model of the process. First, an observer is designed that converges, asymptotically, to actual values. Then the controller is constructed using the separation principle for linear systems. A nonlinear control law to the output concentration of the evaporator in the full model is proposed. Thus, the controller-observer scheme performance for reduced model together with the nonlinear controller is verified by means of numerical simulations.

2. State Observer Design

In order to design a controller for the output sugar concentration of the evaporator, an observer to estimate state variables is introduced. Since in a practical application the state vector is not all accessible by direct measurement, the output vector dimension is smaller than that of system state vector, thus, it is necessary to rebuild the state to design a controller. An observer is a set of equations that allows estimating the state of a system from input and output information of the system. A system is observable if it is possible to rebuild the initial state of the system from output measurements and inputs. In [9] it is shown the construction of a multi-output generalized observable canonical form by means of a change of coordinates and the observation problem based on a high gain nonlinear observer [5].

The observer is based on a linear model obtained from a coordinate change which transforms the original non-linear system given by (1) into a linear system:

$$\begin{aligned} \dot{x} &= f(x) + g(x)u \\ y &= h(x) \end{aligned} \tag{1}$$

Then, from system (1) a linear reduced system in state-space form can be obtained as given by:

$$\begin{aligned} \dot{x} &= Ax(t) + Bu(t) \\ y &= Cx(t) \end{aligned} \tag{2}$$

Where the vector $x \in \Re^n$ stands for the state variables, $u \in \Re^m$ is the input vector that represents the manipulated variable, and the measured outputs are represented by vector $y \in \Re^p$. Furthermore, A_{nxn} is the system matrix, B_{nxm} is the distribution matrix of manipulated inputs and C_{pxn} is the matrix of observable outputs. For this system, it is assumed that the control law is given by $u_l = -F\hat{x}$, thus, the system can be written as:

$$\begin{bmatrix} \dot{x} \\ \dot{\varepsilon} \end{bmatrix} = \begin{bmatrix} A - BF & -BF \\ 0 & A - KC \end{bmatrix} \begin{bmatrix} x \\ \varepsilon \end{bmatrix} \tag{3}$$

Where $\varepsilon = (\hat{x} - x)$ is the estimation error. F and K are the gain matrices of controller and observer, respectively, which can be chosen independently. The gains in matrix K are selected so that the matrix $A-KC$, has all its poles located in the left-half of the complex plane, and matrix F is chosen such that $A-BF=0$.

3. Application to the Sugar Evaporation Process

An evaporator is used to concentrate a dilute solution of a single solute in a volatile solvent [3]. Energy and mass balances give the following governing equations [7].

$$\begin{aligned}
\tau_1 \frac{dF_0}{dt} &= -F_0 + F_1 + O_1 \\
\tau_{01} \frac{dP_0}{dt} &= -P_0 + P_1 + \frac{\gamma_1}{\rho_1^2 (T_{S0}, P_0)} S^2 \\
\tau_{e1} \frac{dO_1}{dt} &= -O_1 + k_1 S \\
W_1 \frac{dC_1}{dt} &= -O_1 C_1 - F_0 (C_1 - C_0)
\end{aligned} \tag{4}$$

Where the pressure drop is given by $\Delta P_1 = \frac{\gamma_1}{\rho_1^2 (T_{S0}, P_0)} S^2$, with $\Delta P_1 = P_0 - P_1$, γ_1 is a conversion factor and ρ_1 is the vapour density that leaves the solution. Furthermore, F_0 is the sugar cane solution feed flow rate with concentration C_0 in Brix, S is the heating steam flow rate at temperature T_{S0} and pressure P_0, O_1 is juice vapour flow rate with temperature T_{S1} and pressure P_1. The sugar cane solution in the evaporator is flowing out through the inside tubes of the heating coils with flow rate F_1, concentration C_1 and temperature T_1. The initial conditions of the state variables and parameters are given in tables 1 and 2. The vapour of the sugar cane solution in the evaporator is passed into a condenser. The conditions imposed on the evaporator model are as follows: The time constants τ_1 and τ_{01} describe the vapour flow rate dynamic of the sugar cane solution. All of the time constants are assumed constant. k_1 represent the static gain. Since a compressor or a vacuum pump controls vapour pressure of the sugar cane solution, the sugar cane solution flow rate O_1 is equal to condensed flow rate F_0.

3.1. *The State space model*

Using the standard notation for dynamic systems given by (4), the evaporator mathematical model can be expressed as a state space model in deviation variables as in the canonical form (2), with measured output variables x_1 and x_2.

$$\dot{x}_1 = -\frac{1}{\tau_1}x_1 + \frac{1}{\tau_1}x_3 + \frac{1}{\tau_1}u_1$$
$$\dot{x}_2 = -\frac{1}{\tau_{01}}x_2 + \frac{1}{\tau_{01}}u_2 + \frac{\gamma_1}{\tau_{01}\rho_1^{\,2}(T_{SO}, P_0)}u_3 \tag{5}$$
$$\dot{x}_3 = -\frac{1}{\tau_{el}}x_3 + \frac{k_1}{\tau_{el}}u_4$$
$$\dot{x}_4 = -\frac{1}{W_1}(x_3 + O_{1s} - x_1 - F_{0s})x_4 + \frac{1}{W_1}(x_1 + F_{0s})u_5$$
$$y_1 = x_1$$
$$y_2 = x_2$$

Where: $x_1 = F_0 - F_{0s}, x_2 = P_0 - P_{0s}, x_3 = 0_1 - 0_{1s}, u_1 = F_1 - F_{1s}, u_2 = P_1 - P_{1s}$, $u_3 = S^2 - S_s^{\,2}, u_4 = S - S_s, x_4 = C_1 - C_{1s}$ and $u_5 = C_0 - C_{0s}$, The sub index s stands for steady state conditions

3.2. *Construction of the observer*

Now, from another perspective of the systems, it is possible to break up it, and to design an observer from the reduced system that includes the variables x_1, x_2 and x_3. The following vector is obtained.

$$\begin{bmatrix} \xi^1{}_1 \\ \xi^1{}_2 \\ \xi^2{}_1 \end{bmatrix} = \begin{bmatrix} x_1 \\ \frac{1}{\tau_1}(-x_1 + x_3) \\ x_2 \end{bmatrix} \tag{6}$$

This constitutes a change of coordinates that transform the system's model in the form of equation (2). Thus the observer in original coordinates is given by

$$\dot{\hat{x}}_1 = \frac{1}{\tau_1}(-\hat{x}_1 + \hat{x}_3 + u_1) + 2\theta(x_1 - \hat{x}_1)$$
$$\dot{\hat{x}}_2 = \frac{1}{\tau_{01}}\left(-\hat{x}_2 + u_2 + \frac{\gamma_1}{\tau_{01}\rho_1^{\,2}(T_{SO}, P_0)}u_3\right) + \theta(x_2 - \hat{x}_2) \tag{7}$$
$$\dot{\hat{x}}_3 = \frac{1}{\tau_{el}}(\hat{x}_3 + k_1 u_4) + (\theta^2\tau_1 + 2\theta)(x_1 - \hat{x}_1)$$

3.3. *Controller construction*

To design the controller, the separation principle is used as a basis, and then it is possible to elect the gains of both the controller and the observer independently. Thus the gain matrices are given by.

$$F=\begin{bmatrix} -1 & 0 & 1 \\ -\frac{\gamma 1}{\rho_1^{\ 2}} & 0 & \frac{\gamma 1}{\rho_1^{\ 2}} \\ 1 & -\frac{\rho_1^{\ 2}}{\gamma_1} & 1 \\ 0 & 0 & -\frac{1}{k_1} \end{bmatrix} \qquad K=\begin{bmatrix} 2\theta & 0 \\ 0 & \theta \\ (\theta^2\tau_1+2\theta) & 0 \end{bmatrix} \tag{8}$$

For the controller and observer respectively.

Considering that the controller for reduced system has the gain F. Then, a nonlinear controller for the dynamic behaviour of the state x_4 (output concentration of the evaporator), can be proposed, the control law is given by

$$u_5=\left(\frac{(x_3+O_{1s})-(y_1+F_{Os})}{(y_1+F_{Os})}\right)\hat{x}_4 \tag{9}$$

4. Numerical Results

In order to evaluate the observer performance, the system was simulated assuming the values of the parameters and variables given in Tables 1, 2, & 3.

Table 1. Initial conditions.

$F_0 = 2561$ kg/min	$O_1 = 708$ kg/min
$C_0 = 15.2$ %	$F_1 = 1901$ kg/min
$S = 933$ kg/min	$C_1 = 22.5$ %
$T_{S0} = 130$ °C	$W_1 = 16607$ kg
$P_0 = 2.7$ bar	$P_1 = 1.73$ bar

Table 2. Steady State values and parameters.

$F_{0s} = 2656.7$ kg/min	$t_1 = 1.10$ min
$P_{0s} = 2.6345$ bar	$\tau_{01} = 1.22$ min
$O_{1s} = 755.73$ kg/min	$\tau_{e1} = 0.82$ min
$C_{1s} = 20.988$ %	$k_1 = 0.81$

Table 3. Control Inputs and Physical Properties.

u_{1s}=1901 kg/min	u_{4s}=933 kg/min
u_{2s}=1.73 bar	u_{5s}=15.2 %
u_{3s}=870489 kg^2/min^2	$\rho(T_{s0},P_0) = 1.4935$ kg/m^3

In evaporation process model, given by equation (5) one can observe that x_4 is uncoupled, i.e., it hasn't an influence over the dynamic performance of full system, then the reduced model that includes the states x_1, x_2 and x_3 can be write as (7). The initial conditions are set to: $x_1(0) = 2651$, $x_2(0) = 2.7$, $x_3(0) = 708$, $x_4(0) = 0.152$, $\hat{x}_1(0)=\hat{x}_2(0)=\hat{x}_3(0)=\hat{x}_4(0)=0$. The used value of parameter θ is 1, in all simulations.

The observer performance is first analyzed. The estimation error results are shown in Fig. 1, estimation error dynamic of feed flow rate.

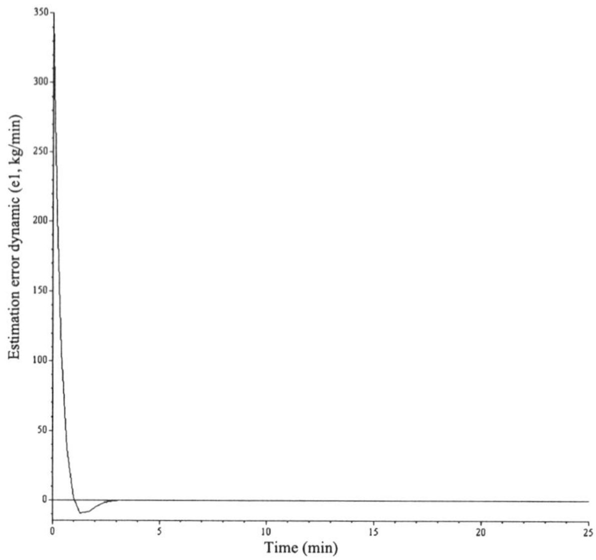

Figure 1. Estimation error dynamic, e_1, for the steam pressure, F_0.

In order to test the behaviour of the observer-controller scheme and the nonlinear control proposed in disadvantageous conditions, simulations were carried out based on noise-corrupted measurements. The outputs x_1 and x_2 were perturbed with white noise signals of mean zero and standard deviation of 0.5. The sample time was 0.2 s. The results of this numerical simulations are depicted in Fig. 2, both estimate output concentration and vapour flow rate, rapidly converges to the actual values. In Fig. 3, actual and estimate feed flow rate and measured output with noise (feed flow rate) are shown, and Fig. 4, depicts the actual and estimate steam pressure and measured output with noise (steam pressure). This last figure shows how the observer behaves efficiently.

The observer-controller scheme shows good convergence properties and the gain can be arbitrarily assigned in order to speed the convergence.

Conclusion

We present in this paper a study of and observer controller of a single-effect sugar evaporation process. Since there is not model uncertainty, the estimation error converges to zero and the estimator presents good stability properties. The estimated states by the observer were used in a controller-observer scheme to regulate the output concentration of the evaporated solution by means of a single nonlinear control. Numerical results show the performance of the nonlinear controller and the effective use of the controller-observer applied to the sugar cane evaporation process.

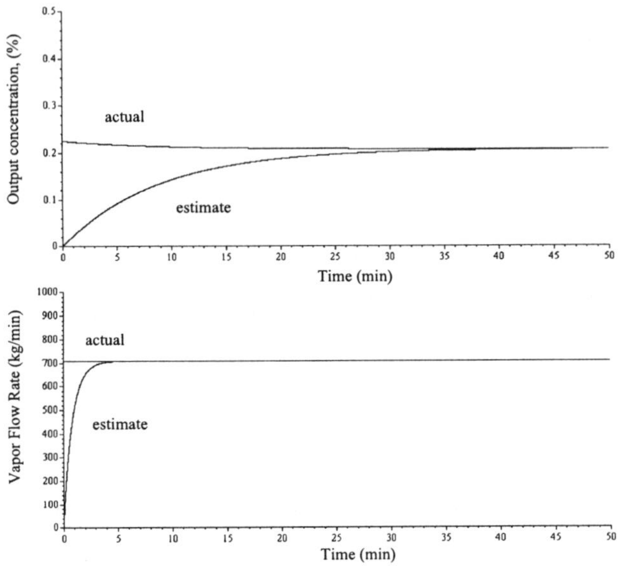

Figure 2. a) Output concentration, C_1, of the evaporator. b) vapour flow rate, O_1, from sugar solution.

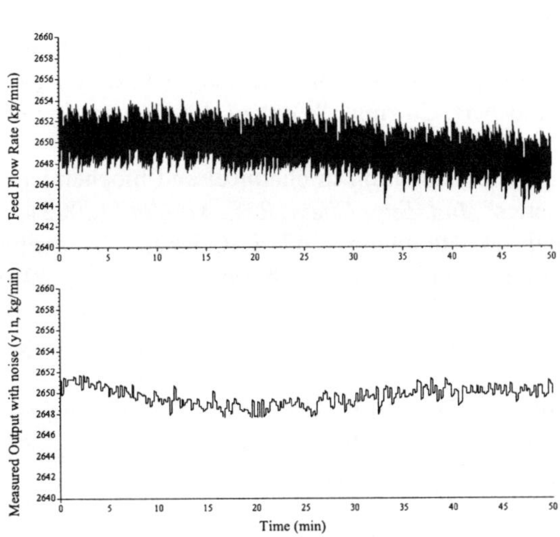

Figure 3. a) Feed flow rate, F0, with noise in the measurement. b) measured output with noise, y_1

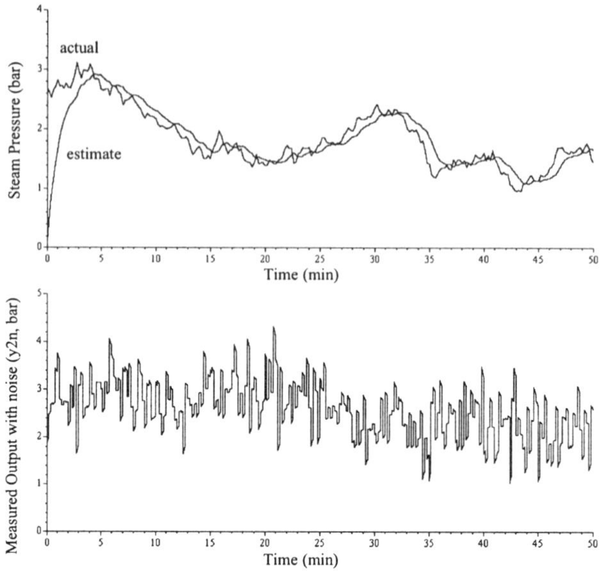

Figure 4. a) Steam pressure, P0, b) Measured output with noise, y_2 (steam pressure).

References

1. S.I. Bigiola and J.L. Figueroa. "A high gain nonlinear observer: application to the control o fan instable nonlinear processes", *Comp. and Chem. Eng.*, **28**,1881 (2004)
2. S.I. Bigiola and J.L. Figueroa. "State estimation in nonlinear processes. Application to pH process control". *Ind. Eng. Chem. Res.*, **41**, 4777 (2002).
3. C. A. Smith and A.B. Corripio. "*Control Automático de Procesos. Teoría y Práctica*", Limusa, México, (2004).
4. D. Dochain. "State estimation in chemical and biochemical processes with uncertain kinetics", *Ind. Eng. Chem. Res.*, **41**, 4777 (2002).
5. J. P. Gauthier, H. Hammouri and S. Othman. "A simple observer for nonlinear systems: applications to bioreactors", *IEEE Trans. On Autom. Control.* **36**, 875, (1992).
6. A. Isidori, "*Nonlinear Control Systems*", Springer Verlag, London, Great Britain, 1995.
7. S. Lisane Elhaq. F. Giri and H. Unbehauen. "Modelling, identification and control of sugar evaporation: theorical design and experimental evaluation", *Control. Eng. Practice,* **7**, 931 (1999)
8. R. Martínez-Guerra, R. Garrido and A. Osorio-Mirón. "The fault detection problem in nonlinear systems using residual generators", IMA *J. Math. Control Info.* **22**, 119, (2005).

SENSOR FUSION- SONAR AND STEREO VISION, USING OCCUPANCY GRIDS AND SIFT

ALFREDO CHAVEZ PLASCENCIA and JAN DIMON BENDTSEN

Department of Control Engineering, Institute of Electronic Systems, Faculty of Engineering and Science, Aalborg University, Frederik Bajers Vej 7C, 9220 Aalborg East, Denmark, e-mail: acp@control.aau.dk, dimon@control.aau.dk, Phone: +4596358747, *Fax:* +4598151739

The main contribution of this paper is to present a sensor fusion approach to scene environment mapping as part of a SDF (Sensor Data Fusion) architecture. This approach involves combined sonar and stereo vision readings. Sonar readings are interpreted using probability density functions to the occupied and empty regions. SIFT (Scale Invariant Feature Transform) feature descriptors are interpreted using gaussian probabilistic error models. The use of occupancy grids is proposed for representing the sonar as well as the features descriptors readings. The Bayesian estimation approach is applied to update the sonar and the SIFT descriptors' uncertainty grids. The sensor fusion yields a significant reduction in the uncertainty of the occupancy grid compared to the individual sensor readings.

Keywords: Sensor fusion, mobile robots, stereo vision, occupancy grids, SIFT

1. Introduction

In order for a Wheeled Mobile Robot (WMR) to achieve full autonomy and consequently widen the range of its applicability, it is necessary to develop more reliable systems which can operate in structured and unstructured environments. To achieve full autonomy the WMR has to incorporate sensors to gather information from the environment. Without sensing the WMR cannot react to its surroundings and all the tasks the robot has to perform must be programmed *a priori*. By combining and fusing different readings from a variety of different sensors, on the other hand, the WMR can potentially achieve full autonomy. The result of the fusing process from different sensors can be used to reconstruct the robot's world environment, and the robot can plan its own path and avoid obstacles. The robot can also adapt by itself to unexpected environments. However, it is almost impossible for the mobile robot to fully reconstruct its world environment and adapt by itself to the former by means of one single sensor. Therefore fusion of different sensor readings must be applied in the hierarchical architecture of the robot.

When dealing with Sensor Data Fusion (SDF) architecture, one of the requirements to take into account is the choice of the internal representation. This internal representation must be chosen such that it is common to all sensors.

One internal representation that fulfill the above requirements is the occupancy grid introduced by Elfes in Elfes;[1] Elfes;[2] Elfes;[3] and Elfes.[4] An occupancy grid is a tessellation of the robot's environment into cells defined over a discrete spatial lattice. The parameters in this random field are the stochastic random variables or cells C_i, $i = 1, ..., n$. The status of these cells being occupied and empty is denoted *occ* and *emp* respectively. The state of each cell is exhaustive and exclusive, meaning that $P(C_i = occ) + P(C_i = emp) \equiv 1$, where $P(\cdot)$ is the probability of a random variable being in a particular state.

This paper is focused on the problem of fusing range readings from a sonar with landmarks produced from stereo vision using the SIFT (Singular Invariant Feature Transform) algorithm. This method was originally introduced by David G. Lowe[5] and later refined by the same author Lowe.[6] The use of occupancy grids is proposed in this paper to map the readings from sonar and stereo vision.
There are several approaches to solve the sensor data fusion problem. In Ren C. Luo[7] and Ren C. Luo[8] a classification is presented. Alberto Elfes[4] proposed to use the recursive Bayes formula to update the occupancy grid for multiple sensor observations $(s_1, ..., s_n)$. This approach is used in this paper.

In Daniel Pagac and Durrant-Whyte,[9] C.A.[10] and Mathies and Elfes[11] some of the drawbacks of the sonar are mentioned. Sonar readings are more uncertain perpendicular to the main axis of the beam and more accurate along the axes; this is due to the width of the beam. Specular readings do not give direct information on the distance to the nearest surface. The contribution of this paper is to reduce these drawbacks by fusing the sensor readings with the landmarks produced by the stereo vision.

Section 2 is concerned with the sonar operation as well as the sonar model. Section 3 deals with a brief description of the SIFT. A background of the pinhole camera model is provided. This section also shows how to get and model the stereo triangulation and the error in the stereo matcher.

Section 4 describes the SDF architecture and how the Bayes recursive formula can be applied to the occupancy grids in order to update the maps. Section 5 shows the results of the reduction of the uncertainty by fusing the SIFT descriptors with the sonar readings using occupancy grids in the SDF architecture. Section 6 gives the conclusion and directions for future work.

2. Sonar

2.1. *Theory of Operation*

It was mentioned previously how important it is for the robot to sense the environment. A common sensor to measure distance is the ultrasonic range finder or sonar. The sonar used is the polaroid 6500 ranging module. The sonar transmits a series of 16 pulses at 49.9 KHz. The transmitted pulses reflect off the object and are received at the transducer. The reflection time is proportional to the distance from the object to the source. The sonar can thus measure the distance from the transducer to the object accurately, but it cannot estimate at what angle within the sonar cone the pulse was reflected. Hence the angle at which the obstacle was measured is uncertain.

2.2. *Sonar Model*

Experiments have shown that the energy is stronger close to the transducer and along the axis of the beam. To reflect this, probabilistic sensor models have been formulated to represent the behavior of the sonar beam. A wide range of sonar models have been developed in the past years by various researchers, Elfes,[1] Elfes,[2] Konolige[12] and C. A..[10] The sensor model proposed by Elfes[1] has been chosen due to its success in applying gaussian probabilities to the echo beam.

The model of the sonar beam is formulated as two probability density functions, f_E and f_O. These functions measure the confidence and uncertainty of an empty and occupied region in the cone beam of the sonar respectively. They are defined based on the geometrical aspect and the spatial sensitivity of the sonar beam. Figure 1 shows the occupied and empty probability distributions for a sonar beam that returns a reading R. In this model the following variables and parameters are used:

R is the range measurement return by the sensor.
ϵ is the mean sonar deviation error.
ω is the width of the cone beam.

S is the sonar sensor.
P is the cell being updated.
δ is the distance from P to S.
θ is the angle between the main axis of the beam and the line SP.

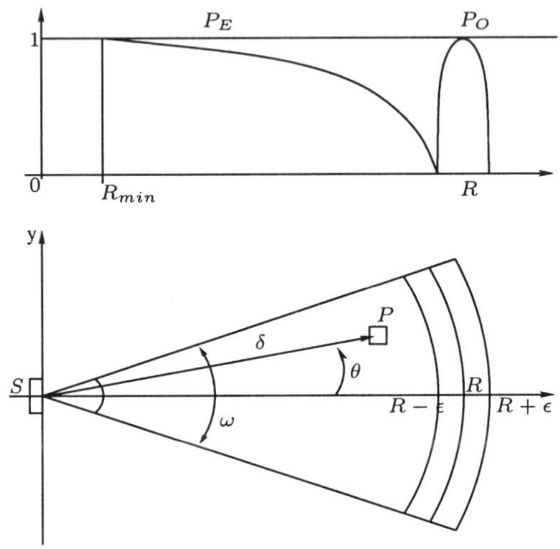

Fig. 1. Morevec-Elfes probabilistic sensor model

The beam is divided in two regions: *a*) The free space area or empty probability region which is the part of the beam between the sensor and the range where the obstacle was detected. This includes cells $P(x, y)$ inside the sonar beam. Each cell has an empty probability $P_E(x, y) = E_r(\delta) \cdot E_a(\theta)$, where $E_r(\delta)$ is the estimation of the free space cell based on the range measurement from the sonar. $E_a(\theta)$ is the estimation that the cell is free based on the angle of the cone beam.

These two probabilities are defined as follows:

$$E_r(\delta) = \begin{cases} 1 - \left(\frac{\delta - R_{min}}{R - \epsilon - R_{min}}\right)^2 & \text{for} \quad R_{min} \leqslant \delta \leqslant R - \epsilon \\ 0 & \text{otherwise} \end{cases} \tag{1}$$

$$E_a(\theta) = 1 - \left(\frac{2\theta}{\omega}\right)^2 \quad \text{for} \quad \frac{-\omega}{2} \leqslant \theta \leqslant \frac{\omega}{2} \tag{2}$$

b) The occupied area or probability occupied region. This is the area where the obstacle was detected. In this region the uncertainty of the exact distance to the obstacle (ϵ) has to be taken into account. The probability of a cell being inside the occupied region is $P_O(x,y) = O_r(\delta) \cdot O_a(\theta)$. $O_r(\delta)$ is the probability that a cell is occupied based on the range reading. $O_a(\theta)$ is the probability that the cell is occupied based on the difference of angle between the obstacle and the beam axis. These two probabilities are defined as follows:

$$O_r(\delta) = \begin{cases} 1 - \left(\frac{\delta - R}{\epsilon}\right)^2 & \text{for} \quad (R-\epsilon) \leqslant \delta \leqslant (R+\epsilon) \\ 0 & \text{otherwise} \end{cases} \tag{3}$$

$$O_a(\theta) = 1 - \left(\frac{2\theta}{\omega}\right)^2 \text{ for } \quad \frac{-\omega}{2} \leqslant \theta \leqslant \frac{\omega}{2} \tag{4}$$

The gaussian probability distribution of a single echo beam is shown in fig 2.2.

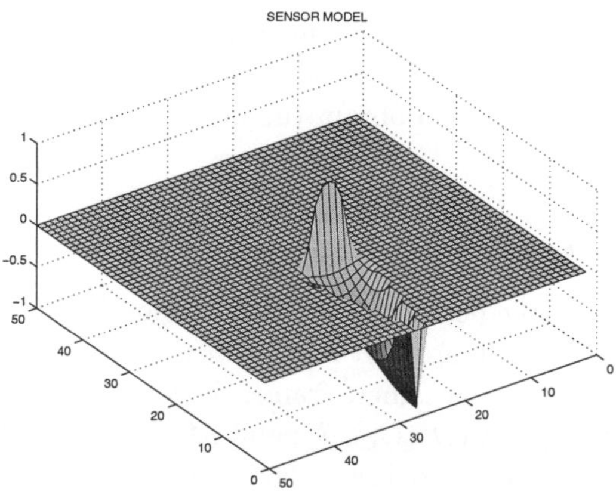

Fig. 2. Occupancy grid for the Moravec Elfes probabilistic sensor model. The sensor is placed at (30,0), and the obstacle is detected at (30,20)

3. Vision

3.1. *SIFT*

The Scale Invariant Feature Transform is a method for extracting distinctive invariant features from digital images. The features are invariant to scaling and rotation. They also provide a robust matching across a substantial range of affine distortion, change in 3D view point, addition of noise, and change in illumination. Furthermore, the features are distinctive, i.e. they can be matched with a high probability with other features in a large database with many images.

The SIFT algorithm consists of the following mayor steps.

- Scale-space peak detection: The aim of this step is to find locations in the image that are invariant to scale change in the same image.
- Accurate key-point localization: In this step the position of each point candidate is determined; points with low contrast and poor localization along the edge are removed.
- Majority orientation assignment: This step makes the rotation descriptor invariant. This is done by assigning a consistent orientation to each key-point.
- Computation of the local image descriptor: This step associates each feature point with a 128-element feature vector or interest point descriptor that uniquely identifies that point.

Once the descriptors are found in each image, i.e., left and right images, a robust matching algorithm is applied in both images. Stereo triangulation is implemented in order to obtain the depth from the pair of cameras to the features.

3.2. *Camera Model*

The model used is the perspective or pinhole model. This model can be seen in the left part of the figure 3. It represents the camera by its optical center C_l, an image plane π, the camera frame, the focal length f, the optical axes and the principal point (O_x, O_y). A point $M' = [X, Y, Z]$ in a $3D$ real world coordinate frame is projected into the image plane or camera frame as the vector $m_l = [x, y, z]$.

3.3. *Triangulation*

The triangulation algorithm outlined in Trucco and Verri[13] has been implemented in order to obtain the depth of the matching SIFT descriptors. The triangulation is straightforward since the intrinsic and extrinsic parameters are known in advance. The problem of this algorithm is to determine the midpoint of the vector M' of the segment parallel to the vector n that joins $l = am_l$, and $r = T + bR_{stereo}^T m_r$,where $a \in \mathbb{R}$, and $b \in \mathbb{R}$, as it is shown in fig 3. The $3D$ coordinates of each descriptor can be obtained by solving the equation (5), and the coefficients a, b, c can be obtained by solving the linear system (6). The Z component in equation (5) represents the depth to the point M'.

$$M' = [X, Y, Z]^T = am_l + \frac{1}{2}c(m_r \times m_r) \tag{5}$$

$$am_l - bR^T_{stereo} + c(m_l \times R^T_{stereo}m_r) = T \tag{6}$$

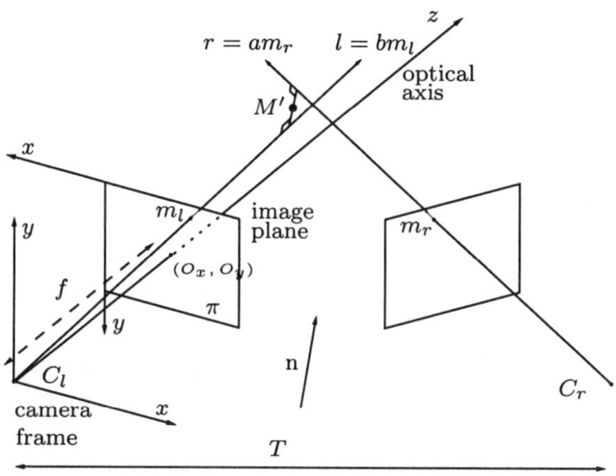

Fig. 3. The figure shows in the left part the perspective or pinhole model and also the triangulation

R_{stereo} and T are defined as the intrinsic and extrinsic parameters of the stereo system, where, $R_{stereo} = R_l R_r^T$ and $T = T_r - R^T_{stereo}T_l$. ($T_l$, R_l) and (T_r, R_r) are the intrinsic and extrinsic parameters of the two cameras in the world reference frame, the former parameters have been obtained from the matlab Camera Calibration Toolbox using single camera calibration. m_l and m_r are the projection vectors of the $3D$ point into the left and right image planes respectively.

3.4. *Stereo Triangulation Error*

Due to the factors of quantification and calibration errors, a certain degree of uncertainty must be expected in the triangulation. Mathies and Shafer[14]

shows how to model and calculate the triangulation error in stereo matched with $3D$ normal distributions.

In the following, the details of the triangulation error modelling are shown. Assuming a $3D$ point $M' = [X, Y, Z]$, which is projected onto the left and right image planes respectively as $m_l = [x_l, y_l]$ and $m_r = [x_r, y_r]$ as they can be depicted in figure 3. These vectors are considered to be normally distributed with means μ_l and μ_r and covariance matrices V_l and V_r. The covariance matrix V_p of the point M' is shown in equation 7.

$$V_p = J \begin{bmatrix} V_l & 0 \\ 0 & V_r \end{bmatrix} J^T \tag{7}$$

Where J is the jacobian of first partial derivatives of $f(m_l, m_r) = M'$. These equations can be obtained by solving the equation (5).

4. Sensor Fusion

4.1. *Sensor Data Fusion Architecture*

Joris W.M., Krose J.A. and Groen C.A.[10] define a SDF (Sensor Data Fusion) architecture, which is similar to the one proposed by Elfes,[2] Whyte,[15] Mayica and Durrant-Whyte.[16] In this architecture the WMR should be able to fuse different readings from different modalities of sensors. Fusion can be applied at any stage in the hierarchy architecture of the fusion process. In this way, the readings of the different kind of sensors must be converted to the internal representation. This internal representation is common to all the sensors. The fusion of the data of all the sensors is performed in this internal representation. The use of the SDF architecture is proposed to fuse sonar readings with features from stereo vision.

4.2. *Bayes Update Formula*

Elfes and Matthies[1] have proposed in their previous work the use of a recursive Bayes formula to update the occupancy grid for multiple sensor observations $(s_1, ..., s_n)$. When the former is transferred to the occupancy grid framework, the following is obtained:

$$P(occ|s_r) = \frac{P(s_r|occ)P(occ)}{P(s_r|occ)P(occ) + (1 - P(s_r|occ))(1 - P(occ))} \tag{8}$$

In the above formula.
s_r is the relevant evidence giving by the sensor reading. In the case of the

sonar it is the sonar reading R and in the case of the stereo system it is the depth Z from the triangulation algorithm.
$P(occ)$ and $1 - P(occ)$ are the prior probabilities that a cell is occupied or empty respectively, obtained from an existing map.
$P(s_r|occ)$ is a conditional probability given by the probabilistic sensor model.

5. Results

Some simulations of a WMR are shown in this section, with the SDF architecture implemented. Two sonars as well as a stereo system have been placed on the front of it, and two occupancy grids were defined, one for the sonar readings and one for the stereo system. In these experiments it is assumed that the robot is in static position. In the first experiment the two sonars measure and fuse two different readings. Figure 4 shows the two sonars readings which have been fused using the recursive Bayes formula and it also shows the occupied, empty and unknown probabilities. In the second experiment, a pair of images have been taken from the stereo system set up. Matching feature descriptors from stereo image pair have been identified; this can be seen in Figure 5. The green lines indicate the connection between the descriptor matches from the SIFT algorithm. $3D$ reconstruction by triangulation has also been implemented on the matching descriptors in order to obtain the depth measurements. Stereo triangulation error wich estimates the uncertainty of each descriptor was fused by using the recursive Bayes formula. Fig 6 shows the result of fusing the sift descriptors uncertainties.

The decision rule as shown in the formula 9 has been implemented between the two sets -the sonar reading and the sift reading descriptor sets- in order to reduce the uncertainty.

$$
\begin{gathered}
if(r_{sonar} > 0.5)\ then \\
0.5 + \frac{(2(r_{sonar}-1))(2(r_{sift}-1))}{2} \\
if(r_{sonar} < 0.5)\ then \\
0.5 + \frac{(r_{sonar})(r_{sift}-1)}{(r_{sonar})(r_{sift})+(r_{sonar}-1)(r_{sift}-1)} \\
if(r_{sonar} = 0.5)\ then \\
r_{sift}
\end{gathered}
\tag{9}
$$

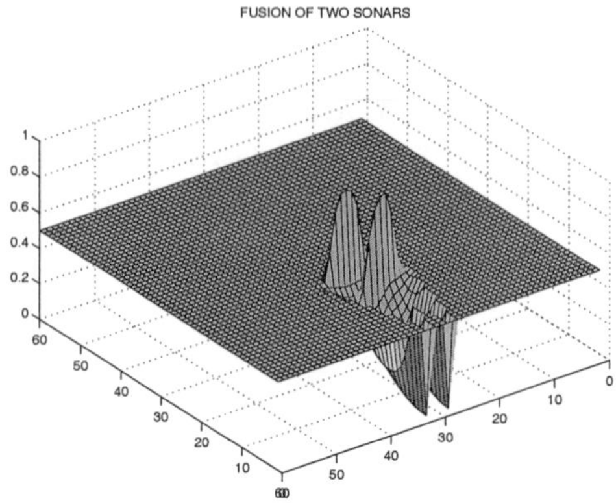

Fig. 4. Fusion of two sonars

Fig 7 shows the results. It can be seen that this fusion has indeed led to reduction of uncertainty in the estimated proposition.

6. Conclusion and Future Work

This article presents an application of sensor fusion in a SDF architecture and the use of the Bayes recursive theorem for updating the occupancy grid. This work also considers the uncertainties produced by the sonar as well as the ones produced by the $3D$ triangulation from the SIFT descriptors. Sonar readings were fused with the features produced by the SIFT algorithm. It has also been shown that under this fusion, the uncertainty as well as the specular reflexions have been reduced. The fusion process has been implemented using two sonars and a single stereo picture. It was demonstrated that the uncertainty can be reduced by fusing sonar readings with SIFT descriptors. Further research could be to fuse readings from a sonar ring with several pictures of the same environment in order to construct a precise world model. This fusion can also be combined with other sensors e.g. laser range finding equipment.

Fig. 5. Descriptor matches left and right images

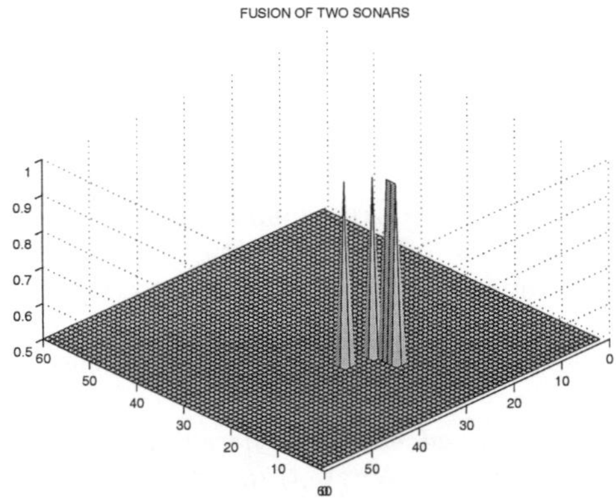

Fig. 6. Fusion of the SIFT descriptors using the bayes recursive formula

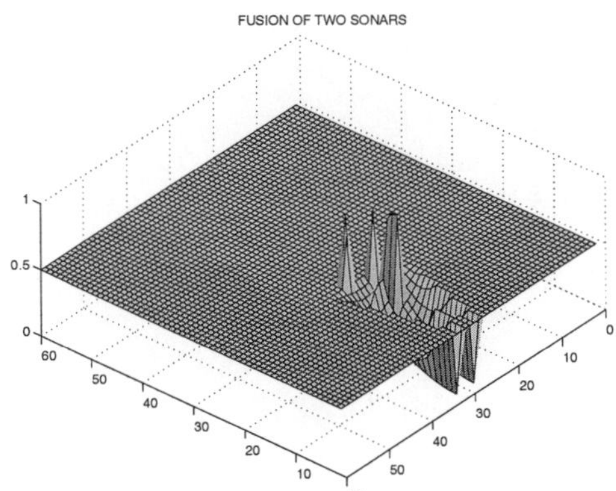

Fig. 7. Uncertainty decrement

References

1. H. P. M. A. Elfes, *IEEE International Conference on Robotics and Automation* , 116 (1985).
2. A. Elfes, *IEEE Journal of Robotics and Automation* (1989).
3. A. Elfes, *Prceedings Nasa Publications* **N90**, 341 (1989).
4. L. M. Alberto Elfes, *IEEE Conference on Decision and Control* **3**, 1802 (1987).
5. D. G. Lowe, *Computer Science Department, University of British Colombia, Vancouver, B.C., Canada* **21**, 1150 (1999).
6. D. G. Lowe, *Computer Science Department, University of British Colombia, Vancouver, B.C., Canada* **2**, 91 (2004).
7. K. L. S. Ren C. Luo, Chin-Chen Yih, *IEEE Sensors Journal* **19**, 107 (2002).
8. M. G. K. Ren C. Luo, *IEEE Transactions on Systems, Man, and Cybernetics* **19**, 901 (1989).
9. E. M. N. Daniel Pagac and H. Durrant-Whyte, *IEEE Transactions on Robotics and Automation* **14**, 623 (1998).
10. J. W. K. J. G. C.A., *Proceedings of the IEEE* (1996).
11. L. Matthies and A. Elfes, *IEE Autonomous Robots* (1988).
12. K. Konolige, *Autonomous Robots* **4**, 351 (1997).
13. E. Trucco and A. Verri, *Introductory Techniques for 3-D Computer Vision* (Prentice Hall.
14. L. Matthies and S. A. Shafer, *IEEE Journal of Robotics and Automation* **3**, 239 (1987).
15. H. F. D. Whyte, *Kluer Academic Publishers* (1988).
16. J. Mayika and H. F. Durrant-Whyte, *Ellis Horwood publishers* (1994).

MODELING AND ANALYSIS OF A MICROMACHINED TACTILE SENSOR FOR MINIMALLY INVASIVE SURGERY

MOHAMMAD AMEEN QASAIMEH, ION STIHARU AND JAVAD DARGAHI

CONCAVE Research Centre, CONCORDIA University 1455 de Maisonneuve Blvd., Montreal, QC. H3G 1M8, Canada

m_qasai@encs.concordia.ca, istih@vax2.concordia.ca, jdargahi@alcor.concordia.ca

In this paper the design and simulation of a new tactile array sensor for minimally invasive surgery (MIS) is presented. Using MEMS technology, this sensor can be integrated with the current commercial endoscopes tips of the grasper. The designed sensor can detect magnitude and the position of applied forces on the endoscope jaw. Current available commercial endoscopes are not equipped with tactile feedbacks. The presented sensor could be used as object imaging, where a data processing of the feedback signal could detect the mechanical properties and the shape of the grasped tissues. The designed sensor is built up from three layers. The upper layer is fabricated from monocrystalline silicon to form the teeth shape, the middle layer is the sensing element which is fabricated from Polyvinylidene Fluoride (PVDF) film, and the lower layer is a supporting substrate to support the PVDF layer. The designed sensor consists of eighteen sensing elements distributed as a matrix of six by three on the area of one jaw, where the length and the width of this sensor are two millimeters and one millimeter respectively. In other array sensors, the regions between the neighboring sensing elements are not active. In this sensor, all surface points of this designed sensor are practically active. Simulation results show that for any applied force, the magnitude and the position of this force can be detected; also good linearity between voltages on the sensing layer with respect to the applied force on the teeth layer is reached. This sensor as proposed is a good candidate for batch micromachining, which is yet another commercial advantage for this design and, because of its cheap manufacturing costs; the surgeon can use it as a disposal part of the endoscope tool. Thus re-sterilization is not required leading to reduced in cost and safer surgery.

1. Introduction

Recently, many open surgical procedures have been replaced by minimally invasive surgery (MIS) which requires only a small incision to gain access to the internal organs [1, 4, 6, 9]. Through the very small incisions associated with the micro-surgery, the endoscope and the mechanism of surgery tools are inserted to manipulate the tissues and viewing the surgery site on a screen. MIS usually results in less pain and strain on the organism and the patient, less scarring due to smaller incisions. The other benefits are the short recovery time for the patient, as well as reduced health-care costs due to shorter hospitalization time. Some of disadvantages currently associated with MIS is the loss of tactile force and location perception , the reduced dexterity of surgeon due to restricted vision, difficult handling of the instruments and the very restricted mobility [4, 6, 9].

In most of the MIS procedures, the surgery is performed on very small tissues at micro scale and these tissues are very soft and sensitive to any excessive force; may yield to undesired results. Moreover, hidden and hard to access locations make the surgery more difficult and highlights the need for readiness of tactile sensing [13]. To make MIS and microsurgery more efficient, a feedback signal must be provided to the surgeon to enable him to feel the applied force, to know the position of force, to detect the stiffness of tissues, to predict the shape of the grasped tissues, and to sense the presence of the blood vessels and ducts during the surgery procedure [6, 7, 9, 22]. Sensing is most of the time associated with an electric output due to the portability of the signal.

Piezoelectric materials are mostly ceramics which generates a voltage when it is stressed by a force and the materials for making this type of sensor exhibits high stiffness comparable to steel. Therefore, the deformation of the sensing element in a piezoelectric force sensor will be much smaller than in other measuring systems. The high rigidity of piezoelectric force sensors also provides an inherently high natural frequency and short rise time. This permits the measurement of extremely fast events. Some of the positive properties of the piezoceramics are found in piezo-electric polymers. Largely used for low dynamic forces is PVDF which is an ideal sensing material because of its responsiveness to a wide range of frequencies, relatively high mechanical strength, and high sensitivity [5, 25].

The endoscope graspers should be set with teeth to be able to grasp slippery and soft tissues associated with endoscope operations [6, 9, 24].

Many papers have presented design of force sensors limited just to μN measurements and therefore, not adequate for use in MIS application [19-21, 23]. Also, several tactile sensor designs were addressed but were limited to measuring only tens of mN of forces [14-15, 18]. Some tactile imagers were fabricated using complex structures associated with combining several technologies [16]. There has been an attempt to realize tactile sensor capable to find the position of applied force, but was limited to measuring the position in one direction only [17].

A number of publications were concerned with designing a tactile sensor especially for use in MIS. Other built devices were able to find the force and the compliance of the sensed tissues but were poor at finding the exact location in two dimensions of applied force and were limited to milligram force detection [7, 8, 10, 11]. Some of the work was to design a membrane based tactile sensor for MIS to measure the magnitude of applied force and the polar direction of position [12]. However, this design cannot be used for grasping tissues because it does not possess a teeth-like structure. Interest in the tactile sensor ability to find the magnitude of the applied force, in addition to the position of applied force and the pressure distribution for MIS tactile imagers were early developed by researchers. Finite element methods were applied in parallel development works [6, 9, 22, 24], but none of the proposed design has the ability to measure the position in two direction of the applied force.

This present paper presents a new concept of a tactile sensor for micro-surgery and MIS that can be integrated with the endoscopes grasper tools as shown in Figure 1., this design is able to find the force of maximum 1N, the X-Y position of the applied force along the area of the grasper in increments, and the pressure distribution created by forces. Moreover, there are no dead points on this design such that all of points are active and any applied force at any location on the jaw can be detected and measured.

The designed sensor has eighteen teeth, micro-machined from monocrystalline silicon by bulk etching.

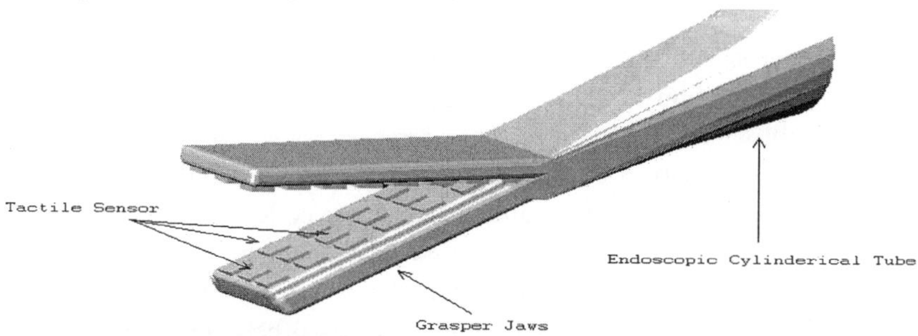

Figure 1. The endoscopic tool with integrated tactile sensor.

Under each tooth there is an active region of the PVDF layer, the applied force on the silicon layer transfer to PVDF film. Further, the charges are collected via conductive strips to feedback its voltage output in order to measure the location and the magnitude of the applied forces.

2. The Sensor Design

The sensor design comprises three layers. The upper layer is micro-machined from silicon wafer (100 orientation) using MEMS technology. It is anisotropically etched from both sides to form the rigid teeth-like structure. These rigid teeth are able to grasp the slippery tissues within micro-surgery operation while the mirror at the bottom is to make sure that the pressure transmitted to the sensing layer is focused in the specific region defined by the tooth and to minimize the effect of this transmitted pressure to the other regions beside. The micro-machined silicon layer is etched to form eighteen teeth on each side; each tooth has a trapezoidal prism shape, squares at the top and bottom. The dimension of its top is 134 μm, and the bottom is 268 μm. Its height is 95 μm due to the anisotropic etching angle (54.74°). The silicon wafer thickness is 475 μm, which is the most commercially available thickness which is another important factor in the cheap manufacturing cost of this sensor. The interval between tooth and the other in length side of layer is 78.4 μm, and in the width side is 98 μm, all of these dimensions have been calculated based on the assumption that the maximum force that can be detected is 1 Newton for a safety factor of 1.5. The structure of the silicon is shown in Figure 2.

The lower layer is a supporting substrate to the middle layer and prevents it from moving. It is made from rigid polymer material, and has the same dimensions of the silicon layer with a thickness 250 microns. All of the force is carried out and resisted by the upper layer and transmitted to the middle layer to be sensed.

The middle sensitive layer is sandwiched between the two structures. It consists of a PVDF film of 25μm thickness and it is this that creates a charge when it is exposed to normal and shear force stresses. The top and bottom surface of the PVDF film is covered by aluminum during manufacturing and is patterned on each side to create aluminum strips for collecting created charges. The top and bottom side of the aluminum cover is patterned to form squares of 134μm length and which are spread out on both sides as a matrix of six columns and three rows, the rows are spaced by 98μm interval between each other and columns are spaced by 78.8μm, each row and column located at edges are shifted by 67 μm from edge to locate strips exactly under teeth. Each PVDF region covered by this squares of aluminum from both side is active and located exactly under each tooth. So under each tooth there is an active PVDF region covered by aluminum coating from each side, and the active region is presented by two patterned parallel aluminum squares on the top and bottom of the PVDF layer of which each is 134μm in length. The connecting wires are patterned also from coatings; each connecting wire on both sides has its own path without any intersection. Care is taken to avoid any parallel connections on top and bottom to avoid forming other active regions rather than squares active regions. The width and length dimensions of the PVDF layer is little larger than other layers to enable the integration of the wires and the contact pads as illustrated in Figure 2., the other regions of the PVDF layer are not active because the created charges on it cannot be collected since no conductor covers it. The complete structure of the designed sensor and the dimensions is shown in Figure 2.

Any applied force at any point on the silicon layer is transmitted to the PVDF layer which creates charges at the active regions. These are collected by the aluminum strips. The magnitude of the measured voltage at various active regions is proportional to magnitude of the applied force. By comparing these magnitudes of the created voltages at various regions and the slope for each created voltage, the position of the applied force can be detected.

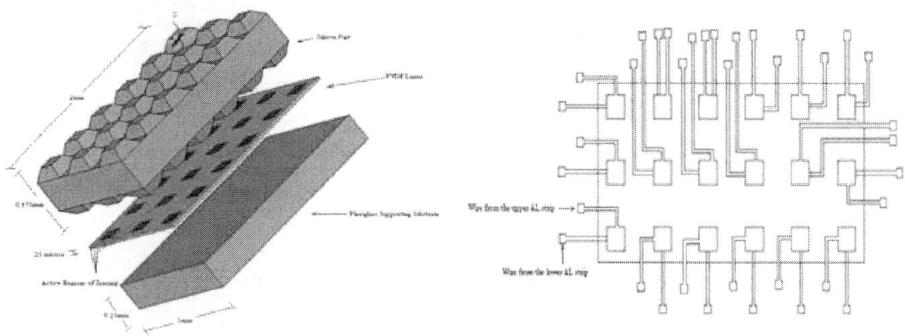

Figure 2. The structure of the designed sensor (to the left), the PVDF film, patterned aluminum strips, connecting wires, and the connecting pads (to the right).

3. Finite Element Modeling

The performance of the proposed sensor was modeled and further simulations were carried out using ANSYS10.

The silicon part, PVDF layer and Plexiglass layer are meshed as shown in Figure 3. , the silicon and rigid polymer parts are added to PVDF element as super element layers.

For the silicon and plexiglass parts, SOLID95 element was used, the advantages of SOLID 95 are: plasticity, creep, swelling, stress stiffening, large deflection, large strain capabilities, and a high-order version. It can tolerate irregular shapes without much loss of accuracy.

For the PVDF part, a coupled field SOLID227 was used. Coupled-field 10-node tetrahedron is a coupling element between structural and electrical fields and its properties are very appropriate for this model.

The top layer of the grasper jaw was made from silicon, while the bottom is made form plexiglass. Their properties are given in the literature [9, 26].

The beta phase of the PVDF layer forms crystal symmetry C_{2v}, and the d-form of its piezoelectricity properties is denoted by the following matrix [9]:

$$d_{ij} = \begin{bmatrix} 0 & 0 & 0 & 0 & d_{15} & 0 \\ 0 & 0 & 0 & d_{24} & 0 & 0 \\ d_{31} & d_{32} & d_{33} & 0 & 0 & 0 \end{bmatrix}.$$

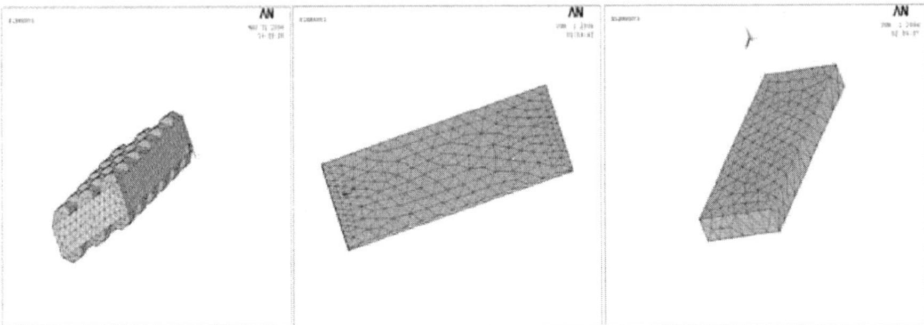

Figure 3. The meshed layers of silicon, PVDF and plexiglass, respectively.

The output charge of the PVDF film is dependent on the applied force on the sensing element and proportional to the electrode area of the sensing elements [9] as shown in Eq. (1):

$$q = F(\phi_1 d_{31} + \phi_2 d_{32} + d_{33}) \qquad (1)$$

Where q is the output charge, F is the applied force on PVDF film (transmitted from upper layer), Φ_1 and Φ_2 are constants proportional to the electrode area of the sensing elements, d_{31}, d_{32} and d_{33} is the piezoelectricity properties of the PVDF in the drawn, transverse, and thickness directions, respectively [9]. The piezoelectric coefficient and the mechanical proprieties are given in the literature [26].

4. Simulation Results

The maximum load used for examining the output is (1 N); this load was examined as acting at different teeth, individually. The force applies pressure to the top of the tooth and transmitted to press on the PVDF and create charges. Maximum pressure was applied just on one tooth creating an average peak voltage of 2 Volts as shown in Figure 4., another case was examined when the maximum pressure was divided between two teeth located on the middle, each one carried 1/2 N, result show that under each tooth an average voltage of one volt was created, as shown in Figure 5.

Figure 6. show the Von Mises stress distribution on the PVDF layer when the maximum load is divided between two teeth. The Von Mises stress for the silicon part when maximum load was applied on one tooth is shown in Figure 7.

The created voltage was verified to see the linearity of the PVDF film output with respect to the applied force. Various loads were applied at the same point and the maximum created voltage was recorded. Table 1. shows the magnitude of the applied load and the maximum created voltages. The linearity of the PVDF layer is shown in Figure 8.

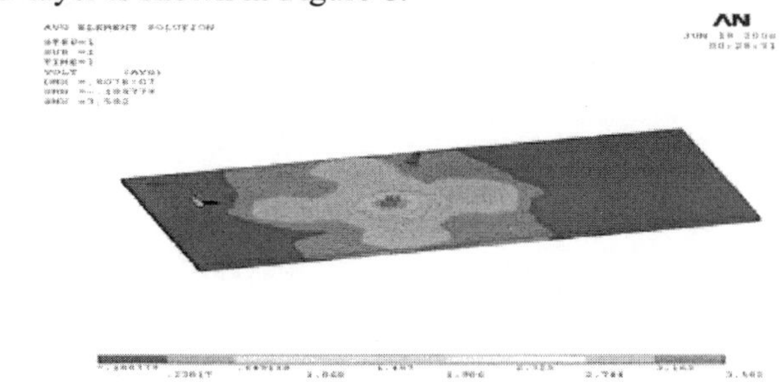

Figure 4. The voltage distribution when pressure is applied on one tooth located in the middle.

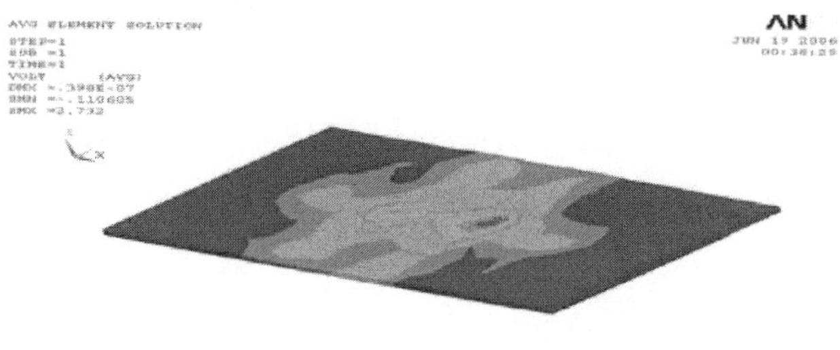

Figure 5. The voltage distribution when two teeth carried the maximum pressure.

From the figures shown, an average value of created voltage is about two volts when maximum force is applied on one tooth, and when the maximum load is carried by more than one tooth, each sensing region under each tooth creates an output, the sum of which is two volts.

Table 1. The Maximum created voltage on PVDF versus different loads

Force (N)	1	0.75	0.5	0.25	0.125	0.0625	0.01	0.001	0.0005
Max. voltage (Volts)	3.929	2.947	1.965	0.98233	0.49116	0.245583	0.0392	0.003929	0.002222

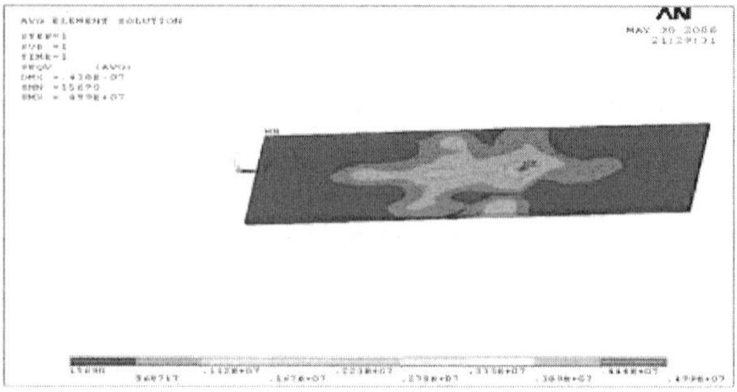

Figure 6. The Von Mises stress distr. on PVDF when the $load_{Max}$ is divided between two teeth.

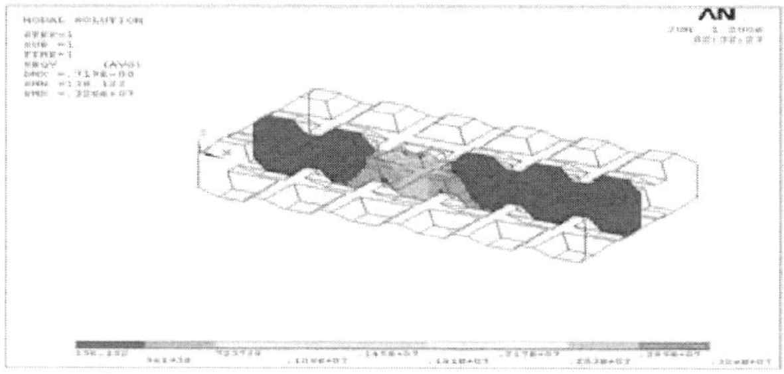

Figure 7. The Von Mises stress dist. on the Si layer when the $load_{Max}$ is applied on one tooth.

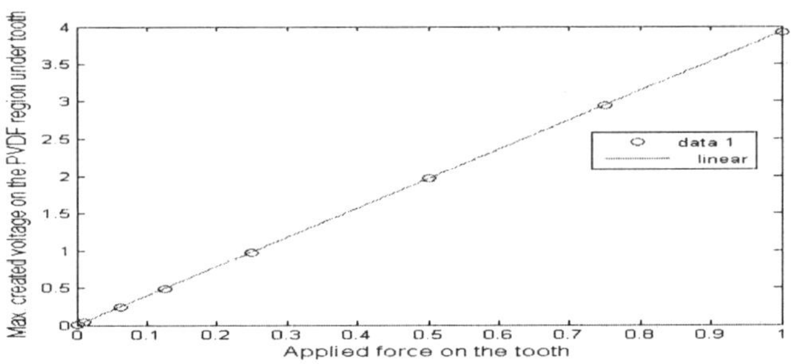

Figure 8. The created voltage versus different magnitude of applied forces.

5. Conclusion

The presented sensor is designed using MEMS technology that makes it relatively simple to fabricate and assemble. Packaging issues are still to be addressed. The process is carried out for the same wafer to produce large batches, which greatly reduces manufacturing costs. The designed sensor could be integrated with the commercial endoscope tips to feedback the applied force information to the surgeon to complete the surgery procedure in safe mode without any exceeded force on the soft tissues associated with the MIS surgery.

The presented sensor, unlike other designs, can measure the magnitude and the position of applied force in a two dimensional jaw plane at any point (no dead-areas). By comparing the voltage outputs of the active regions of the middle layer, the location of application can be detected. Also, since the pressure distribution of the applied forces can be found, this sensor could be used to provide the shape image of the grasped tissues. The designed sensor represents one jaw of the grasper, the other jaw is just one layer fabricated from silicon to form the like-teeth shape. It is this shape that plays such an important role in grasping soft and slippery tissues associated with MIS surgery procedures.

Therefore, both parts of the jaws could be used as the disposal part of endoscopic tips due to low manufacturing costs, and also reduce the cleaning problems. The sensing material improves sensor performance with high durability. It supports a wide range of frequencies, provides an inherently high natural frequency. It has very good mechanical proprieties, high sensitivity, and high signal-to-noise ratio is expected.

Simulation results show excellent ability of the designed sensor to detect the magnitude and 2-D position of the applied force, and show the ability to detect the pressure distribution of the applied forces. Also the simulation show very good linearity of the created voltage with respect to the applied forces. Feedback of this sensor could be displaced or be sensed by a surgeon via haptic interface.

References

1. F. Tendick, S. Sastry, R. Fearing, and M. Cohn, IEEE/ASME Transactions on Mechatronics **3**, 34 (1998).
2. H. Melzer, M. Schurr, W. Kunert, G. Buess, U. Voges, and J. Meyer, J. Endoscop. Surgery **1**, 165 (1993).
3. F. Tendick, T. Mori, and L. Way, PA Churchill-Livingston, (1995).
4. J. Wendlandt, UC Berkeley Electronics Research Laboratory, Memo No. UCB/ERL 94/7, February, (1994).
5. Z. Lu, P. Chen, and W Lin, IEEE Transactions on Systems, Man, and Cybernetics, Part C. In press, (2005).
6. J. Dargahi, M. Parameswaran and S. Payandeh, Journal of MEMS IEEE **9**, 329 (2000).
7. R. Sedaghati, J. Dargahi, and S. Harpiyar, International Journal of Solids and Structures **42**, 5872 (2005).
8. J. Dargahi and S. Najarian, Can. J. Elect. Comput. Eng. **28,** 155 (2003).
9. J. Dargahi, S. Najarian, and K. Najarian, Can. J. Elect. Comput. Eng. **30**, 225 (2005).

10. J. Dargahi, S. Najarian, and X. Zheng, Int J Med Robotics Comput Assist Surg. **2,** 84 (2006).
11. M. Tanimoto, F. Arai, T. Fukuda, H. Iwata, K. Itoigawa, Y. Gotoh, M. Hashimoto, and M. Negoro, The 11th Annual International Workshop on MEMS, Heidelberg, Germany, January 25-29 (1998).
12. J. Dargahi and S. Najarian, Sensor Review **24,** 284, (2004).
13. T. Mustufa, M.Sc. thesis, The Johns Hopkins University in conformity, Baltimore-Maryland, September (2005).
14. E. Kolesar, C. Dyson, R. Reston, R. Fitch, D. Ford and S. Nelms, 8th Annual IEEE International Conference on Innovative Systems in Silicon, Austin, Texas, USA, October 9-11. (1996).
15. E. Kolesar, R. Reston, D. Ford, and R. Fitch, Jr., IEEE International Conference on Systems Engineering, Dayton, Ohio, August 1-3 (1991).
16. Z. Wen, Y. Wu, Z. Zhang, Sh. Xu, Sh. Huang, Y. Li, Sensors and Actuators A: Physical **103**, 301 (2003).
17. J. Son, E. Monteverde, and R. Howe, 1994 IEEE International Conference on Robotics and Automation, San Diego, California, May 8-13 (1994) .
18. B. Gray, R. Fearing, 1996 IEEE International Conference on Robotics and Automation, Minneapolis, Minnesota, April (1996).
19. E. Winder, Y. Shen, N. Xi, W. Sheng, U. Wejinya, and C. Pomeroy, 2005 IEEE/ASME International Conference on Advanced Intelligent Mechatronics Monterey, California, July 24-28 (2005).
20. Y. Shent, N. Xi, W. Li, 5th IEEE International Symposium on Assembly and Task Planning, Besançon, France, July 11 (2003).
21. C. Fung, I. Elhaj ,W. Lil, and N. Xi, IEEE International Conference on Robotics & Automation Washington, DC, May (2002).
22. J. Dargahi and S. Najarian, Industrial Robot: An International Journal **32,** 268, (2005).
23. Y. Sun', D. Potasek, D. Piyabongkam, R. Rajamani, and B. Nelsont, 2003 IEEE International Conference on Robotics & Automation Taipei, Tairsn, September 14-19 (2003).
24. H. Singh, R. Sedaghati and J. Dargahi, 2nd IEEE International Conference on Sensors, Toronto, Canada, October 22-24 (2003).
25. R. Chen and B. Wang, INSTITUTE OF PHYSICS PUBLISHING, Smart Mater. Struct. **13**, 791 (2004).
26. Good Fellow company - USA, retrieved May (2006) from the web: http://www.goodfellow.com/csp/active/gfHome.csp.
27. I. Lee, H. Sung, Experiments in Fluids **26,** 27 (1999).
28. R. Sedaghati, J. Dargahi, H. Singh, International Journal of Solids and Structures **42,** 5872 (2005).

CHARACTERIZATION OF FINGERPRINTS USING TWO NEW DIRECTIONAL MORPHOLOGICAL APPROACHES

LUIS A. MORALES-HERNÁNDEZ*
*Facultad de Ingeniería, Universidad Autónoma de Querétaro, Río Moctezuma 249
San Juan del Río, Qro., México, 76807

IVÁN R. TEROL-VILLALOBOS**, AURELIO DOMÍNGUEZ-GONZÁLEZ*, GILBERTO HERRERA-RUIZ*
**CIDETEQ, SC. Parque Tecnológico Querétaro S/N Sanfandila
Pedro Escobedo, Qro., México, 76700

Abstract

The present paper is focused on the characterization of the fingerprints using directional morphological transformations. The interest of the proposed approaches to compute orientations field is not only useful in computing fingerprint orientation patterns but also for characterizing other structures containing anysotropies (for example, some microstructures in material images). Two approaches are investigated in this paper. The first one is a global approach based on the directional granulometries using line segments as structuring elements. Then, the notion of quatree structure is used to go from a global approach to a local one. The second method considers a local approach by using the concept of distance function. The distance function is computed by the supremum of directional erosions. It begins with a small structuring element by taking into account all orientations. Then the methodology increments the size of the structuring element until the structure is completely removed. The maximum of the distance function contains the sizes of the greater lines that can be included in the structure. To know the direction of the lines, a second image containing the orientations is built when the distance function is computed. Finally, both images, the distance function and the orientation function are used to estimate the lines at different orientations to characterize the fingerprint.

Key words: Fingerprints, identification, image, morphology granulometries, quadtre, , multiscale.

1 Introduction

Fingerprints are today the most widely used in biometric features for personal identification. With the increasing emphasis on identity management, automatic fingerprint recognition has commercially received wide attention. Nevertheless there still exist critical research issues such as a long processing time in large database and dealing with a required improvement to fingerprint classification and identification. In both topics, orientation patterns, which can be defined by a local direction of ridge-valley structures, this plays an important role [1-4,17]. Several methods have been proposed to improve the estimation of the orientation field, which can broadly be categorized as filtering-based [1,2, 18] and model-based [5-9]. The present paper is focused on the use of the mathematical morphology methodology to model orientation fields. Particularly, the study will focus on the characterization of the fingerprints by using the directional morphological transformations. In fact, since the fingerprints can be considered as a structure composed by a set of line segments, a bank of filters which is composed by directional morphological transformations permit to extract the main orientations of the image. This is similar to the perception of the orientation of line segments by the human brain. In order to achieve an image processing oriented to image structures, two approaches are investigated in this paper. The first one is a global approach based on the directional granulometries, computed by morphological openings using directional structuring elements also called line segments. This approach allows to determine the main directions of the structures by studying the minima of the granulometric density function. In order to define a local approach, a quadtree structure is used to decompose the image with a different resolution that is according to the levels of the tree. Thus, it is possible to obtain a multi-scale local approach to define descriptors that take into account size (scale) and orientation. The second method considers a local approach by using the concept of distance function. In our case, the distance function is computed by the supremum of directional erosions. After that, the maxima of the distance function contains the information of the biggest structuring elements (line segment) that can be placed inside the structure. In order to know the orientations, a second image is computed by detecting the orientation of the supremum of directional erosions. These local descriptors for the size and orientation provide an excellent reconstruction of the orientation and allow to describe the model of the fingerprint orientation pattern in a piecewise manner. Compared to the prior works, this algorithm is able to predict orientation in a more efficiently way because the proposed approach does not need to determine the core of the image. In previous works, more complex algorithms that use the geometric theory of differential equations or constrained nonlinear models that require optimal estimators have been proposed.

2 Some basic concepts of mathematical morphology

Morphological filters are increasing and idempotent transformations [10-12,16]. While the increasing property expresses that the order is preserved, one says that a transformation ψ is idempotent if and only if for all function f, $\psi(\psi(f)) = \psi(f)$. The basic morphological filters are the morphological opening $\gamma_{\mu B}$ and the morphological closing $\varphi_{\mu B}$ with a given structuring element μB, where in this work, B is an elementary structuring element (3×3 pixels) that contains its origin. $\breve{B}$ is the transposed set ($\breve{B} = \{-x : x \in B\}$) and μ is an homothetic (scale) parameter. In this work, the homothetic parameter only takes integer values. The morphological opening is an anti-extensive filter and the morphological closing is an extensive filter. These transformations are expressed by means of the morphological dilation $\delta_{\mu B}$ and morphological erosion $\varepsilon_{\mu B}$. Thus,

$$\gamma_{\mu B}(f)(x) = \delta_{\mu B}(\varepsilon_{\mu B}(f)) \quad \text{and} \quad \varphi_{\mu B}(f)(x) = \varepsilon_{\mu \breve{B}}(\delta_{\mu B}(f)) \quad (1)$$

The morphological erosion and dilation are respectively expressed by: $\varepsilon_{\mu B}(f(x)) = \bigwedge\{f(y); y \in \mu\breve{B}_x\}$ and $\delta_{\mu B}(f(x)) = \bigvee\{f(y); y \in \mu\breve{B}_x\}$, where $\bigwedge$ is the infimum operator ($\bigvee$ is the supremum operator). Another interesting class of filters is composed by the openings and closings and closings by reconstruction. When filters by reconstruction are built, the basic geodesic transformations, the geodesic dilation and geodesic erosion of size 1 are iterated until idempotence is reached [13,18]. Where the geodesic dilation and the geodesic erosion of size one are given by $\delta_f^1(g) = f \wedge \delta_B(g)$ with $f \geq g$ and $\varepsilon_f^1(g) = f \vee \varepsilon_B(g)$ with $f \leq g$, respectively. When the function g (the marker) is equal to the dilation or to the erosion of the original function by a given structuring element, the closing and the opening by reconstruction are obtained.

$$\tilde{\varphi}_{\lambda B}(f) = \lim_{n\to\infty} \varepsilon_f^n(\delta_{\lambda B}(f)) = \varepsilon_f^1 \varepsilon_f^1 \cdots \varepsilon_f^1(\delta_{\lambda B}(f))$$

$$\tilde{\gamma}_{\lambda B}(f) = \lim_{n\to\infty} \delta_f^n(\varepsilon_{\lambda B}(f)) = \delta_f^1 \delta_f^1 \cdots \delta_f^1(\varepsilon_{\lambda B}(f)) \quad (2)$$

3 Directional granulometry and supremum of directional erosions

In mathematical morphology there are some operators based on the detection of the residues of parametric transformations. Some examples are the ultimate erosion, the skeleton by maxima balls, the granulometry function and the distance function. These two last transformations are computed by means of the difference of successive openings or erosions, respectively. Generally, an associated function is linked to these transformations; for example in the granulometry residues the density and distribution functions are related to these residues. Here, the directional granulometry and a function associated to the directional erosions are introduced. Granulometry (anti-granulometry) was formalized by Matheron[16] for binary images and extended to complete lattices by Serra[10].

Definition 1. A family of openings $\{\gamma_\lambda\}$ (or respectively of closings $\{\varphi_\lambda\}$) where $\lambda \in \{1,...,n\}$, is a granulometry (respectively anti-granulometry) if for all $\lambda, \mu \in \{1,...,n\}$ and all function f, $\lambda \le \mu \Rightarrow \gamma_\lambda(f) \ge \gamma_\mu(f)$ (respectively $\varphi_\lambda(f) \le \varphi_\mu(f)$).

The granulometric curves can be computed from the granulometric residues between two different scales $\gamma_{\mu_i}(f) - \gamma_{\mu_j}(f)$ with $\mu_i \le \mu_j$. Then, one says that $\gamma_{\mu_i}(f) - \gamma_{\mu_j}(f)$ contains features of f that are larger than the scale μ_i, but smaller than μ_j. Since the interest in this paper is to introduce the notions of directional granulometry and a function computed by means of the supremum of directional erosions, two parameters are required to characterize them, the size of the structuring element λ and the angle α (direction) of this element. Thus, the set of points of a line segment is computed by:

$$\text{If } \alpha \le 45° \text{ then, } y_i = x_i \tan\alpha \text{ for } x_i = 0,1,...,\lambda\cos\alpha$$

$$\text{If } \alpha > 45° \text{ then } x_i = y_i \cot\alpha \text{ for } y_i = 0,1,...,\lambda\sin\alpha$$

Then, morphological opening and closing are given by:

$$\gamma_{L_{\alpha,\lambda}}(f) = \delta_{\check{L}_{\alpha,\lambda}} \varepsilon_{L_{\alpha,\lambda}}(f), \quad \varphi_{L_{\alpha,\lambda}}(f) = \varepsilon_{L_{\alpha,\lambda}} \delta_{\check{L}_{\alpha,\lambda}}(f) \tag{3}$$

where the morphological erosion and dilation are given by:

$$\varepsilon_{L_{\alpha,\lambda}}(f)(x) = \min\{f(y) : y \in L_{\alpha,\lambda}(x)\} \quad \delta_{L_{\alpha,\lambda}}(f)(x) = \max\{f(y) : y \in L_{\alpha,\lambda}(x)\} \tag{4}$$

When a directional erosion size λ is applied to an image at all directions, some output images computed for some directions contain more information of the original image than other ones, depending of the anisotropies presented in the image. For example, if the erosion is applied at all directions of the image in Fig. 1(a), the erosion at angle 112 removes less information than others. Then, the idea of applying directional erosions consists in preserving the maximum of information for all directions. To achieve this objective, the supremum of the erosions, as expressed by the following relationship, is computed.

$$\varepsilon_\lambda^{sup}(X) = \bigcup_{\alpha \in [0,180]} \varepsilon_{\alpha,\lambda}(X) \tag{5}$$

4 Fingerprint orientation field based on directional granulometry and quadtree structure

Recently, a work for fingerprint orientation field, based on constrained nonlinear models that require optimal estimators has been proposed [9,8,5,6]. While in mathematical morphology, an interesting work for directional filtering was proposed in [15]. Here, a morphological approach that uses the notion of directional granulometries and the quadtree structure is proposed in order to characterize anisotropies in the images. The Fig.1 (b) illustrates the distribution function of the image in Fig.1 (a) for $\lambda = 80$ and $0 \leq \alpha < 180$. The minimum of this function permits to determine the direction of the main structures. The minimum in Fig.1 (c) was computed from the function in Fig. 1(b) using morphological transformations in one-dimensional case. To carry out the minimum detection, the distribution function was transformed into the interval [0,255] in integer numbers and then, the traditional morphological tools for detecting minima in mathematical morphology was applied [12]. In fact, the minima of the image will enable to have a criterion to go from a global approach to a local one by means of the quadtree structure. In the quadtree approach, the coding by regions is made by an homogeneity criterion that enables to discriminate whether a square region can be considered a connected component. One starts with a square of 2^n pixels that is divided in four square zones. Each square zone is studied on the original image using one or several homogeneity criteria (variance, max-min values, ...). If the homogeneity criterion is verified, a function value is given at all points of the square region; for instance, the average of the intensity values in the square. For any square that does not verify the homogeneity criterion, a similar procedure is performed in a recursive way by dividing the square region by four. For orientation fields, it is clear that a homogeneity criterion is given by a directional one and in our case, the minima of the distribution function is used as criterion. If the distribution function in a square region presents only a principal minimum, then the region is considered homogeneous. In this case, the pixel values of the region are affected by the angle of the minimum where the minimum was found. Otherwise, if the distribution function of a square region has several representative minima, then the region is devided by four. The Fig 1(b) illustrates the approach to determine the orientation field in the image. After devising the image by four, their distribution functions were computed as illustrated in Fig 3. In particular, observe that the distribution functions in Figs. 3(c) and 3(d), correspond to the bottom right and left squares, both containing only one principal minimum, while the other two squares contain several representative minima. Thus, these two squares were divided by four and their distribution functions were computed to know their directional homogeneity. The image in Fig.2 shows the orientation fields (gray level). Figure 2(a) shows the orientations of the first hierarchy of the quadtree (four square zones), and Figs. 2(b) and 2(c) illustrate the up square regions divided by four squares regions. Finally, Fig. 2(d) illustrates the final hierarchy, while in Fig.2 (e), the regions were affected by a color in order to better illustrate the orientation fields. The color representation is the classical hue definition of the perceptual color spaces.

(a)

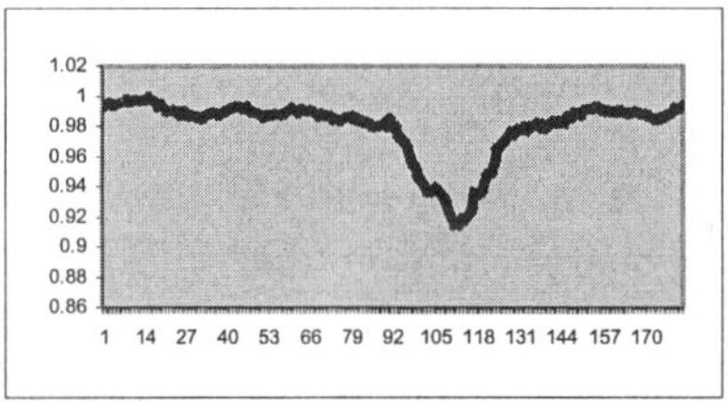

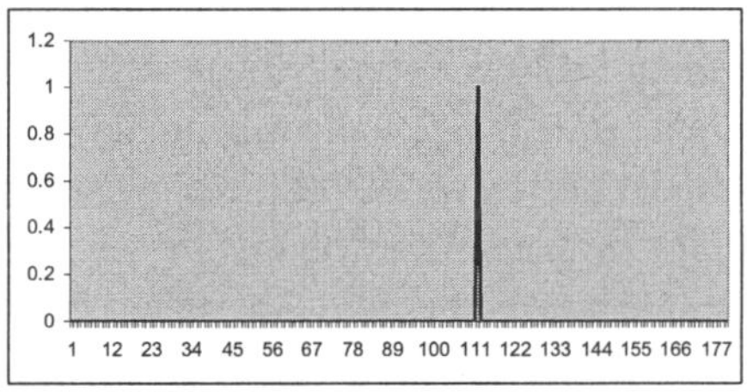

(b) (c)

Figure 1. a) Original image, b) Directional granulometry size 60, c) Minimum detection (112 degrees)

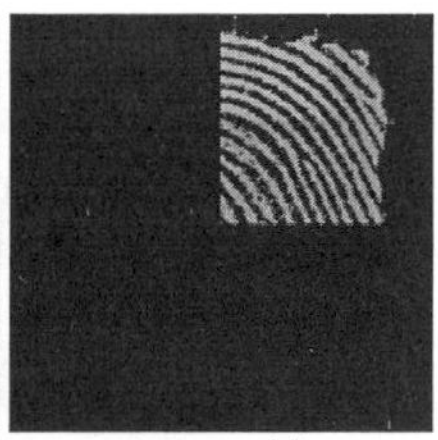

(a) (b) (c)

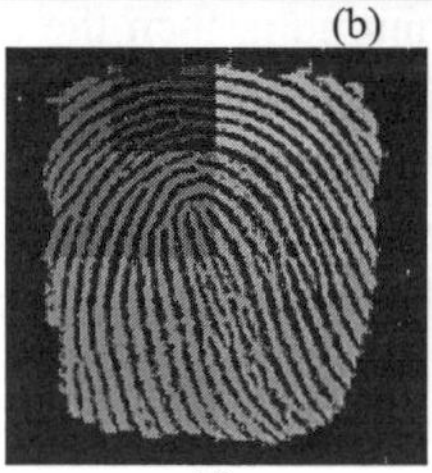

(d) (e)

Figure 2 a) First hierarchy of the quatree, b) and c) Second hierarchy of the quadtre, d) Final Quadtree structure, e) Color representation of the Quadtree

In summary, the directional granulometry or pattern spectrum notion allows to know the size and the predominant orientation in the quadtree at a given scale. By computing the minima of the pattern spectrum at each level of the hierarchy of the quadtree, one stops the process of the image division in square blocks when only one minimum is found. The image is divided in subimages in order to find the predominant orientation. Then, we can reconstruct the fingerprint image according to the quadtree since each one has the size information and orientation.

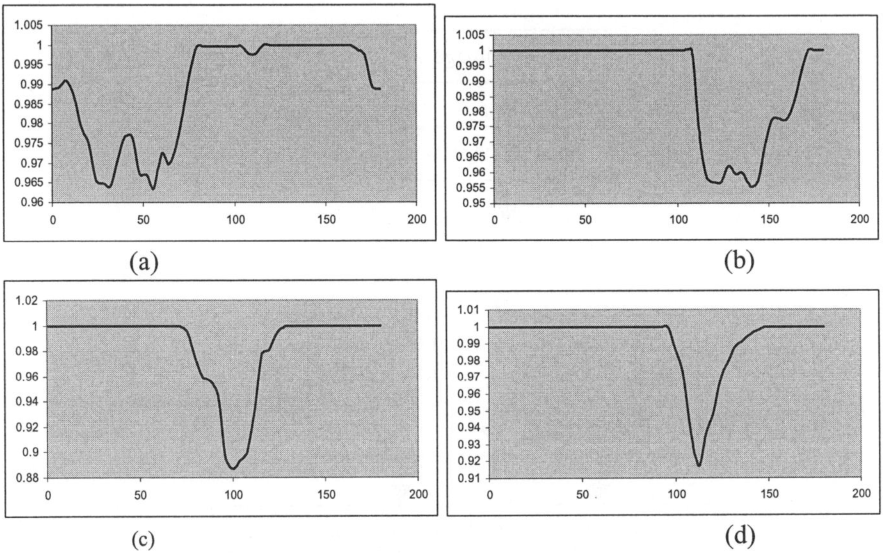

(a) (b) (c) (d)

Figure 3 Directional granulometries at first level of the hierarchy: a) and b) Top squares regions, c and d) bottom square regions

5 Fingerprint orientation field based on the supremum of directional erosions

As expressed above, the idea of applying directional erosions consists in preserving the maximum of information for all directions. To stock the information for all λ value, a gray level image is used in a similar way that the distance function is built. This means that the procedure is begun for $\lambda = 1$ by increasing by one all point x belonging to $\varepsilon_{\lambda}^{\sup}(X)$. Then, the procedure continue until ψ has a λ_{max} value such that $\varepsilon_{\lambda_{max}}^{\sup}(X) = \varnothing$. The maxima of this function is the locus of maximal structuring elements (ultimate erosions). Thus, one knows the position of the greatest structuring elements that can be completely included in the structure. However, the angle of these structuring elements is not known. To overcome this problem, a second image containing the angle of the structuring elements is built. The scheme in Fig. 4 illustrates the procedure for extracting the directional information of the image. First, the tophat size 4 was computed on the original image (Fig. 1(a)) for obtaining a binary one Fig. 5(a). Then, at each size of the structuring element, the supremum of directional erosions for all orientations is computed in order to obtain the distance function and the orientation function illustrated in Fig. 5(b). Thus, these functions are used to compute the segment of lines that characterize the structures. The maxima of distance function is computed for obtaining the locus of maximal structuring elements (ultimate erosions) and the values of the orientation function enable us to know the orientation lines (see Fig. 5(d)).

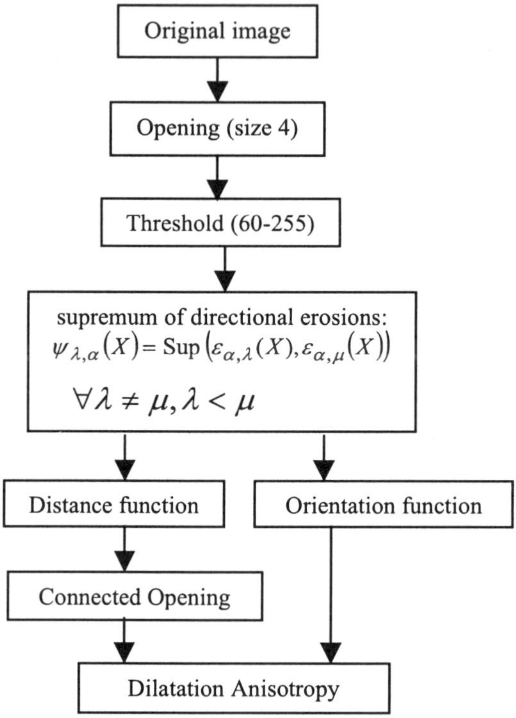

Figure 4 Procedure scheme for computing the orientation field

Since, the number of maxima is too much, a filtering process using a class of rank-max connected filters was carried out on the distance function. This enables us to have a better description of the image by using the segments of lines as illustrated in Figs. 5(f)-(g).

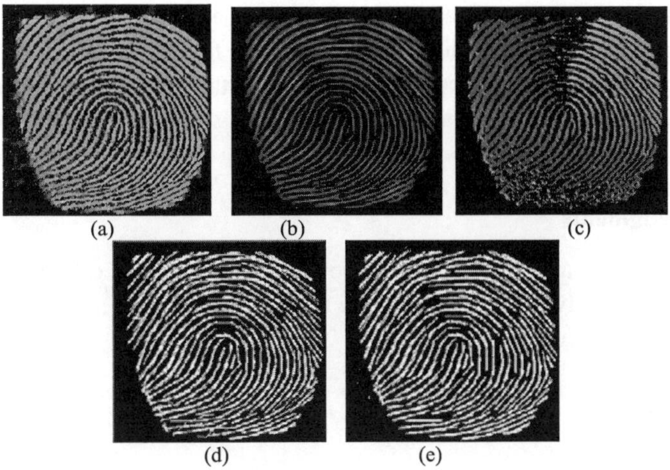

Figure 5 a) Tophat 4, b) Image distance funtion, c) Image orientation function, d)Image dilation anisotropy connectec 4, e)Image dilation anisotropy connectec 6

In conclusion, this connected filter allows to leave alone the structures that have certain orientation and size, those that are better adapted to the structures. The fingerprint is reconstructed starting from 2 images, the first image gives the information of the preferential orientations, while the second one gives the information of the predominant sizes. This enable us to compute an image as result, containing a set on line segments that characterize the orientation fields of the image.

4 Conclusion

In this paper, a study of orientation fields based on directional morphological transformations was made. The study was focused on the characterization of fingerprints which is a typical practical application. Two approaches were investigated to characterize orientation fields. The first one is based on the directional granulometries and the notion of quadtree structure. The quadtree is used to describe a class of hierarchical data structures, thus it permits to classify the orientation fields at different scales . The proposed homogeneity criterion for selecting the fields at a given hierarchy of the quadtree is specified by the principal minimum (or minima) of the distribution function of the granulometry. The second one considers a local approach by using the concept of distance function. The maxima of distance function was used for computing the locus of maximal structuring elements and a second function (orientation function) was used to obtain the angles of the line segments. This pair of local parameters enables us to have a good description of the image orientation fields by means of line segments. Other methods seek the core of the image to begin to work with the image, however, the disadvantage is that in a poor quality image is very difficult to find the core. Additional methods look for adapting lines to he image but the procedures become complex. Moreover, the filter-bank only compares with eight directions and these do not consider the size of the lines.

Acknowledgments

The first author acknowledges the government agency CONACyT for the financial support. Iván Terol would like to thank Diego Rodrigo and Darío T.G. for their great encouragement.

References

1. A.K. Jain, L. Hong, S. Pankanti, R. Bolle, Proc. IEEE, 85 (9) 1365, (1997).
2. D. Maio, D. Maltoni, IEEE Trans. Pattern Anal. Mach. Intell., 19(1), 27, (1997)
3. X.D. Jiang, W.Y. Yau, Proc. of the 15th International Conference on Pattern Recognition, 2, 1042, (2000).
4. R. Cappelli, A. Lumini, D. Maio, D. Maltoni, IEEE Trans. Pattern Anal. Mach. Intell. 21(5), 402, (1999).
5. B.G. Sherlock, D.M. Monro, Pattern Recognition 26(7), 1047, (1993).
6. P.R. Vizcaya, L.A. Gerhardt, Pattern Recognition, 29(7), 1221, (1996).
7. J. Gu, J. Zhou, D. Zhang, Pattern Recognition, 37(3), 543, (2004).
8. J. Zhou, J. Gu, Pattern Recognition, 37(2), 389, (2004).
9. J., Li, W. Y., Yau, H., Wang , Pattern Recognition, 39, 102, (2006).
10. J. Serra, Image Analysis and Mathematical Morphology, J. Serra, Ed., Vol. II, Academic Press, New York, (1988).
11. H. J. A. M. Heijmans, Morphological Image Operators, Acad. Press, (1994).
12. P. Soille, Morphological image analysis, Springer-Verlag, Heidelberg(2003.)
13. P. T. Jackway and M. Deriche, IEEE Trans. Pattern Anal. Mach. Intell. 18(1), 38, (1996)
14. F. Meyer and P. Maragos, J. Visual Comm. Image Represent., 11(3), 245, (2000).
15. P.Soille and H. Talbot, IEEE Trans. on Pattern Anal. Machine Intell., 23(11), 1, (2001).
16. G. Matheron, Random Sets and Integral Geometry, Wiley, New York, 1975.
17.S. Greenberg and D. Kogan, Pattern Recognition Letters, 27(1), 59, (2006).
18. A. U. Khan, M. K. Khan and M. A. Khan, International Technology Journal 4(1), 16,(2005).

POLARIZATION BASED MODIFIED MIRAU INTERFEROMETRY WITH INSTANTANEOUS PHASE SHIFTING FOR SURFACE PROFILING

N. R. SIVAKUMAR

Department of Mechanical and Industrial Engineering, Concordia University, Montreal, Canada H3G 1M8

In-line metrology, though has been the requirement of the precision industries, it has been difficult to achieve with high precision, primarily due to vibrations while manufacturing. This is even more important in phase shifting interferometry which is highly precise and hence sensitive to vibrations. In this work, a modified Mirau interferometer is proposed to work with polarization optics, to enable use of instantaneous phase shifting for working in vibrating environments. Profile measurements were carried out on a smooth mirror surface and the results are described.

Keywords: Surface profile measurement, Instantaneous Phase shifting, Modified Mirau interferometer.

1. Introduction

Several phase-measurement techniques can be used in optical profilometry to characterize object surface[1,2]. The basic advantage of Mirau interferometry over other profile measurement techniques is that, interference fringes are formed only when the reference and the measurement paths are almost equal and this reduces the ambiguity in object positioning. Moreover, Mirau optical setup is quite close to a common path configuration, which reduces the effects of vibration and environment compared to other profilometric techniques. Though this technique has so many advantages over the contemporary profilometers, the measurement is not polarization dependent. Essentially a phase shifting technique, Mirau configuration forces one to use conventional phase shifters using a piezoelectric device[3], which have inherent error sources due to mechanical movement[4]. Much of the works reported until now have been on the algorithms to compensate for the errors due to vibration and mechanical movement.

Not much work has been done to eliminate these errors. One method to eliminate the errors induced by vibration was first conceived by Smyth and Moore in their work of instantaneous phase shifting system[5]. Other techniques based on the use of gratings and polarization optics followed to generate phase-shifted interferograms simultaneously[6-9].

2. Design of the optical layout

To utilize the advantage of non-mechanical phase shifting using polarization based phase shifters, a modification on Mirau interferometry to suit polarization optics is designed in this work. The important component of the design is the beamsplitter, shown in figure 1. A glass plate was coated with birefringent coating on its bottom surface so that the incoming light is split according to its polarization, and is focused on the object and the reference surface. The top surface of the glass is coated with a fully reflective surface for a small area. This reflective surface is designed not only to reflect all the light that is deflected from the polarization coating, but also to act as the reference surface. Due to such a design, a near common optical path is achieved and the effect of environment is reduced. Additionally, interference is effected only when the object and reference beam traverse almost equal path length, which reduces the ambiguity of the optical setup.

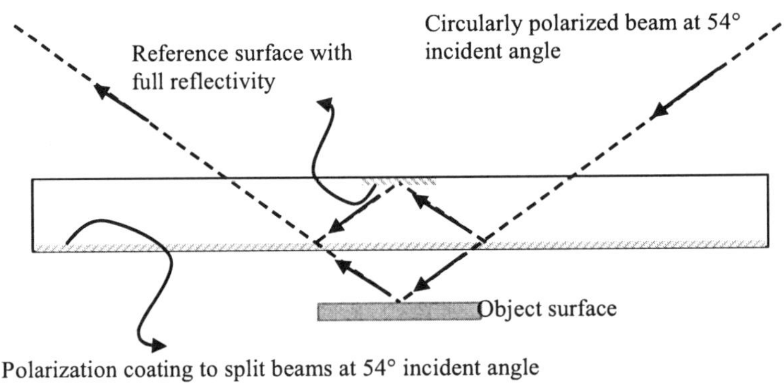

Figure 1 Design of Beamsplitter for modified Mirau interferometry

Figure 2 shows the experimental setup for surface profiling using polarization based on the new optical layout. Laser from a 1mW, 632nm source is made to pass through a spatial filter to increase the diameter of the beam. The beam is then made to travel through a quarter waveplate to make it circularly polarized so that the beam proceeding to the object and the reference have equal intensity. The beam is inclined at 54° to the beamsplitter by a mirror and is focused using a lens of 300mm focal length. The requirement of the mirror to steer the beam at 54° is to split the beam in to orthogonal polarizations once they pass through the beamsplitter designed specifically for this purpose. The beams reflected from the object and the reference is combined by the same birefringent material.

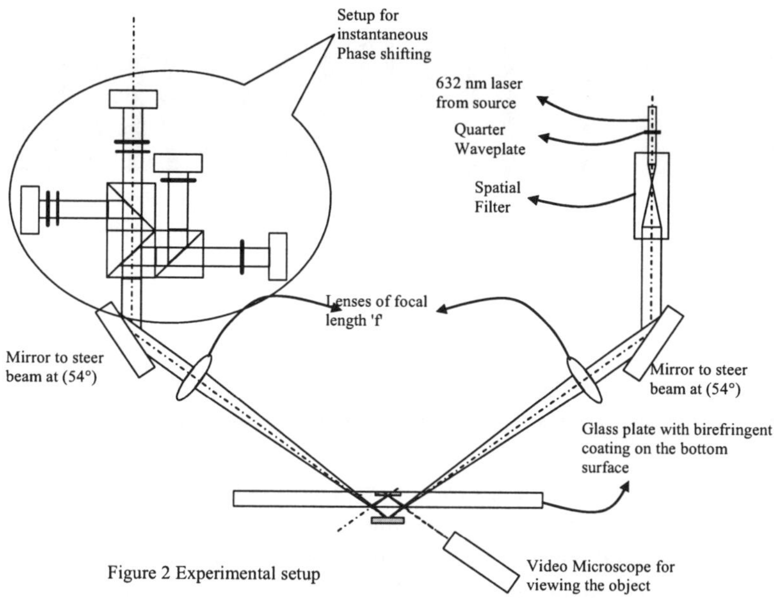

Figure 2 Experimental setup

This combined beam is collimated using a collimating lens of focal length 300mm and is made to travel through the three non-polarization beamsplitters pasted together to form the four arms of the phase shifting arrangement. The phase of the four output beams are same as that of the incoming combined beam as it has gone through the same path length and there are no polarization elements in the path. For phase shifting interferometry, the main requirement is that there should be three or more equally phase shifted images, so that the phase value of the image can be calculated. The phase shifting is proposed to be instantaneous and non-mechanical. In this technique, the required phase shift is achieved by quarter waveplates and polarizers as described in our earlier works[9-11].

3. Experiments and results

Experiment was carried out on a mirror surface. The object was positioned and the tilt and focus adjusted to get a few fringes on the CCD detectors. The four phase shifted images formed on the respective CCD's are digitized instantaneously through a frame grabber and are input to phase shifting software for further analysis of the surface characteristics. The four phase shifted images are shown in figure 3. However, it is important to note that, due to the arrangement of cameras, images from two of the cameras need to be mirrored to get proper orientation. The intensity values of corresponding pixels in the cameras are calculated using the four phase shift algorithm and phase values are calculated using the equation 1.

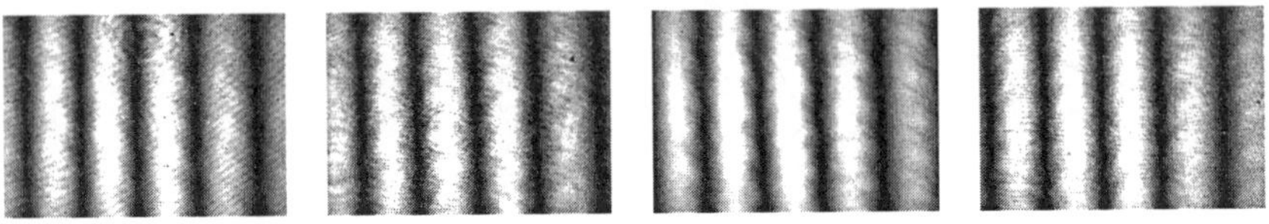

Figure 3 Four phase shifted images

$$\phi(x,y) = Tan^{-1}\{\frac{[I_4(x,y) - I_2(x,y)]}{[I_1(x,y) - I_3(c,y)]}\} \tag{1}$$

where the $I_i(x,y)$ is the intensity at a particular pixel in a camera and $\phi(x,y)$ is the phase value due to path length variation between the object and reference at that particular pixel.

The phase image of the object will have a saw tooth profile owing to the 2π discontinuities. The process of generating continuous phase image from the modulo 2π image and removal of phase discontinuities by adding or subtracting a value of π is called unwrapping[12,13]. The unwrapped image in figure 4, consists of the three dimensional phase information about the object. The unwrapped image is processed to convert the phase information into height image in the corresponding X, Y points. The relative surface height can be determined from the phase data by the equation 2[14].

$$h(x,y) = \frac{\lambda[\phi(x,y)]}{2\pi(\cos\theta_1 + \cos\theta_2)} \tag{2}$$

where $h(x,y)$ is the height, $\phi(x,y)$ is the phase information and θ_1 and θ_2 are the angle of light incidence and the angle of light detection respectively.

4. Discussion of results

As can be seen from the unwrapped image, the continuous phase image is tilted, and this tilt can be removed by using a first order least square fitting, in which the tilt plane is identified and subtracted from the image. Figure 5 shows the image after the removal of tilt. The figure and the profile below show that the surface is curved on the X-axis. Even though the mirror is a flat surface, the reason for the curvature is due the fact that the light incidents at an angle. Hence, the focused spot on the object forms an ellipse instead of a circle on the point of incidence. This implies that the distance imaged in the X-axis is considerably more than that being imaged on the Y-axis. After being reflected from the object, the light again becomes circular. This shrinks the scale of X-axis while maintaining the scale in the Y-axis, which is the reason for the curvature.

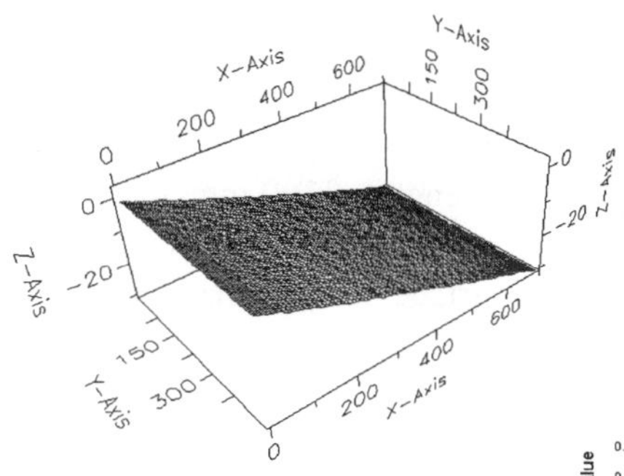

Figure 4 Unwrapped image

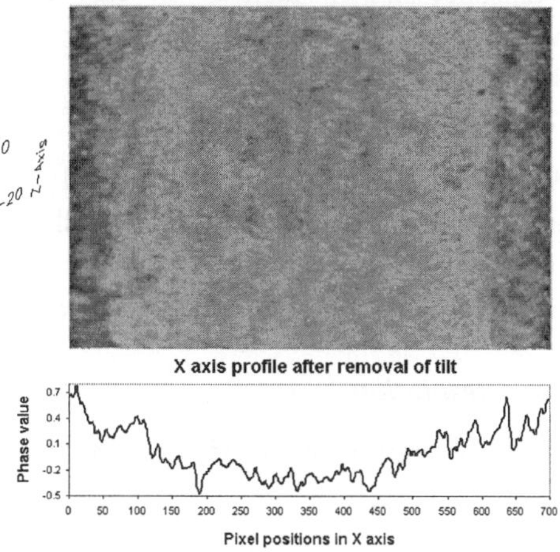

Figure 5 Image after tilt removal

Since this curvature is a constant feature, as the spot size and the angle of inclination remain the same, it can be removed by using a second order least square fitting. The final image after removal of curvature is shown in figure 6. The average roughness and the root mean square value of the surface was found to be 8.5 nm and 10.5nm respectively.

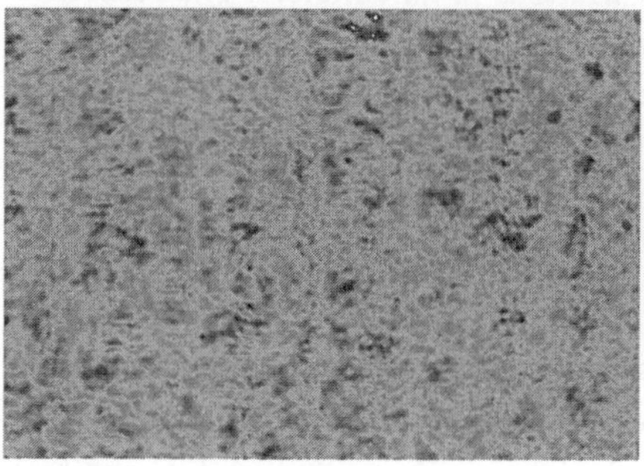

Figure 6 Final Image of Mirror Surface

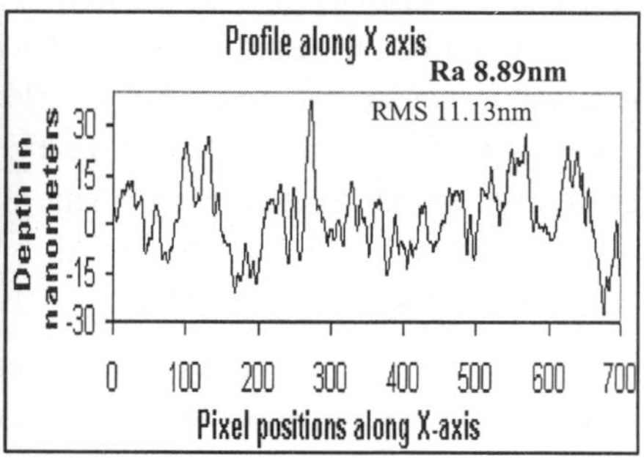

Figure 7 X-axis profile of surface

4.1. *Object Positioning*

To ensure that the incident laser light hits the object where the slot is to be measured, a video microscope was employed. As shown in the optical setup (figure 2), the video microscope with high magnification was initially focused on to the plane of incidence of the laser light on the object. The object is then moved in the X and Y directions to position the incident laser beam on the slot that is to be measured. Then the focusing lens is shifted forward to increase the beam spot size in such a way that the beam covers the required analytical surface area.

4.2. *Beam spot size*

With the beam diameter of 10mm and the focal length of 300 mm, the spot size on the object is determined theoretically to be around 25microns. This value can be increased to accommodate the required surface feature that needs to be analyzed by moving the focusing lenses.

4.3. *Resolution*

The horizontal resolution depends on the size of the spot on the surface, and the number of pixels on the CCD camera. The vertical resolution however needs to be discussed in detail. It has been discussed earlier that the phase information obtained from the pixels of the CCD camera can be converted to height information, if the angle of incidence and angle of viewing is known. In our case, the angle of incidence and angle of viewing is 54°. With this, we arrive at the sensitivity factor which is multiplied to the phase unwrapped image so that the deformation in the Z-axis can be calculated. The line profile along the X-axis is shown in figure 7 where the horizontal-axis indicates the pixel numbers of the corresponding points on the line where the profile is taken. The vertical-axis indicates the displacement in the height variations in nanometers. The reason for noisy measurement result is due to the fact that the surface used as a reference was not a standard reference but a reflective surface coated inside the designed plate. Further experiments to standardize the reference and to calibrate the same is required to reduce the measurement noise.

5. Conclusion

In summary, we have used a new polarization based Mirau interferometric layout for micro profiling of flat surfaces with sub-nanometer resolution. With this setup, need for separate reference had been eliminated and a common optical path configuration achieved. Moreover there is no ambiguity in object positioning as interference happens only when the measurement and reference beam travel equal path Additionally, the phase shifting does not rely on any moving objects and the interference is registered only when the object and the reference beam traverse almost equal path length. Since there are no mechanical devices and no ambiguity, the system once setup need not be calibrated with time, thereby increasing the accuracy and reliability of the phase shift. Moreover, the measurement of phase is instantaneous which increases the versatility of this technique to measure vibrating objects.

References

1. B. K. A. Ngoi, K. Venkatakrishnan, N. R. Sivakumar and T. Bo, Opt Commun. 190 (2000) 109-116
2. R. Windecker, H. J. Tiziani, Opt. Eng. 38(6) (1999) 1081-1087
3. P. J. Groot, Journal of OSA A, 12(2) (1995) 354-365
4. K. Kinnstaetter, App Opt. 27(24) (1988) 5082-5089
5. R. Smythe, R. Moore, Opt Engg. 23(4) (1984) 361-365
6. C. L. Koliopoulos, Proc of SPIE 2861 (1996) 86-93
7. C. L. Koliopoulos, Proc of SPIE 1531 (1991) 119-127
8. Y. K. Osuk, Opt Lett 9(2) (1984) 59-61
9. B. K. A. Ngoi, K. Venkatakrishnan, and N. R. Sivakumar, App Opt 40(19) (2001) 3211-3214.
10. N. R. Sivakumar, K. Venkatakrishnan, and B. K. A. Ngoi, Lasers in Engg. 12(1) (2002) 43-52
11. N. R. Sivakumar, B. Tan, K. Venkatakrishnan, Optics Communications 257 (2006) 217-224
12. Onodera. R, and Ishii. Y, Optical Engineering, 38(12) (1999) 2045-2049
13. Jingang. Z, Optical Engineering, 38(12) (1999) 2075-2080.
14. K. Creath., Proc of SPIE 4101A-06 (2000) 1-10

FUZZY CONTROL OF A HYDRODESULPHURIZATION REACTOR

SALVADOR CRUZ DEL CAMINO, FABIAN S. MEDEROS NIETO, ENRIQUE ARCE MEDINA[†]

Instituto Politécnico Nacional, Edif. 7, Unid. Prof. A.L.M., México 07738, DF, Mex.
[1]e-mail: earcem@yahoo.com.mx

A. MORALES SÁNCHEZ

Instituto Mexicano del Petróleo, Eje Lázaro Cárdens 152, 07730, DF, Méx.

This paper discusses the design and MATLAB simulation of a Fuzzy Logic Controller (FLC) for the hydrodesulphurization (HDS) process. HDS is a key process in existing petroleum refining operations. It occurs in a hydrotreating reactor in which sulfur is removed when a catalyst, of small particle size, is contacted with atmospheric gas oil and hydrogen. A fuzzy controller of the hydrogen recycle stream is proposed in this paper. It also provides an overview of the mathematical algorithm to solve the partial differential equations that constitutes the model. Verification of the model was conducted using data from pilot plant experiments. The proposed adaptive FLC controller is shown to be capable of compensating disturbances that affect the systems dynamics and to provide good overall system performance.

1. Introduction

The need to meet more stringent standards limiting the sulfur content of fuels urges a deeper understanding of the mechanism by which sulfur-containing compounds are removed using hydrodesulphurization (HDS) catalysts. The modeling and simulation of a process are necessary for a better understanding of the operation control. Examples for applications of fuzzy control to chemical reactors can be found in the literature [1]. In this paper, we compare the application of the Takagi – Sugeno [2] and a self-tuning [10] fuzzy controller to a model of a HDS pilot plant.

Extensive studies have been conducted on modeling pseudo homogeneous plug flow reactors on the HDS process using a steady state model, Andrigo et al. [3], Chen and Ring [4].

Korsten and Hoffman [5] have presented a three phase heterogeneous model for HDS of vacuum gas oil in a trickle bed reactor. Further explorations of the idea were conducted by Bhaskar et al. [6]. We use this model as a basis to develop our model for the HDS of atmospheric gas oil, which is less heavy than the vacuum gas oil. The model was used to perform studies on sensitivity to feed variable changes and process control.

[†] Work partially supported by grant 2-4570.5 of the Swiss National Science Foundation.

2. Formulation of Mathematical Model

The HDS reaction of atmospheric gas oil can be written as:

$$\nu_1 \bullet A_1 \text{ (liquid)} + \nu_4 \bullet A_4 \text{ (Gas)} \rightarrow \nu_2 \bullet A_2 \text{ (liquid)} + \nu_3 \bullet A_3 \text{ (Gas)}$$

Where: A_1 is hydrogen, A_2 is hydrocarbon, A_3 is hydrogen sulfide, and A_4 is organic sulfur compound. ν_1, ν_2, ν_3 and ν_4 are the stoichiometric coefficients of A_1, A_2, A_3, and A_4 respectively. In this work the feed stock used is atmospheric gas oil.

The mathematical model equations are indicated in Table 1. The superscripts G, L and S stand for gas, liquid and solid respectively. The mass balance of H_2 and H_2S are given by equations 1 and 2. $k_i^L \cdot a_L$, describe the mass transfer between the gas and the liquid phase. P_i and C_i are partial pressures and concentrations, u_G is the superficial gas velocity, ε is the catalytic bed void fraction, Hi is the Henry's Law constant, R is the gas law constant and T is the absolute temperature. Equations 3 and 4 describe the mass balance of the volatile compounds H_2 and H_2S using the two-film (gas-liquid and liquid-solid) approach, and equation 5 describes the mass balance of the organic sulfur compound at the liquid-solid interface. Equations 6 to 8 represent relationships of the components transported between the liquid phase and the surface of the catalyst $k_i^S \bullet a_S \bullet (C_i^L - C_i^S)$ and the extent of chemical reaction, $\nu_i \cdot \rho_B \cdot \varsigma \cdot r_C$, . ρ_B is the catalyst bulk density; r_c is the reaction rate over the catalyst.

Kinetics of HDS reaction is given in equation 9, four parameters have to be estimated from the pilot plant data in this equation, namely, the orders of reaction, m_1 and m_2, the absorption equilibrium constant K_3 , and the apparent rate constant, k_{app}. Equation 10 is an empirical correlation as used by Korsten and Hoffman [5], it determines k_{app} as a function of the superficial mass-flow velocity G_L.

In the pilot plant reactor liquid oil and H_2 move downstream through a fixed bed reactor *(trickle flow reactor)*. The reaction products leave the reactor and after having been cooled to a low temperature, enter a liquid/gas separation stage, see Fig. 3. The bottoms liquid is the desulfurized gas oil and constituted the principal product.

2.1. *Solution of the Mathematical Model*

Concentration profiles are obtained by means of the discritization of the model in space, with proper boundary conditions. The discritization for the entire length of the reactor is performed by the splines lines approach; a set of ordinary differential equations is obtained [7].

Table 1. Equations of the mathematical model.

$\frac{\in}{RT}\frac{\partial p_1^G}{\partial t}=-\frac{u_G}{RT}\frac{\partial p_1^G}{\partial z}-k_1^L\cdot a_L\cdot\left(\frac{p_1^G}{H_1}-C_1^L\right)$	(1)
$\frac{\in}{RT}\frac{\partial p_3^G}{\partial t}=-\frac{u_G}{RT}\frac{\partial p_3^G}{\partial z}-k_3^L\cdot a_L\cdot\left(\frac{p_3^G}{H_3}-C_3^L\right)$	(2)
$\frac{\partial C_1^L}{\partial t}=-u_L\bullet\frac{\partial C_1^L}{\partial z}+k_1^L\bullet a_L\bullet\left(\frac{p_1^G}{H_1}-C_1^L\right)-k_1^S\bullet a_S\bullet(C_1^L-C_1^S)$	(3)
$\frac{\partial C_3^L}{\partial t}=-u_L\bullet\frac{\partial C_3^L}{\partial z}+k_3^L\bullet a_L\bullet\left(\frac{p_3^G}{H_3}-C_3^L\right)-k_3^S\bullet a_S\bullet(C_3^L-C_3^S)$	(4)
$\frac{\partial C_4^L}{\partial t}=-u_L\cdot\frac{\partial C_4^L}{\partial z}-k_4^S\cdot a_S\cdot\left(C_4^L-C_4^S\right)$	(5)
$\frac{dC_1^S}{dt}=k_1^S\bullet a_S\bullet\left(C_1^L-C_1^S\right)+\nu_1\cdot\rho_B\cdot\varsigma\cdot r_C$	(6)
$\frac{dC_3^S}{dt}=k_3^S\bullet a_S\bullet\left(C_3^L-C_3^S\right)+\nu_3\cdot\rho_B\cdot\varsigma\cdot r_C$	(7)
$\frac{dC_4^S}{dt}=k_4^S\bullet a_S\bullet\left(C_4^L-C_4^S\right)+\nu_4\cdot\rho_B\cdot\varsigma\cdot r_C$	(8)
$r_C=k_{app}\frac{(C_4^S)^{m_1}\cdot(C_1^S)^{m_2}}{(1+K_3\cdot C_3^S)^2}$	(9)
$\frac{1}{k_{app}}-\frac{1}{k_{in}}=\frac{A}{G_L^B}$	(10)

2.2. *Open Loop Simulations of the Model*

Results from the simulation produce good agreement with those reported in the literature [5]. Figure 1 depicts the concentrations profiles of H_2S in the liquid phase, at different times, which increases sharply at the reactor inlet, then comes to a maximum and then gradually decreases down through the reactor until a steady state is reached, around 31,000 seconds (solid line). The plot in figure 2 depicts how the concentration of the organosulfur compound, in the liquid phase, decreases through the reactor at different times. The dynamic simulations results of figures 1 and 2, using the model of the reactor reproduced the observed concentration profiles, at steady state, found in the literature [5].

3. The Reactor Control System

The control objective involves maintaining the hydrogen feed flow rate at a desired quality by manipulating the flow of the recycled stream. A ratio control scheme is to be used to maintain the hydrogen feed flow rate to the reactor regulated, as recommended by Smith and Corripio [11]. The hydrogen feed

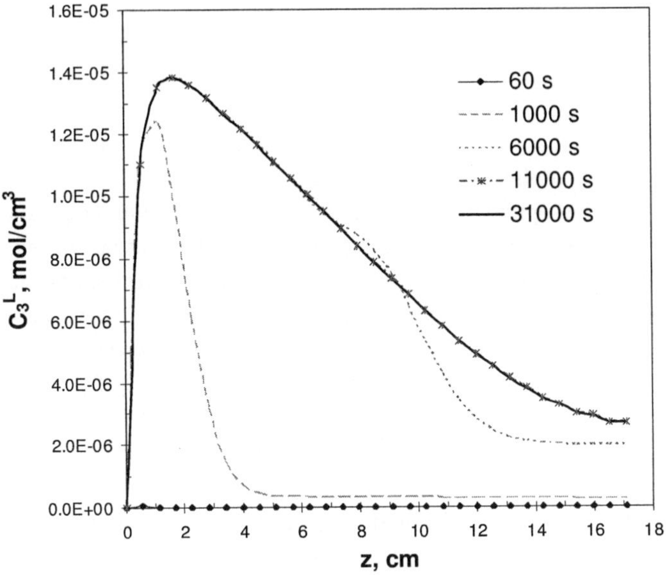

Figure 1. Concentration profiles of H_2S in the liquid phase through the reactor.

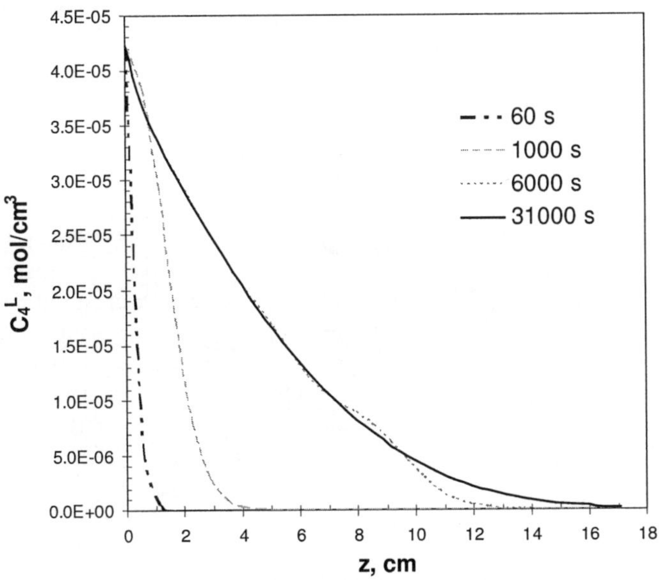

Figure 2. Concentration Profiles of organosulfur compound in the liquid phase through the reactor.

flow rate to the reactor is equal to the sum of the recycled hydrogen and the fresh feed hydrogen. The diagram for the control scheme is shown in Fig. 3.

3.1. *Fuzzy-PI Controller*

In a FLC the system behavior is described by means of fuzzy relational equations. A typical linguistic model is expressed as `IF x is A THEN y is B', where A and B are fuzzy sets. The fuzzy sets are determined from experience.

Since the first application of fuzzy logic control to industrial processes was developed [8], the applications of fuzzy controllers have been successfully installed in the process industries [9]. Two approaches can be followed for the fuzzy controller implementation. 1) The fuzzy PI controller can be designed to mimic traditional PI control action, adjusting the manipulated variable directly or 2) it can be designed to provide controller parameters for the PI controller as given by Hyung-Soo this is known as the self-tuning FLC [10]. Application of the first approach uses the Tagaki-Sugeno model. Which consists of a number of fuzzy implications, each one is composed of a set of premises in the IF part and a set of consequences in the THEN part. The IF part provides a logic-based guidance to the use of regression models in the THEN part, as in Equation 11.

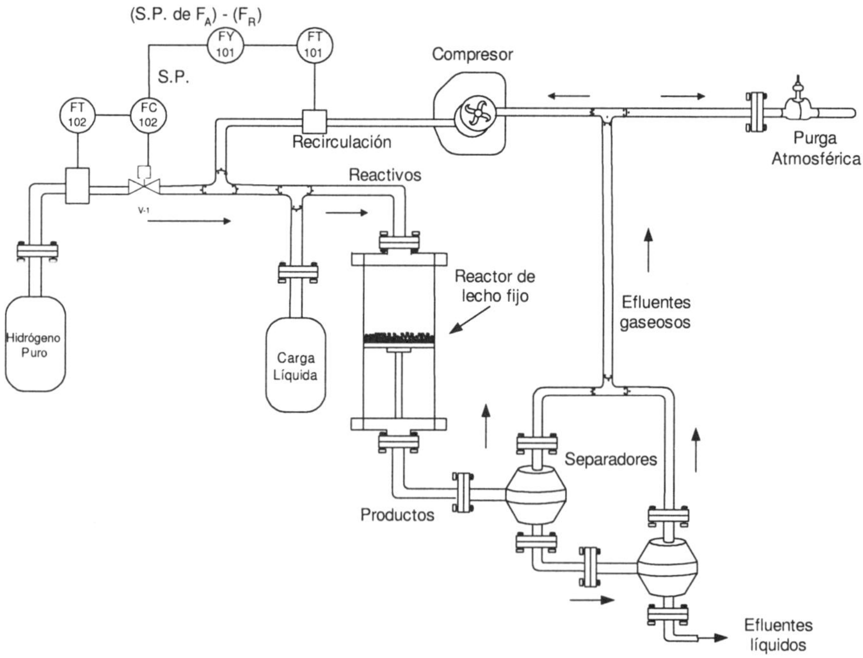

Figure 3. Block diagram of the control system of the recycle stream.

$$L^i : \quad IF\ e(k)\ is\ A_1^i\ and\ \Delta e(k)\ is\ A_2^i \qquad (11)$$
$$THEN\ u(k)^i = a_1^i e(k) + a_2^i \Delta e(k)$$

Where $u(k)^i$ denotes the output from the i-th implication, $a_p^i (p = 1,2)$ are consequent parameters of the pseudo PI controller. A_1^i *and* A_2^i are fuzzy sets on e(k) and $\Delta e(k)$.

In the second approach the velocity form of the PI control law is used in its discretisized form, Eq. 12, where the manipulated variable change, Δm_n, is a function of the error and the change in error:

$$\Delta m_n = K_c(\Delta E_n + \frac{K}{\tau_i} E_n) \qquad (12)$$

Calculation of the tuning parameters is done by the following relationships.

$$K_c(k) = K_c(k-1) + \Delta K_c(k) \qquad \tau_i(k) = \tau_i(k-1) + \Delta \tau_i(k) \qquad (13)$$

The e and Δe inputs are sampled once in every cycle of integration. The manipulated variable is the recycled hydrogen. We assume that a fuzzy logic controller is given for a single input in such a way that the fuzzy sets used for e and Δe, are triangular functions defined on equally spaced points, the fuzzy control rules are linear. Three fuzzy sets are chosen and defined by the following rule base of fuzzy-set values for the error e and change in error Δe: N – Negative, ZE – Zero, and P - Positive

Values for ΔK_c and $\Delta \tau_i$ are obtained through the fuzzy rules, a simple direct matrix mapping can be used to represent the rules of the controller, and this matrix is called the look-up table or the decision table, see tables 2 and 3. In these tables the column expresses the fuzzy subsets of the input error and the rows the change in error. Entries are the consequences of all rules. For example rule 1 can be stated as IF e is P AND Δe is N THEN u is P, which correspond to the entries shown in the Table 2 for column 1 and row 3.

Although the usual approach in developing a knowledge based system starts with the knowledge acquisition through the interaction with an identified expert in the field, in this work the knowledge acquisition is accomplished through the simulation of the reactor model equations.

4. Simulation Results

The closed loop responses of recycled stream are obtained by simulation using MATLAB software, see figure 4. To demonstrate the performance of the Self-tuning Fuzzy controller compared with the Takagi-Sugeno FLC, experiments were conducted by simulation. In all cases dealing with servo controller or regulatory controller, simulations results show that the self-tuning FLC performance were better than that of the Takagi-Sugeno approach. The error is used in computing some cost function like the ISE for comparison purpose.

Table 2. Decision Table for Kc.

e(k) \ Δe(k)	N	Z	P
N	N	P	N
Z	P	Z	N
P	P	N	P

Table 3. Decision Table for τ_I.

e(k) \ Δe(k)	N	Z	P
N	P	N	P
Z	N	Z	P
P	N	P	N

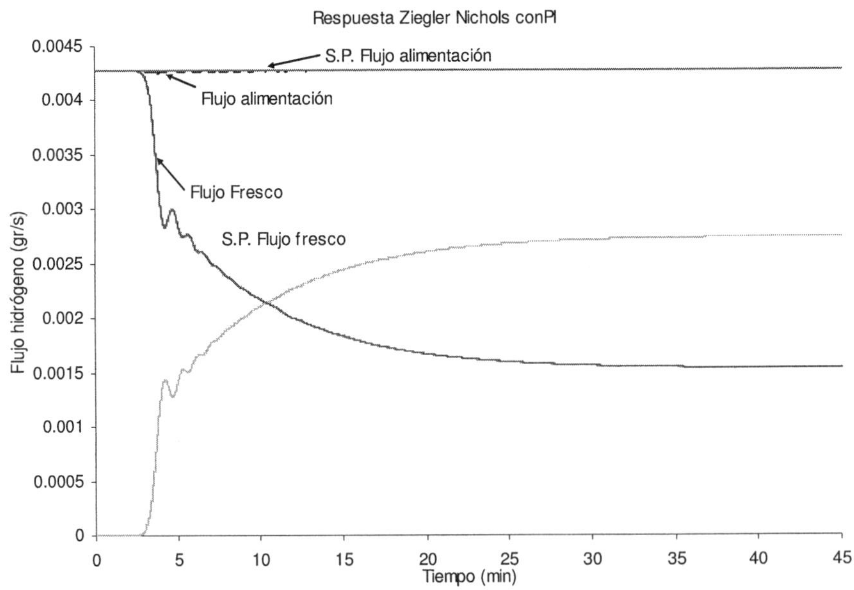

Figure 4. Controlled response of hydrogen flow rate.

Conclusions

A model has been obtained that could be a useful tool that can be employed in the analyses of the influence of variables and parameters in the HDS process. The model adequately predicts the concentration at the reactor outlet. The self-tuning FLC controller outperforms the Takagi-Sugeno FLC both models are capable of compensating disturbances that affect the systems dynamics and to provide good overall system performance.

References

1. P. J. King and E. H. Mamdani, "The application of fuzzy control system to industrial processes," *Automatica*, **13**, pp. 235-242 (1977).
2. Takagi, T., Sugeno, M., Fuzzy identification of systems and its applications to modeling and control. *IEEE Trans. Syst., Man, and Cybernt.* **15** , 116-132.(1985).
3. Andrigo, P., Bagatin, R. and Pagani, G., Fixed Bed Reactors, *Catalysis Today*, **52**, 197-221, (1999).
4. Chen, J. and Ring, Z., Modeling and Simulation of a Fixed-Bed Pilot-Plant Hydrotreater, *Ind. Eng. Chem. Res.* , **40**, 3294-3300, (2001).
5. Korsten, H. and Hoffmann, U., Three-phase Reactor Model for Hydrotreating in Pilot Trickle-Bed Reactors, *AIChEJ* , **42**, (1996),1350-1360, (1996).
6. Bhaskar, M., G. Valavarasu, A., Meenakshisundaram and K.S. Balaraman., Application of a Three Phase Heterogeneous Model to Analyze the Performance of a Pilot Plant Trickle Bed Reactor. *Petroleum Science and Tech.* **20**, (3&4) 251-268. (2002).
7. Mederos Nieto, F. S., M. S. Thesis, ESIQIE, IPN, México, 2004.
8. Mamdani, E.H. and S. Assilian, A fuzzy Logic Controller for a Dynamic Plant. *Inter. J. Man-Machine Studies,* **7**, pp 1-13, (1975).
9. Schwartz T. J., Fuzzy Systems in the Real World. *AI Expert*, pp 29-36, August 1990
10. Hyung-Soo, Hwang. A tuning algorithm for the PID Controller Utilizing Fuzzy Theory. IEEE (1999).
11. B. Smith C. A. and Corripio, A. B., Principles and Practice of Automatic Process Control. John Wiley & Sons, 1985.

DISSOLVED OXYGEN CONTROL IN AN AEROBIC SEQUENCING BATCH REACTOR FOR TOXIC WASTEWATER TREATMENT

ALEJANDRO VARGAS*, DANIEL GONZÁLEZ, ALEJANDRO NÚÑEZ, and FRANCISCO VELARDE

Bioprocesos Ambientales, Instituto de Ingeniería, UNAM
Circuito Escolar s/n, Ciudad Universitaria, Coyoacán, D.F.,04510 Mexico
**E-mail: avargasc@ii.unam.mx*
www.ii.unam.mx

Three different controllers for dissolved oxygen setpoint regulation were implemented in a sequencing batch reactor used to treat toxic wastewater. In particular, these were a PID controller with anti-windup, a linearizing plus PID controller, and a fuzzy PD controller. They were tested and designed in simulation using a nonlinear fourth order model and implemented in real time in a pilot scale laboratory bioreactor with promising results.

Keywords: Dissolved oxygen, PID control, exact linearization, fuzzy control, SBR

1. Introduction

Industrial wastewater containing toxic organic compounds is difficult to treat by conventional biological wastewater treatment systems like the activated sludge process. This is due to several factors that include the variability of the wastewater flow and composition, as well as the presence of compounds that may be inhibitors to the microorganisms under relatively low concentrations. However, sequencing batch reactors (SBR) have been shown to be a viable alternative, especially under suitable monitoring and control.[1]

In particular, for toxic wastewaters containing phenols, it has been shown that by controlling the feed rate to a SBR-type biological reactor, such that the substrate concentration remains near the value where the growth rate is maximal—and thus also the degradation rate—, it is possible to make it robust against influent variations and furthermore reduce the degradation time.[2] Such a control strategy uses dissolved oxygen measurements to estimate the degradation rate,[3] but it requires a continuous airflow to the bioreactor, since it relies on monitoring the changes in dissolved oxygen concentration, which occur because the respiration rate is practically proportional to the degradation rate.

Much of the energy consumption in an aerobic wastewater treatment system is due to aerators that supply oxygen for the microorganisms' metabolism. Enough air must be pumped so that the dissolved oxygen concentration stays above the limiting value, below which the operation of the system is compromised. On the other hand, too much air increases the energy consumption without any significant increase in performance. It is therefore of interest to regulate the dissolved oxygen concentration at an optimal setpoint.

It has been reported that a simple PID controller performs adequately for activated sludge processes, and there have been several attempts to increase robustness or reliability by using adaptive or model predictive schemes.[4,5] However, for batch reactors they may not be adequate, since the wide range of substrate concentration values during the reaction make the respiration rate highly variable.

The control challenge reported here was to regulate the dissolved oxygen at a setpoint, despite the unknown (unmeasured) respiration rate. Three strategies were tested: a PID controller with anti-windup reset, a linearizing controller coupled with a PID controller, and a fuzzy PD controller.

The paper is divided as follows. The next section presents the mathematical model used, both for dissolved oxygen and for the oxygen mass transfer dynamics. Following this, the various controllers used are explained. The practical implementation and testing of the control algorithms in a laboratory pilot bioreactor are then explained and shown, and finally, some conclusions are given.

2. Mathematical models

A bioreactor in the sense considered here consists of a tank of a certain volume, where a consortium of microorganisms that are specialized in biodegrading certain substrates is present. These are diluted in the wastewater that enters the bioreactor (the influent). The reactor is assumed to be perfectly mixed, so that concentration of any component is the same within the reactor. In an aerobic reactor the microorganisms require oxygen to perform their metabolic functions, and this is supplied by pumping air to the mixed liquor, usually through diffuser at the bottom.

2.1. *Dissolved oxygen model*

Considering perfect mixing within the reactor, a mass balance on the dissolved oxygen yields the following model

$$\frac{\mathrm{d}S_O}{\mathrm{d}t} = -r + (S_O^{\mathrm{sat}} - S_O)k_L a - D_{\mathrm{in}} S_O, \tag{1}$$

where S_O represents the dissolved oxygen concentration, $D_{\mathrm{in}} = Q_{\mathrm{in}}/V$ the dilution rate with Q_{in} the inflow rate and V the volume, ${S_O}^{\mathrm{sat}}$ the dissolved oxygen saturation constant, and $r(t)$ represents the net respiration rate, accounting for endogenous, as well as metabolic respiration of the microorganisms. In fact, considering the simplest case of only one substrate and one type of biomass, this net respiration rate is given by

$$r = \sigma(S_O)\left(b + \frac{1}{Y_{XO}}\mu(S_S)\right)X. \tag{2}$$

In this model bX corresponds to the endogenous respiration rate and is of first order with respect to biomass concentration X. The term $\sigma(S_O)$ is an activation function for dissolved oxygen, *i.e* $\sigma \to 0$ as $S_0 \to 0$ (no respiration if there is no oxygen) and $\sigma \to 1$ for values of S_O above some critical (limiting) value; this is usually modeled as $\sigma(S_O) = S_O/(K_O + S_O)$. Under non-limiting conditions the greatest contributor to r is the metabolic respiration rate $\mu(S_S)X/Y_{XO}$ (*i.e.* biological oxidation of substrate), which is proportional to the biomass concentration X and to the specific biomass growth rate μ, which depends on the substrate concentration S_S, such that $\mu \to 0$ as $S_S \to 0$.

In a continuous reactor the flow rates of the influent and effluent are the same, and therefore, it operates under a pseudo steady state condition. In the widely used activated sludge wastewater treatment plant, even though there are variations in dilution rate and influent substrate concentration, the substrate concentration in the reactor is more or less constant. Therefore, under non-limiting conditions, all other terms being constant, the respiration rate r may be modeled as constant with a possibly unknown value, or as a slowly varying parameter. This is the philosophy behind many proposed dissolved oxygen control strategies. For example, there have been successful reports on the use of PID,[6] gain-scheduling,[7] adaptive,[4] and even nonlinear model predictive control.[5]

In a sequencing batch reactor (SBR), the tank is initially at a low volume with microorganisms and almost no substrate. It is then filled with the influent up to a maximal volume and let react for some time, after which the biomass is sedimented and the clarified effluent is removed, in order to begin another cycle. The reaction phase therefore starts with a high concentration of substrate that decreases in time as it is biodegraded until its concentration is again near zero. This implies that the specific growth rate $\mu(t)$, which is proportional to the specific degradation rate, is variable, and therefore the respiration rate $r(t)$ too. For non-toxic substances $\mu(S_S)$ is modeled by the Monod law $\mu(S_S) = \mu_{\max}/(K_S + S_S)$, which is an increasing function that basically states that under no substrate, there is no growth, *i.e.* $\mu(0) = 0$, and $\mu \rightarrow \mu_{\max}$ as S_S increases. If the half-saturation constant K_S is small with respect to the initial substrate concentration, then during most of the reaction $\mu \approx \mu_{\max}$ and thus a controller that considers $r(t)$ as a constant may be adequate if convergence occurs fast enough.

For toxic substrates, on the other hand, the specific biomass growth rate is usually modeled by the Haldane law, being zero at $S_S = 0$, attaining a maximum μ^* at some substrate concentration value S_S^* and then decreasing for higher values $S_S > S_S^*$, because the toxicity inhibits the growth of microorganisms. In this case, there is no way to circumvent the problem that $r(t)$ is time variable, if the substrate concentration is not measured and the parameters of the Haldane model are unknown.

2.2. *Oxygen mass transfer model*

The oxygen mass transfer coefficient indicates the rate at which oxygen is transferred from the air bubbles to the liquid medium per unit volume. From the control viewpoint, $k_L a$ depends nonlinearly on the airflow Q_{air} entering the diffusers, mainly because in a certain range, this affects both the size and the number of bubbles. Therefore, Q_{air} may be regarded as the input to the system for control purposes.

A commonly used model for $k_L a$ is[4]

$$k_L a(Q_{\text{air}}) = k_1 \left(1 - \exp(-k_2 Q_{\text{air}})\right), \tag{3}$$

which in some cases may be approximated by a linear one (if k_2 is large with respect to the maximum airflow allowed Q_{air}^*). This exponential model being heuristic, it may as well be approximated by a Monod-type curve:

$$k_L a(Q_{\text{air}}) = \frac{(m+1) k_L^* a Q_{\text{air}}}{Q_{\text{air}}^* + m Q_{\text{air}}}, \tag{4}$$

such that $k_L a(Q_{\text{air}}^*) = k_L^* a$ and m is a parameter that indicates the nonlinearity of the relationship (more linear if $m < 1$). Because oxygen transfer occurs at a higher rate than biodegradation, such an algebraic relationship is sufficient in most cases. However, for dissolved oxygen control, it may be convenient to consider the response dynamics of k_La with respect to Q_{air}, *i.e.* k_La does not change immediately as Q_{air} changes. A proposed model, that has the same steady-state relationship (4) is

$$\frac{\mathrm{d}k_La}{\mathrm{d}t} = \frac{1}{\tau_k}\left((m+1)\,k_L^*a\left(\frac{Q_{\text{air}}}{Q^*_{\text{air}}}\right) - \left(m\left(\frac{Q_{\text{air}}}{Q^*_{\text{air}}}\right)+1\right)k_La\right), \quad (5)$$

with τ_k some time constant.

2.3. *Dissolved oxygen sensor model*

A linear model for the dissolved oxygen sensor —obtained through experimentation on the pilot biorreactor— is given by the following second order differential equation, where S_O^{m} is the measured dissolved oxygen:

$$\tau_1\tau_2\frac{\mathrm{d}^2 S_O^{\text{m}}}{\mathrm{d}t^2} + (\tau_1+\tau_2)\frac{\mathrm{d}S_O^{\text{m}}}{\mathrm{d}t} + S_O^{\text{m}} = S_O \quad (6)$$

Eqs. (1), (5), and (6) constitute the nonlinear time-varying fourth order dynamic model that was used for simulation and controller design.

3. Control strategies

Three controller design strategies were tested. These are reported in the next subsections.

3.1. *PID control with anti-windup*

The PID controller is perhaps the most widely used in process control due to its reliability and simplicity.[8] Its design is straightforward and does not necessarily require a model of the process. The control signal $u(t)$, which in this case is $u = Q_{\text{air}}$, is calculated using the expression

$$u(t) = K_p\left(e(t) + \frac{1}{T_i}\int_0^t e(s)\mathrm{d}s + T_d\frac{\mathrm{d}e(t)}{\mathrm{d}t}\right) \quad (7)$$

where $e(t) = y_{\text{ref}}(t) - y(t)$, with y the output of the system and y_{ref} the reference output or setpoint; in the bioreactor y was the measured dissolved oxygen S_O. The PID controller parameters K_p, T_i and T_d need tuning, which can be done manually or using specialized software.[8]

Two problems arise during its practical implementation. The first one is the derivative term $\mathrm{d}e/\mathrm{d}t$, which cannot be implemented exactly. However, since the signals are discretized and ultimately processed in a computer, the derivative is computed with a differentiator followed by a median filter to eliminate some of the noise amplification.

The second problem is actuator saturation. The airflow is limited to be between 0 and Q^*_{air}. If the controller demands the actuator to operate beyond its limits, the integrator may suffer from windup. A standard anti-windup technique of integrator reset was implemented using a feedback signal of the difference between the theoretical input $u(t)$ and the actual control signal capped by the actuator limits $\bar{u}(t)$ as follows:

$$u(t) = K_p \left(e(t) + \frac{1}{T_i} \int_0^t \left(e(s) + \frac{u(s) - \bar{u}(s)}{\gamma K_p} \right) \mathrm{d}s + T_d \frac{\mathrm{d}e(t)}{\mathrm{d}t} \right). \tag{8}$$

The design parameter γ is used for tuning the anti-windup and a value between 0.1 and 1 is usually enough.

3.2. *Linearizing+PID control*

For the design of this controller, a reduced order model was considered, namely the first order model nonlinear model of Eq. (1) and the algebraic $k_L a$ relationship in steady-state of Eq. (4). Assume that the respiration rate $r(t)$ is measured. Then it is possible to exactly linearize the system. Simply set as desired model reference the system

$$\frac{\mathrm{d}S_O}{\mathrm{d}t} = \frac{1}{\tau_d} \left(S_O^{\text{ref}} - S_O \right), \tag{9}$$

with τ_d some desired time constant (a design parameter) and equate both right hand sides of Eqs. (1) and (9) to solve for $k_L a$, *i.e*

$$k_L a = \frac{\left(S_O^{\text{ref}} - S_O \right) / \tau_d + D_{\text{in}} S_0 + r}{S_O^{\text{sat}} - S_O}. \tag{10}$$

The input to the system, *i.e.* the airflow Q_{air}, is computed by solving Eq. (4), such that

$$Q_{\text{air}} = \frac{k_L a Q^*_{\text{air}}}{k_L^* a + m \left(k_L^* a - k_L a \right)}. \tag{11}$$

Of course, the implementation of such a strategy depends on knowing the respiration rate $r(t)$ on-line. Eq. (1) can be used for this purpose if *approximations* of the dissolved oxygen time derivative and of the true $k_L a$ used are available. The first one can be computed using a differentiator and a filter, while the second one can be computed by passing the implemented airflow signal through the dynamic $k_L a$ model of Eq. (5). In this sense one may estimate $r(t)$ on-line with

$$\hat{r} = \left(S_O^{\text{sat}} - S_O^m \right) \widehat{k_L a} - D_{\text{in}} S_O^m - \widehat{\dot{S_O}}, \tag{12}$$

where $\widehat{k_L a}$ and $\widehat{\dot{S_O}}$ are the estimated values of $k_L a$ and $\frac{dS_O}{dt}$, respectively, while S_O^m is the measured dissolved oxygen. This estimator, together with Eqs. (10) and (11) constitute an implementable linearizing controller. However, it is still prone to plant-model mismatches and time delays due to the estimators. Therefore, for better performance, a PID controller with anti-windup was implemented as an outer loop in a cascade control, such that the airflow Q_{air} was computed using Eq. (8), adding the linearizing term of Eq. (11), computed using Eqs. (10) and (12).

3.3. *Fuzzy PD control*

Another controller that is becoming popular for process control is the fuzzy controller. Heuristic *if-then* rules expressed linguistically are converted to a numerically implementable algorithm. A fuzzy controller consists of three parts:[9] a *fuzzifier*, which converts the crisp (numerical) inputs to fuzzy membership values in fuzzy sets, an inference engine which determines the fuzzy output sets that are activated according to the rule set, and a *defuzzifier*, which converts the fuzzy outputs to crisp numerical values of the output.

The fuzzy controller chosen here was of PD type. The two inputs were the setpoint error $e = S_O^{\text{ref}} - S_O$ and the derivative of this error $\text{d}e/\text{d}t$, which was numerically estimated using a differentiator and a filter. These two inputs were fuzzified according to the trapezoidal membership function set shown in Fig. 1 and defuzzified using the method of center of gravity using the singleton membership functions for the output at the following set of values: $\{0, 0.25, 0.5, 1, 2.5, 3\}$. The rule base is summarized in Table 1.

Table 1. If-then rule base for the fuzzy controller.

$\frac{de}{dt} \setminus e$	NMG	NG	NM	NP	Z	PMP	PP	PM	PG	PMG
DNG	sV0	sV0	sV0	sV0	sV0	sV1	sV1	sV2	sV3	sV4
DN	sV0	sV0	sV0	sV0	sV0	sV1	sV1	sV3	sV4	sV5
DZ	sV0	sV0	sV0	sV0	sV0	sV1	sV3	sV4	sV5	sV5
DP	sV0	sV1	sV1	sV2	sV2	sV2	sV3	sV5	sV5	sV5
DPG	sV1	sV2	sV2	sV3	sV3	sV3	sV4	sV5	sV5	sV5

4. Results

The three controllers were tested first under simulation using *Simulink* and *Matlab* (The Mathworks, Inc.). The parameters for the model were obtained

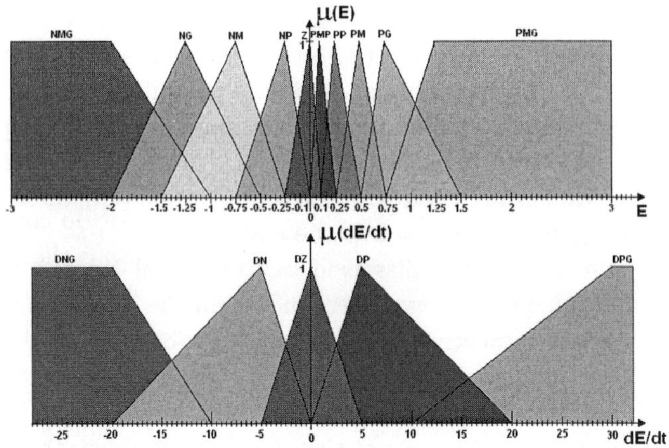

Fig. 1. Fuzzy membership functions for the inputs $e(t)$ and $\frac{de}{dt}$.

experimentally, resulting those shown in Table 2. The respiration rate was not calculated using a model for the bioreactor. Instead, it was calculated from off-line data of dissolved oxygen measurements from experiments that were run with $Q_{\text{air}}(t) = Q^*_{\text{air}}$. The resulting signal was used as an external input to the fourth order dissolved oxygen model for simulation.

Table 2. Parameters of the bioreactor in simulation

$\tau_k = 25$ [s]	$k^*_L a = 15.2[1/h]$	$m = 2.1$	
$S_O^{\text{sat}} = 6.8[mg/L]$	$\tau_1 = 36$ [s]	$\tau_2 = 72$ [s]	$Q^*_{\text{air}} = 3.5$ [L/min]

Once the controllers had been tested under simulation using real time data, they were implemented in *LabView* (National Instruments) as a module for the BIOREC program developed for monitoring and control of the bioreactors at the labaroatory.[2]

The experimental setup was an 8 liter laboratory SBR-type bioreactor inoculated with activated sludge from a nearby wastewater treatment plant and acclimated for 4-chlorophenol biodegradation. The temperature was regulated at 27±1°C by a recirculating water bath. Peristaltic feed and draw pumps (MasterFlex, Cole-Parmer), as well as an mixer were digitally controlled by the computer running the BIOREC program and equipped with a PCI-6025E data acquisition card (National Instruments). The air-flow was regulated using a programmable mass flow controller (Aalborg

Instruments, Inc.). The tests for the various controllers were performed in a batch system, *i.e.* $Q_{\text{in}} = 0$ for most of the reaction, except at the beginning when filling up to a working volume. Therefore the substrate concentration went from a high value of approximately 150 mg L^{-1} at the beginning of the reaction to practically zero at the end, and thus the respiration rate started at some value, increased to a maximum value and then decreased to the (low) endogenous rate towards the end.

Figures 2 and 3 show the dissolved oxygen signal for one SBR cycle when the fuzzy PD controller was used and when the linearizing plus PID controller were used, respectively. The setpoint for both cases was $S_O^{\text{ref}} = 4$ [mg/L]. Notice that despite the fact that the respiration rate was highly variable, both controllers perform quite well. The bump at the end for the second controller happens because at this moment the substrate is depleted and thus the respiration rate becomes rather small, too fast to be correctly estimated.

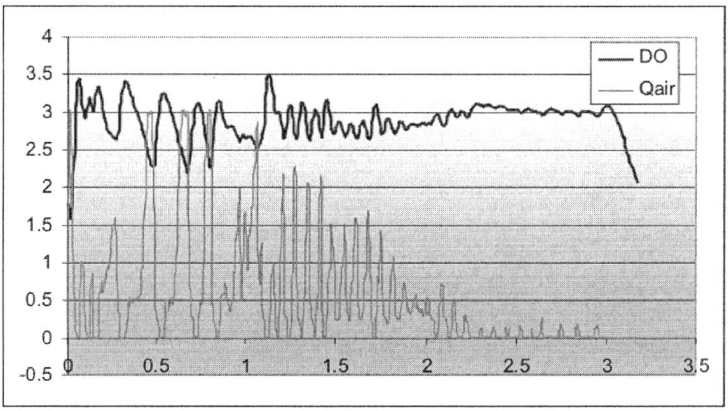

Fig. 2. Dissolved oxygen regulation using the fuzzy PD controller.

5. Conclusions

Three controllers were tested for dissolved oxygen regulation at a setpoint in a sequencing batch reactor used for toxic wastewater treatment. The respiration rate is highly variable during the reaction phase so the system never reaches a steady-state. Despite this, the controllers perform well, both in simulation and in real time for a pilot scale bioreactor used to treat

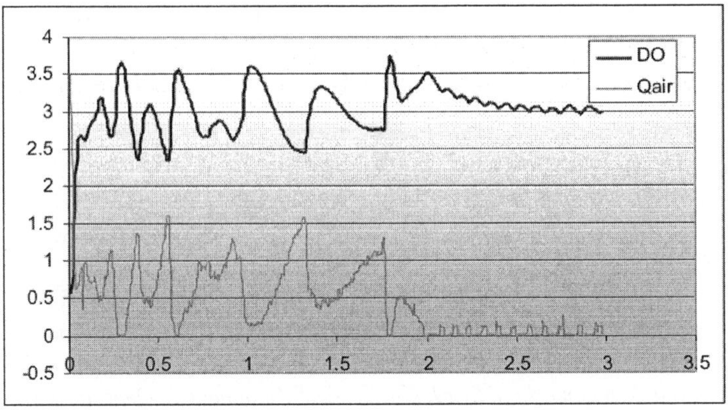

Fig. 3. Dissolved oxygen regulation using the linearizing+PID controller.

synthetic wastewater containing 4-chlorophenol. Currently, the controllers are being further tested and in the future it is expected to use the airflow signal to indirectly estimate the degradation rate in order to implement a time optimal controller.

Acknowledgements

This project was supported by DGAPA-UNAM (IN104805). The fourth author is grateful to CONACYT for the scholarship granted.

References

1. P. Wilderer, R. Irvine and M. Goronszy, *Sequencing Batch Reactor Technology*, Scientific and Technical Reports, Vol. 10 (IWA Publishing, London, 2001).
2. G. Buitrón, M.-E. Schoeb, I. Moreno-Andrade and J. Moreno, *Wat. Res.* **39** (2005).
3. M. Betancur, J. Moreno, I. Moreno-Andrade and G. Buitrón, *Int. J. Robust and Nonlinear Control* **16** (2006).
4. C.-F. Lindberg and B. Carlsson, *Wat. Sci. Tech.* **34** (1996).
5. W. Chotkowski, M. Brdys and K. Konarczak, *Int. J. Syst. Sci.* **36** (2005).
6. G. Olsson, M. Nielsen, Z. Yuan, A. Lynggaard-Jensen and J.-P. Steyer, *Instrumentation, Control and Automation in Wastewater Systems*, Scientific and Technical Reports, Vol. 15 (IWA Publishing, London, 2005).
7. B. Carlsson, C.-F. Lindberg, S. Hasselblad and S. Xu, *Wat. Sci. Tech.* **30**, 255 (1994).
8. K. Ang, G. Chong and Y. Li, *IEEE Trans. Control Syst. Tech.* **13** (2005).
9. D. Driankov, M. Reinfrank and H. Hellendoorn, *An Introduction to Fuzzy Control*, 2 edn. (Springer, New York, 1996).

COMPUTER-DRIVEN CHEMICAL VAPOR DEPOSITION REACTOR FOR THE DEPOSITION OF METALLIC OXIDE LAYERS AND MULTILAYERS

LUIS M. APÁTIGA C., EDGAR MÉNDEZ M., VICTOR M. CASTAÑO M.
Centro de Física Aplicada y Tecnología Avanzada, Universidad Nacional Autónoma de México, Querétaro, A.P. 1-1010, C.P. 76000, México

DOMINGO RANGEL M.
Centro de Física Aplicada y Tecnología Avanzada, Universidad Nacional Autónoma de México, Querétaro, A.P. 1-1010, C.P. 76000, México
Facultad de Ingeniería, DEPFI, Universidad Autónoma de Querétaro, Cerro de las Campanas s/n, Querétaro, C.P. 76010, México

A computer-driven Chemical Vapor Deposition (CVD) reactor for the deposition of metallic oxide layers and multilayers is described. The so-called Pulsed-Injection Metal Organic (PI-MOCVD) reactor is a low-pressure system for the growth of metallic oxide layers from a liquid solution of a metal-organic precursor. An effective pulsed injection mechanism for the precursor supply, normally used for fuel injection in internal-combustion engines, delivers a precise amount of liquid precursor to the reactor through injectors, whose pulse intervals (injection frequency) are controlled by a computer-driven system. Once the liquid precursor is injected, it is instantaneously evaporated (flash evaporation), so there is no time for chemical changes of the precursor. The precursor in vapor phase is transported through a carrier gas (Ar) towards the high temperature reaction chamber, where the oxide layers are deposited onto the hot substrates. The time between the injection and the arrival of the vapor at the substrate is extremely short (below 1 s), so less stable precursors can be used without the typical problems of classical CVD. Some results obtained to grow different metallic oxide coatings, such as cobalt oxide and titanium dioxide using commercially available metal-organic precursors, are presented along with the experimental parameters used in their deposition. In addition, different micrographs taken by Atomic Force Microscopy (AFM) and Scanning Electron Microscopy (SEM) of the coatings surfaces are shown, as well.

1. Introduction

Recent developments in ceramic technology are expanding the use of ceramics for structural applications by increasing their strength and resistance to fracture. The vast majority of commercially important ceramics are chemical compounds between at least one metallic element and one of five nonmetallic elements (C, N, O, P, or S). Oxide ceramics, which include some traditional materials such as magnesia (MgO), alumina (Al_2O_3), titania (TiO_2), etc., are widely used in industry. Solid materials grown as thin films fall into two general classifications, depending on the application: passive as for the majority of optical and mechanical coatings, and active, as for electro-opitical applications. By adding the appropriate impurities, some of the ceramics display semiconducting behavior, such as zinc oxide (ZnO), which is used as phosphor in color television screens.

Thin films of ceramic materials have been prepared from different deposition techniques such as spray pyrolysis, radio frequency sputtering, Chemical Vapor Deposition (CVD), pulsed laser deposition, sol-gel process, etc., onto a variety of substrates [1-5]. Among these techniques, CVD offers many advantages, such as uniform deposition over large area, conformal coverage and selective deposition. The CVD from metal–organic precursor (MOCVD) technique is a rapidly developing method for producing films and coatings of ceramic materials for a variety of applications [6-7]. In our laboratory, a pulsed injection MOCVD reactor, custom-designed and built in Vilnius University was installed recently to produce thin films of ceramic oxides on top of a variety of substrates. The reactor utilizes a computer-controlled interface in open loop mode to inject pulsed liquid-precursor into a low-pressure vertical reaction chamber to accomplish metal-organic CVD on a horizontal, conductively heated substrate [8]. The main goal of this present report is to demonstrate the feasibility of depositing oxide ceramics using this new, simpler and less expensive technology. In addition, the main characteristics of the pulsed injection MOCVD reactor and their principle of operation are presented as well as some results obtained in the deposition of cobalt and titanium oxide layers, along with their characterization made by AFM and SEM, which are the traditional techniques to study such layers.

2. Design and Operation Principle

Figure 1 shows a schematic drawing of the vertical pulsed-injection MOCVD reactor. The system is able to deposit oxide layers and multi-layers by using small micro-doses of precursors mixed in organic solvents [9]. The precise micro-doses of such solution are injected through a computer-driven system into the evaporation zone, where they instantly evaporate (flash evaporation), so there is no time for chemical changes of the precursor. The injected micro-doses in vapor phase are transported towards the deposition chamber by a The injected micro-doses in vapor phase are transported towards the deposition chamber by a carrier gas (Ar), where the oxidation gas source flows (O_2) and the chemical reaction and film growth on the hot substrate occurs.

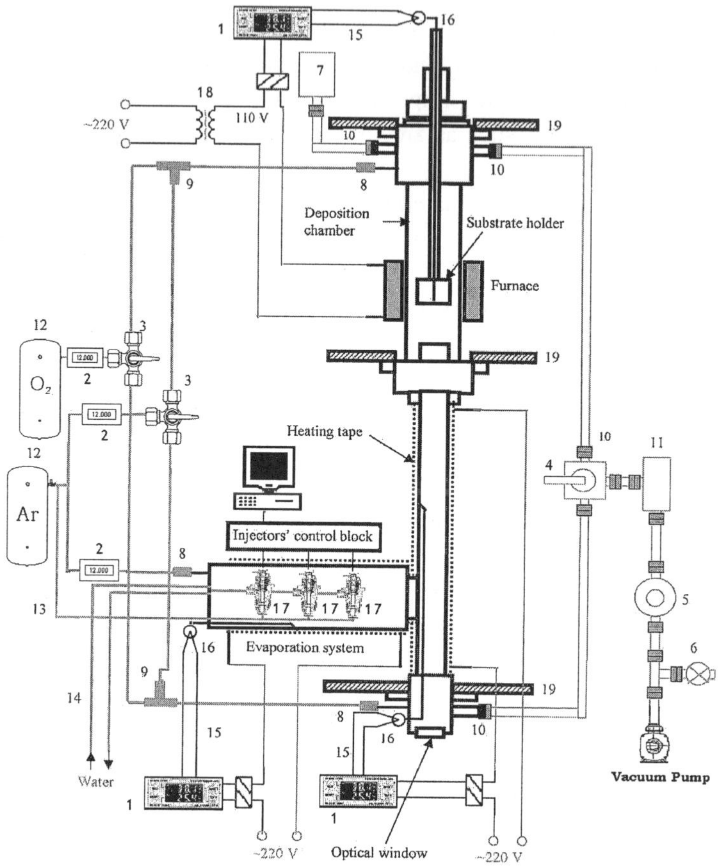

1. Temperature controller (3)
1a. Solid state relay (3)
2. Gas flowcontrollers (3)
3. Three way valve (2)
4. Three way valve (1)
5. Pressure regulation valve (1)
6. Valve (1)
7. Pressure gauge (1)
8. Compression fitting (3)
9. Compression fitting (tee)(2)
10. KF fitting (DN25)(~15)
11. Vapor trap (1)
12. Gas source (2)
13. Compression line for solution container
14. Water line for injectors´cooling
15. Thermocouple extension cable
16. Thermocouple (3)
17. Inyector- solution container system (3)
18. Transformer 220V/110V (1)
19. Shelf (3) for reactor fixation in the rack

Pulsed injection MOCVD reactor

Figure 1. Schematic diagram of the Pulsed-Injection MOCVD reactor.

The reactant and carrier gas that flows through the reaction chamber are regulated by means of digital electronic mass-flow controllers. Temperatures along the evaporation zone, vapor transport and substrate are measured by means of temperature controllers, enabling to set a number of important parameters, such as alarms, ramps, set point, power, etc. [10]. The reaction chamber is a fused quartz tube surrounded by a resistive furnace, which can heat the substrate up to 900 °C. The film thickness, roughness and morphology in general [11], can be adjusted through the injector frequency by means of the computer-driven system, using a computer program operating in open loop mode. This program was developed for the pulsed-injection MOCVD reactor.

The pressure in the reactor chamber is measured by using a digital Pirani manometer. Typical pressure in this reaction zone ranges from 2 to 10 torr, depending on the carrier and reactant gas flow. In addition, the flow of the reactant (O_2) and carrier (Ar) gases are measured and controlled by electronic mass flow controllers, operating typically in the range 0.2-1.4 liters per minute.
On the other hand, the application of several injection sources allows in-situ deposition of multilayered oxide structures. The computer program is able to vary the order of the operation of three injectors, as well as to control the time of the opening and the frequency of the injections, so as to perform many cycles with different injection parameters. The injector frequency ranges from 100 ms up to 10 s, meanwhile the opening time ranges from 2 to 10 ms, the computer program checks out for possible errors in the input of injector parameters. It is important to verify the minimum opening time value or pulse duration, which depends on the injector characteristics. Control data are sent to the system through an interface RS-232 on full duplex mode. The injector parameters are displayed on the computer screen, along with a digital counter that shows the total number of pulses. In general, the features of the computer program make the process very flexible, especially for the deposition of multilayer structures with different parameters and thicknesses [12].

3. Ceramic Oxide Deposition

Some results involvng the use of the pulsed liquid injection MOCVD technology in the deposition metal oxide layers are presented. We have already produced thin films of cobalt oxide (Co_2O_3) and titanium dioxide (namely titania, TiO_2) ceramics on silicon substrates [13]. Cobalt oxide ceramics as magnetic material are widely used in information storage devices basically, whereas the anatase phase of titania is useful as photocatalyst [14].

3.1 Cobalt oxide deposition

The experimental conditions for the synthesis of cobalt oxide films were as follows:

Experimental conditions	
Substrate	Si (100) wafer
Precursor	cobalt (II) acetylacetonate
Substrate temperature	650 °C
Total Pressure	10 Torr
Carrier gas flow (Ar)	0.8 l/min
Reactant gas flow (O_2)	0.4 l/min
Injection frequency	0.5 s (2 Hz)
Opening time	100 ms
Micro-dose per pulse	6.25 x 10^{-3} ml

Figure 2a shows an AFM overview of the cobalt oxide surface. The films were grayish in color, free of pin-holes and strongly adhered to the substrate, as observed in the AFM micrograph. The control of the experimental parameters is important since small variations could produce strong changes in the cobalt oxide morphology and therefore on the desired optical and magnetic properties. The experimental parameters used during the growth process led to a very homogeneous surface, since at the substrate temperature utilized (650 °C), the Ar and O_2 flow rates were high enough to produce smooth films by completely oxidizing the cobalt. Figure 2b show the shape of the injector pulse, which was measured by a Tektronix TDS 3032 digital oscilloscope. The amplitude of the injector pulse is 24V with an opening time of 2 ms in active state. These values are such since the process requires an optimal time in the injection pulses to obtain high performance in the film deposition. The shape of the injector pulse suggests that the injector is working satisfactory within the established optimal conditions.

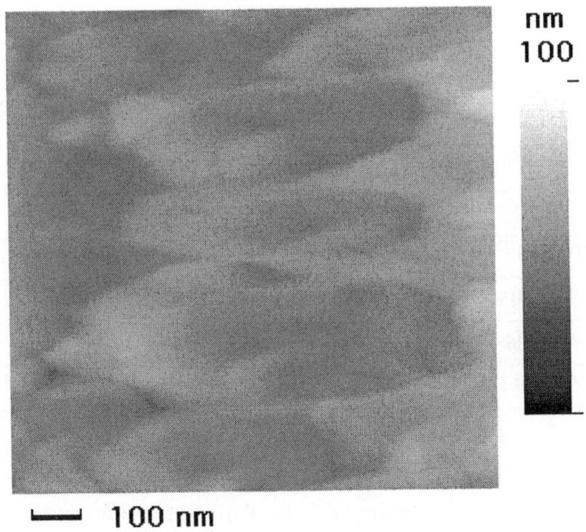

a)

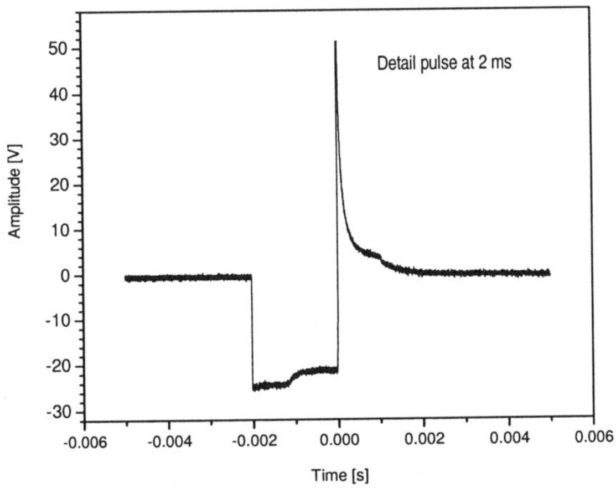

b)

Figure 2. a) AFM image of the cobalt oxide surface. b) Shape of an injector pulse measured by a digital oscilloscope.

3.2 Titanium dioxide deposition

For the synthesis of cobalt oxide films, the experimental parameters were:

Experimental conditions	
Substrate	Si (100) wafer
Precursor	titanium (IV) isopropoxide
Substrate temperature	500 °C
Total Pressure	6 Torr
Carrier gas flow (Ar)	0.8 l/min
Injection frequency	0.5 s (2 Hz)
Opening time	100 ms
Micro-dose per pulse	6.25 x 10^{-3} ml

Figure 3a shows a SEM micrograph of the titania surface. A crack-free nanostructured film surface is observed, forming stepwise nanostructures, which are the result of an overgrowth of nanometer-sized planes along well defined faces. The use of a single molecular titanium (IV) isopropoxide liquid solution at high concentration (without any solvent) as the precursor was sufficient to produce anatase films, without reactant gas. The reactivity of the oxygen present in the single molecular precursor along with the temperature and the experimental conditions provide thermodynamically favorable conditions to produce the stepwise nanostructured anatase thin films. Figure 3b shows the shape of several pulses at 4 Hz. The high reproducibility of the pulses provides an excellent efficiency to the reaction time. The spikes in the original signal of the injections are due to internal artifacts of the injector, which are damped by a power electronic circuit.

a) 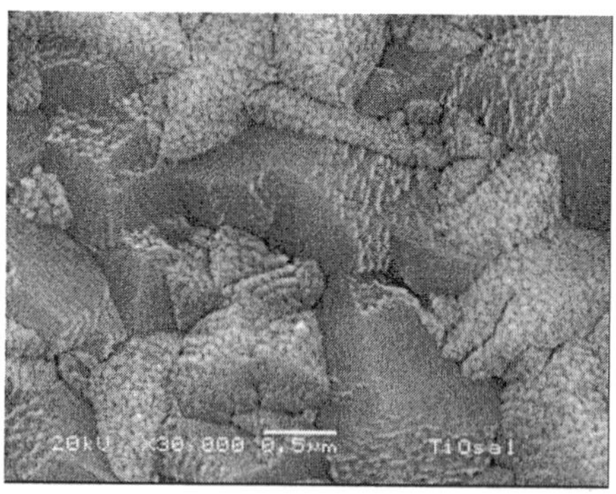

Figure 3. a) SEM image of the titanium dioxide surface and b) four pulses measured by a digital oscilloscope.

b) 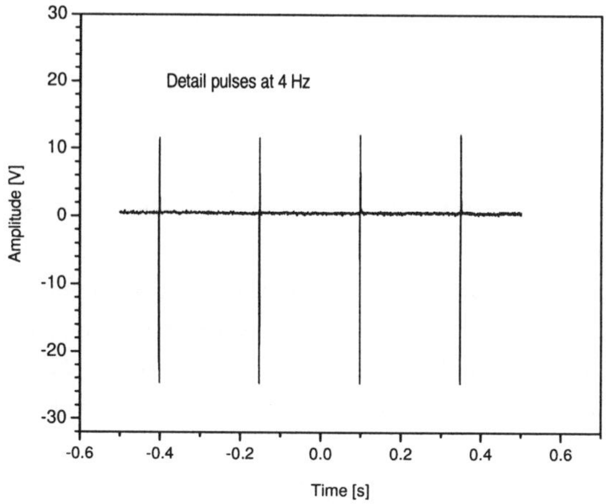

Figure 3. (*Continued*)

4. Conclusions

Although there are several factors to be considered in the nucleation and growth of cobalt and titanium oxide films by the PI-MOCVD technique, its principle of operation, which is based on the flash evaporation of a liquid precursor and its rapid transport in vapor phase to the deposition zone, seems to be an important factor relative to the formation of a solid material in thin film form. The principle for stable generation of precursors in vapor phase along with the computer-controlled mechanism makes the thin film deposition process very flexible and reproducible. The advantages of this novel technique over conventional CVD and other techniques are: significantly higher deposition rates, smoother film surfaces and higher film purity.

References

[1] N. Kovtyukhova, P.J. Olliver, S. Chizhik, A. Dubravin, E. Buzaneva, A. Gorchinskiy, A. Marchenko and N. Smirnova, *Thin Solid Films*, **337,** 166 (1999).

[2] S. Krumdieck and R. Raj, *Surface and Coatings Technology*, **141,** 7 (2001).

[3] M.L. Hitchman and S.E. Alexandrov, *The Electrochemical Society (Interface)*, **40** (2001).

[4] M.J. Henderson, D. King and J.W. White, *Aust. J. Chem.*, **56,** 933 (2003).

[5] T.M. Fuchs, R.C. Hoffmann, T.P. Niesen, H. Tew, J. Bill, F. Aldinger, *J. Mater. Chem.*, **12,** 1597 (2002).

[6] M.P. Moret, R. Zallen, D.P. Vijay and S.B. Desu, *Thin Solid Films*, **366,** 8 (2000).

[7] B.C. Kang, S.B. Lee, and J.H. Boo, *Surface and Coatings Technology*, **131,** 88 (2000).

[8] F.H. Kaatz, J.Y. Dai, P.R. Markworth, D.B. Buchholz, and R.P.H. Chang, *J. of Crystal Growth*, **247** , 509 (2003).

[9] L. Gao, P. Härter, Ch. Linsmeier, J. Gstöttner, R. Emling, and D. Schmitt-Landsiedel, *Materials Science in Semiconductor Processing*, **7** , 331 (2004).

[10] V. Woods, and N. Dietz, *Materials Science Engineering B*, **127** , 239 (2006).

[11] J. Achard, A. Tallaire, R. Sussmann, F. Silva, and A. Gicquel, *J. of Crystal Growth*, **284,** 396 (2005).

[12] V. Hopfe, D.W. Sheel, C.I.M.A. Spee, R. Tell, P. Martin, A. Beil, M. Pemble, R. Weiss, U. Vogt, and W. Graehlert, *Thin Solid Films*, 442, 60 (2005).

[13] L.M. Apátiga and V.M Castaño, *Thin Solid Films*, **496,** 576 (2005).

[14] A. Mills, G. Hill, S. Bhopal, I.P. Parkin, S.A. O'Neill, *J. of Photochemistry and Photobiology A; Chemistry*, **160,** 185 (2003).

NOISE CANCELLATION USING ADAPTIVE NEURAL NETWORKS

JESSE ELÍ QUIJANO AND CARLOS RAMÍREZ
Departamento de Electrónica y Computación, ITESM Campus Querétaro
Epigmenio González 500, Frac. San Pablo, Querétaro, Qro., México
jquijanoc@gmail.com, cramireg@itesm.mx

This paper describes ongoing work on a method for active noise cancellation. The method is based on adaptive neural networks (ANN). It can be particularly useful to clean a particular signal, such as voice, from any kind of noise in a combined signal, as far as a sample of the noise signal can be obtained independently. The idea is to actively, through the feedback ANN, subtract the noise signal from the combined one. Experiments were designed to clean a voice from a second one, and also to clean a voice from various sources of noise, coming from domestic appliances, such as a vacuum cleaner and a hair dryer. In all the study cases, the results showed an important reduction of the noise signal. The objective of this work was the exploration of an ADALINE neural network, as an alternative to least squares algorithms.

Keywords: ANC, active noise control, adaptive neural networks, Adaline

1. Introduction

Active noise cancellation (active noise control) is the name used since the 30's, to the method for the prevention of sounds non-wished in a signal. Nowadays, signal filtering and noise cancellation still are areas of research that offer important challenges [3]. Applications of Active Noise Cancellation (ANC) go from domestic applications, like the cancellation of a vacuum cleaner noise, or a hair dryer noise, to the cancellation of noise in particular locations of work within a factory or within the cabin of an airplane or automobile.

The traditional approach to acoustics control is carried out using passive techniques like isolations, barriers and silencers to attenuate the noise. The passive silencers also use the concept of change of impedance when a combination of speakers and tubes to silence the sound (reactive silencers), or use of energy inflict casualties by the propagation of the sound in a duct aligned with absorbent materials to the sound to provide the noise attenuation (resistive silencers). Most of these approaches are not sufficiently effective or are not practical for certain applications.

Adaptive ANC systems were described in [7], a steepest descent gradient-type algorithm is used to drive the parameters of a FIR filter, in order to reduce a function of the error between the noise to be cancelled and the cancellation´s system output, requiring of a training signal that is not explained in the algorithm. Most ANC systems describen in the literature are mainly applied to ducts or headsets for dynamic cancellation of noise of engine aircrafts, through the use of different techniques, such as genetic algorithms [6] and adaptive feedback LMS algorithms [7][8].

In this work it is proposed a method for cleaning acoustic signals based on Adaline, an adaptive neural network (ANN), by means of the prediction of the interference noise contribution to a combined signal *m*, where *m* is the combined signal of a noise signal *n* and some other signal *v*, such as a human voice. The model consists of two parts: in first part the treatment of the input signals *n* and *m* is carried out, *i.e.*, the signal that transports the noise and the contaminated signal, respectively; the second part is concerned with the acoustic environment, which is related to the output of the neural network. The cleaning process works as follows: from the iterative operation of the ANN it is obtained an error *e* as a result of processing the noise and the contaminated signal, *e* is iteratively used to feedback the network; the model is designed such that the error turns out to be *v*, the voice, not the noise, after subtracting *n*, from *m*.

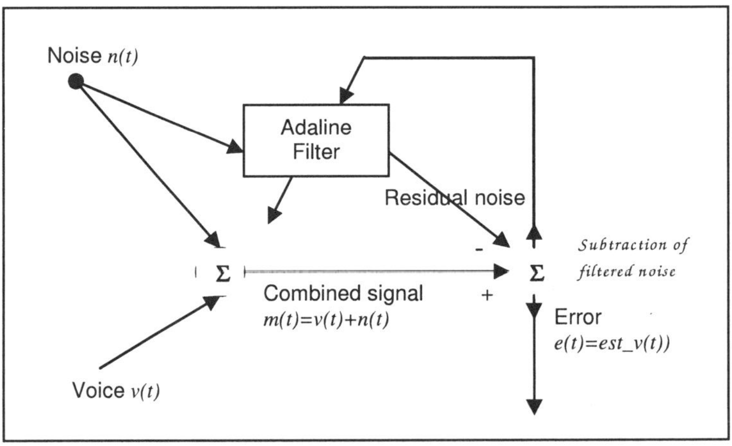

Fig. 1. Schematic of Adaline adaptive feedback ANC system.

2. Adaline Linear Network

The adaptive linear element (Adaline) is a 'learning' system, suggested by Widrow and Hoff [1] that sum up their input signals linearly. An Adaline unit consists basically of a linear combination of input units, and an output as a function of the activation input values, the weights of the connections, and a transfer function, as shown graphically in Fig. 2. The adaline learning algorithm tries to minimize the error of the output adjusting the weights of the connections using the delta rule (described below), in terms of the network output error, but before applying the transfer function.

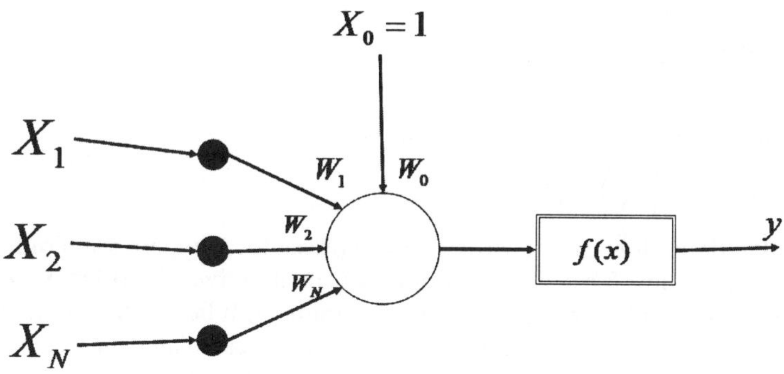

Fig. 2. The Adaline architecture.

2.1 *The Delta Rule*

The *delta rule*[1] is an important generalization of the perceptron training algorithm, presented by Widrow and Hoff as the 'least mean square' (LMS) learning procedure, that extends this technique to continuous inputs and outputs. Given a set of inputs $(\vec{x}_1, \vec{x}_2, \ldots, \vec{x}_n)$ and their corresponding outputs $(d_1, d_2, \ldots, d_n)$, the election of the optimal weights $\vec{w}^*$ depends of what it's understood as optimal. If the equations system has not an exact solution it is going to be chosen as optimal the solution that minimizes the quadratic mean error

$$\left\langle E_k^2 \right\rangle = \frac{1}{L}\sum_{k=1}^{L}(d_k - y_k)^2 = \frac{1}{L}\sum_{k=1}^{L} E_k^2 ,$$

representing L the number of vectors of the training set. Expanding the mean error equation:

$$\left\langle E_k^2 \right\rangle = \left\langle (d_k - \vec{w}^T \vec{x})_k{}^2 \right\rangle = \left\langle d_k{}^2 \right\rangle + \vec{w}^T \left\langle \vec{x}_k \vec{x}_k^T \right\rangle \vec{w} - 2\left\langle d_k \vec{x}_k^T \right\rangle \vec{w}$$

assuming the set is statistically stationary. Then if the correlation matrix of the input is defined as $R = \left\langle \vec{x}_k \cdot \vec{x}_k^T \right\rangle$, the vector $\vec{p} = \left\langle d_k \vec{x}_k \right\rangle$, and $\varepsilon = \left\langle E_k^2 \right\rangle$, then the definitions the previous equation is:

$$\varepsilon = \left\langle d_k \right\rangle + \vec{w}^T R \vec{w} - 2\vec{p}^T \vec{w}$$

This equation shows that *w* is a function of the weights vector *w*. In order to minimize the quadratic mean error it is derived with respect to *w*, and set to zero.

$$\frac{\partial \varepsilon(\vec{w})}{\partial \vec{w}} = \nabla \varepsilon = 2R\vec{w} - 2\vec{p}$$

$$2R\vec{w}^* - 2\vec{p} = 0$$

$$R\vec{w}^* = \vec{p}$$

$$\vec{w}^* = R^{-1}\vec{p}$$

In this case the minimum quadratic error is

$$\varepsilon_{\min} = \left\langle d_k^2 \right\rangle - \vec{p}^T \vec{w}^*$$

Given the definition of the error, this represents a hyperparaboloid with a negative minimum. This gives the idea of an iterative algorithm to find the optimal: the method of the reduction by the gradient. It begins to assign arbitrary values to the weights. The gradient of the error is calculated in this point and then the direction of greater reduction is obtained. The values of the weights in that direction are changed slightly and it is repeated until it reaches the minimum. The rule to update the weights is

$$\vec{w}(t+1) = \vec{w}(t) - \eta \cdot \nabla \varepsilon(\vec{w}(t))$$

The constant η is the learning rate. The problem is that to calculate the gradient it is required to know ***R*** and ***p***, this implies to know the statistics of the signal. To avoid it, the instant quadratic error is used as an approximation of the mean error, and the gradient is calculated from it:

$$E_k^2(t) = (d_k - \vec{w}^t(t)\vec{x}_k)$$

$$\nabla E_k^2(t) \approx \nabla \left\langle E_k^2 \right\rangle$$

$$\nabla E_k^2(t) = -2E_k(t)\vec{x}_k$$

Then, the delta learning rule is

$$\vec{w}(t+1) = \vec{w}(t) + 2\eta \cdot E_k \vec{x}_k$$

In order to generate analog outputs, a sigmoid transfer function is normally applied to the output of the adaline network.

The adaline algorithm works as follows, having as variables the number of input units, weights, and the learning rate.

1) Initialization of weights.
2) An input pattern is applied to the input layer.
3) The linear output is computed from the network.
4) The error is calculated for this pattern, comparing the outputs against the expected output.
5) The connections are updated by means of the delta rule (explained above).
6) If the average quadratic error is acceptable, the process ends, otherwise steps 2 to 5 are repeated for all the training instances.

2.2 *Training Adaline*

In order to train the neural network it was carried out a number of experiments with recorded acoustic signals and also processing the audio signals in real time, through microphones located near the sources. The network received its inputs in groups of 3 frequency bands of the corresponding signals, the signal to clean (voice) and the noise signals. For the case of noise cleaning in real time, two independent inputs from different microphones were used [2], one corresponding to the noise to learn and the other to the combined signal. Recordings of 10 seconds length were made, making 10 different instances of each signal. It was taken 64 samples from each signal, ie., the network received 65 inputs (64 plus 1 bias input). Initial weights are random, a learning rate of 0.001 to 0.03 was used for all samples, then the adaline network was applied to each of the samplings, producing the corresponding mixed attenuated noise. Finally, it was concatenated, in the recording, the 10 output attenuated signals, for analysis of the complete sequence, in order to verify how the network is learning in real time and its corresponding gradual decrease of noise in the contaminated signal.

The speed of adaptation of the algorithm depends on the size N of the weights vector, the learning rate η, and the total power of the input:

$$\tau = \frac{N}{2\eta \cdot tr(R)}$$

where R is the correlation matrix of the adaline inputs as seen above. In order for the adaptation speed to remain close to a constant it is possible to adapt the “learning rate” to the total input power, having in mind the condition of stability. A way to implement this condition is to take

$$\eta \approx \frac{1}{tr(R)}$$

which is the inverse of the total input power, which guarantees the condition of stability, since the trace of a square matrix is equal to the addition of its eigenvalues:

$$\mathrm{tr}(A) = \sum \lambda_i$$

3. Experiments and Results

The goal of this work is to clean a contaminated signal where voice is mixed with noise coming from diverse sources, such as domestic appliances (*e.g.*, hair dryer, vacuum cleaner, etc.), or even a second voice. Psychoacoustics[9] plays an important role here, since the system should be modeled and can also take advantage of some of the characteristics of the behaviour of human hearing.

3.1 *Critical Bands*

Studies concerning the discrimination of frequencies of the human hearing have shown that in low frequencies, tones with a few hertz of separation can be distinguished; meanwhile, at high frequencies it is required a separation of hundreds of hertz. Nevertheless, the ear responds to the most powerful stimuli present at any region of the spectrum. Critical bands is the name given to this behaviour. The critical bands are approximately 100Hz wide for frequencies between 20 and 400Hz; the wide of the bands increases logarithmically according to the frequency:

$$\textit{Critic Band Wide (Hz)} = 24.7\ (4.3F + 1)$$

where F is the central frequency in Hz.

Bark is the unit of perceptual frequency; one bark measures the critic band rate, ie., a critic band has a wide of one bark. The bark scale is a psychoacoustical scale. Eberhard Zwicker[10] defined a scale that ranges from 1 to 24 and corresponds to the first 24 critical bands of hearing. The subsequent band edges are (in Hz) 0, 100, 200, 300, 400, 510, 630, 770, 920, 1080, 1270, 1480, 1720, 2000, 2320, 2700, 3150, 3700, 4400, 5300, 6400, 7700, 9500, 12000, 15500. It is related to, but somewhat less popular than the mel scale.
To convert a frequency f (Hz) into Bark it is used:

$$Bark = 13 \arctan(0.00076f) + 3.5 \arctan((f/7500)^2)$$

3.2 *Experiments*

The system was implemented in Matlab 7 to take advantage to its facilities to deal with signals and audio cards. All signals were sampled during 10 seconds at 44 and 22Khz, 16 bits-data and recorded in wav format.

In figure 3 it is shown graphs of the signals corresponding to an experiment involving a hair dryer as the source of noise, and the voice of one of the authors. A single microphone was used to record the combined signal of voice and noise, locating the microphone in the middle between the sources of noise and voice, 10 cms. apart from each one.

The comparative graphs shown in figure 4 correspond to an experiment carried out with a hair dryer used as the source of noise, and the voice of one of the authors as the signal to clean.

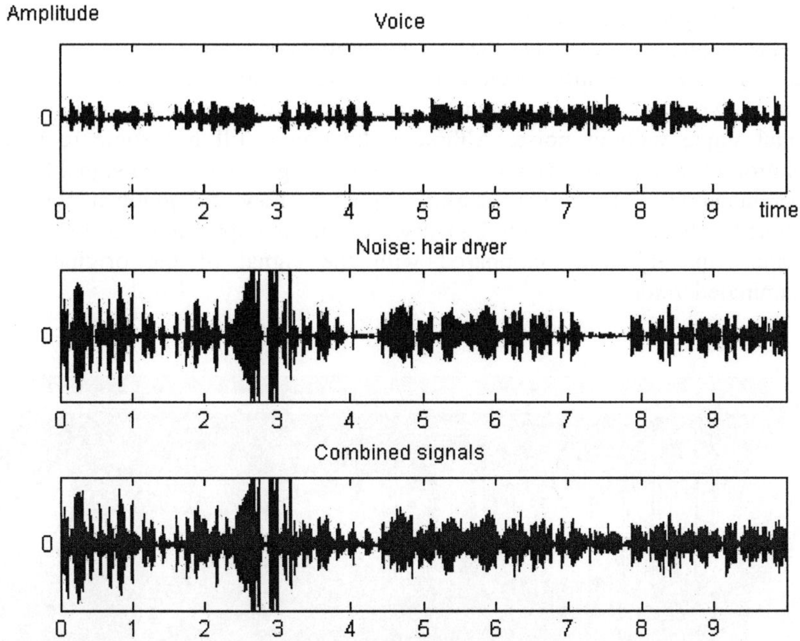

Fig. 3 Input signals.

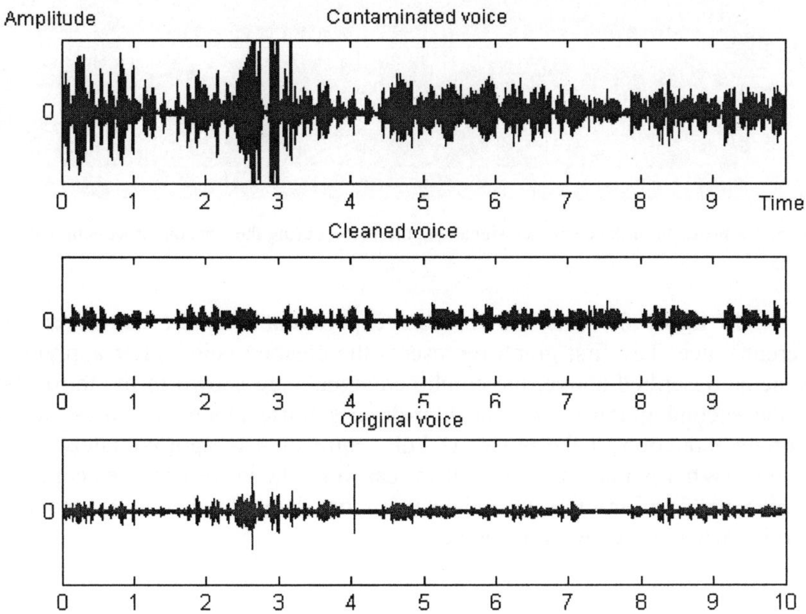

Fig. 4. Sampling of the original signal and the cleaned one after applying Adaline.

The first graph corresponds to the sampled signal of the voice contaminated with the signal of the dryer, the second graph corresponds to the recovered signal, *i.e.*, the voice, after been applied the noise subtraction with the adaline neural network to the mixed signal, and in the third graph its is shown the original voice without noise. Although the voice still has some remaining 'background' noise and suffered some time warping, the resulting signal became comprehensible to the human hearing. Figure 5 shows the graph of the voice combined with the noise of the hair-dryer during the application of the adaline algorithm in real time, compared with the signal of the original, non-contaminated voice.

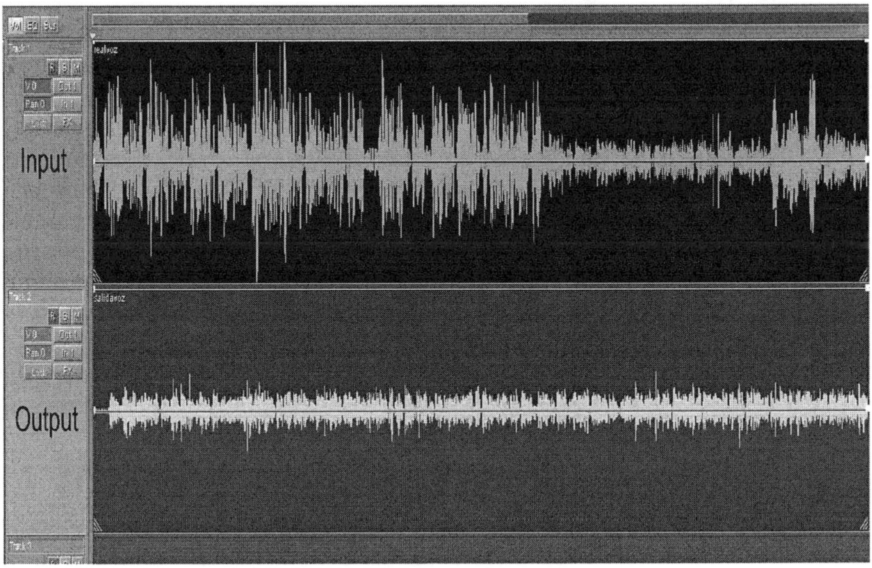

Fig. 5. Comparison of the combined signal and the voice, during the cleaning process in real time.

In figure 6, its is shown other experiment with the hair-dryer and a different voice. The first graph represents the cleaned voice, after applying the adaline as described above, the combined signals are shown in the third graph, and the second graph shows the hair-dryer's noise alone. It can be seen an important reduction of the noise: 18.7dB, equivalent to approximately 86% of the noise. What is interesting is that in real time, the voice becomes completely comprehensible after a few seconds, never more than 10, when it was practically inaudible when mixed with the noise.

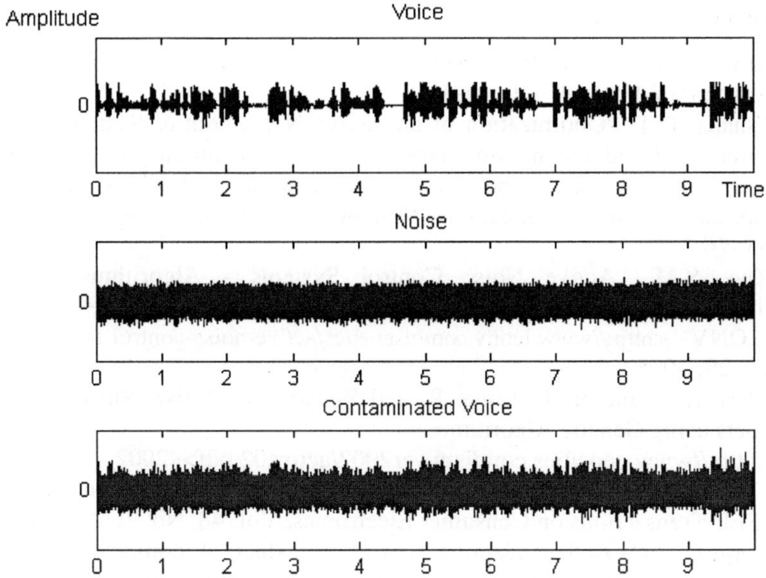

Fig. 6. Noise reduction using adaline algorithm.

4. Conclusion

Experimentally it has been observed an effective reduction of noise, both, on periodic noise, such as the hair-dryer, as with a second more powerful voice took as the noise input signal, and even with white noise. An explanation of this is due to the model: the idea is that the adaline does not work as a filter, it is actually a substraction of noise what is carried out, since the resultant "error" component is the voice, ie., the cleaned signal.

The speed of the signal sampling and the error rate to use in the neural network can be a factor to consider; in the experiments, using a learning error rate of 0.04, the network reached convergence in all the tested cases in a maximum time of 10 seconds, in a Pentium computer to 1200MHz and 256MB RAM, which was enough to carry out the noise cancellation in real time. It is required to carry out more experiments, varying the error rate to a less demanding training condition and trying also other network configurations. Critics bands sampling may also be improved. It is also planned to work on the delays due to sound card interfaces.

References

1. Krose B.J.A., van der Smag, P.P. “Neural Networks”, University of Amsterdam Press, 4th ed., 1991.
2. Tshalis, D.T. “Optimization of an Active Noise control System Inside an Aircraft, Based on the simultaneous optimal positioning of microphones and speakers, with the use of genetic algorithm”. Computational Optimization and Applications. Volume 23, Issue 1, October 2002, Pages: 65 -76.
3. Kuo, S.M. “Active Noise Control Systems - Algorithms and DSP Implementations”, Dennos Morgan. Ed., Wiley Interscience. 1996.
4. iCONV. <http://www.iconv.com/usenet/es/active-noise-control-faq_enes.htm> Jan 29, 2005.
5. Werner, J., Junior, J., Lima, R. and Fogarty, T. Active Noise Control in ducts using Genetic Algorithms. <http://www.geocities.com/jamwer2002/active02.pdf>, 2002.
6. Gan, W.S., Kuo, S.M. An integrated audio and active noise control headset. IEEE Transactions on Consumer Electronics, Vol. 48, No. 2, May 2002.
7. Burgess, J.C. Active adaptive control in a duct: A computer simulation, Journal of Acoustics Soc. Amer., Vol 70, pp 715-726, Sept. 1991.
8. Ericksson L.J., Allie, M.C., Greiner, R.A. “The selection and application of an IIR adaptive filter for use in active sound attenuation.” IEEE Transactions on Acoustics, Speech and Signals Proceedings, Vol 35, No. 4, April 1987.
9. ‘Bandas Criticas’, Universidad de las Américas, Puebla, 17-04-2005, http://ict.pue.udlap.mx/people/raulms/avances/codificacion.html.
10. Principio del formulario
11. Final del formulario
12. Psychoacoustics : facts and models”, Berlin ; New York : Springer, 1999.

ADAPTIVE NEURAL NETWORKS FORECASTING AND ITS ROLE IN IMPROVING A CAMLESS ENGINE CONTROLLER

MOH'D SAMI" S. ASHHAB

Mechanical Engineering Department, The Hashemite University, Zarqa, 13115, Jordan
E-mail: sami@hu.edu.jo, Tel: (962) 5-390-3333, Fax: (962) 5-382-6348

A method for adapting feed-forward neural networks is proposed. The technique handles multi-input multi-output neural networks and is a generalization of previous research results presented in [1] where adaptation of single output feed-forward neural networks was developed. The artificial neural net (ANN) is trained with historical time series input-output process data. Once trained, the ANN forecasts the process outputs in the future. It is assumed that the ANN is linear in the output weight matrix and bias vector which are parameters of the net. This linearity property allows the use of the Kaczmarz's projection algorithm for updating the individual output weight vectors and biases on-line to improve the prediction accuracy. The algorithm uses the errors between the outputs measurements and the predicted outputs values to update the network's parameters recursively. The method's capability is demonstrated through computer simulation on the breathing process in camless internal combustion engines. The adaptive ANN can improve the performance of an ANN based camless engine inverse controller.

1. Introduction

Artificial neural networks (ANN's) have received a lot of attention in recent years due to their attractive capabilities in forecasting, modelling of complex nonlinear systems and control. Applications of neural networks are numerous and include many various fields among which are engineering and business. ANN's have been used for forecasting camless engine torque [1], load [2,3], gasoline consumption [4], energy [5], space weather [6], outdoor sound transmission [7], stream flow [8], wind waves [9] and financial indicators [10-12]. Examples of industrial processes for which modelling and control using neural networks have been investigated include internal combustion engines [13,14], two-stage combustor burning ethylene in air [15] and steel making process [16].

Artificial neural networks are widely used for forecasting. A large number of successful applications have shown that neural networks can be a very useful tool for time series modelling and forecasting [17,18]. The ANN model is trained with historical time series input-output process data or observations and is then used to predict the outputs in the future. Due to gradual degradation of the underlying process, the short-term predictions will be more accurate than the long-term ones. Thus, there is a need to adapt the neural net in order to better accommodate the changing environment and improve the net's prediction accuracy especially when the forecasting horizon is long.

Much effort has been devoted over the past several decades to develop and improve the neural net forecasting models. Training the neural network is a time consuming task and computationally expensive. Thus, adapting the net by subsequent post-training with the updated observations on the item to be forecast is a tedious process. Alternative and better approaches for adapting ANN's have been investigated in the literature. A new time-series forecasting model based on the flexible neural tree was introduced in [19] where the model handles automatically the task of selecting the proper input variables and time-lags to a multi-layer feed-forward neural network. Long-term dependencies can be learnt by neural networks if time-delayed connections are included in the net [20]. The ANN input time delays are updated in [21] to better accommodate the changing process and the simulation results of a time series prediction of Mackey-Glass delay-differential equation and the Korean stock market prediction show that the technique is effective.

In this paper, we propose a method for adapting multi-input multi-output feed-forward neural networks. The proposed algorithm is a generalization of the technique developed in [1] for single-output feed-forward neural networks. Many of the industrial and practical systems have multiple outputs. Thus, the adaptive multi-output neural networks serve a bigger family of processes. The proposed method works for nets that are linear in their output weight matrix and bias vector. The linearity of the neural network in these parameters allows the implementation of the Kaczmarz's adaptive algorithm which updates them recursively. It is assumed that the outputs measurements are available. The adaptive algorithm uses the errors between the measured and predicted (ANN) outputs to update the individual output weight vectors and biases in order to make more accurate forecasts in the future. The computation involved with the proposed method is simple when compared to other available adaptive techniques. The approach responds well to systems that are gradually degrading. Adapting the ANN with the Kaczmarz's algorithm (as proposed in this paper) keeps the weights or importance of the inputs constant.

The method's capability is demonstrated through computer simulation on cylinder air charge/energy loss prediction in camless internal combustion engines. The simulation results show that the adaptive algorithm achieves prediction accuracy and compensates for changes due to speed fast. On the other hand, the static ANN gives large errors in predicting the outputs when the speed changes. The adaptive ANN can improve the performance of the ANN based camless engine inverse controller developed in [14] and eliminates the need to produce engine data at speeds other than 1500 RPM which reduces the memory usage in the engine computer.

2. Problem Formulation

We will consider multi-input multi-output systems that are governed by the following nonlinear discrete-time difference equation

$$Y(t+1) = F(U(t)) \tag{1}$$

or equivalently,

$$\begin{bmatrix} y_1(t+1) \\ y_2(t+1) \\ \vdots \\ y_q(t+1) \end{bmatrix} = \begin{bmatrix} f_1(U(t)) \\ f_2(U(t)) \\ \vdots \\ f_q(U(t)) \end{bmatrix}, \tag{2}$$

where, $U(t)$ is a column vector given as

$$\begin{aligned} U(t) = [&y_1(t)\cdots y_1(t-n_1+1) \;\; y_2(t)\cdots y_2(t-n_2+1)\cdots \\ &y_q(t)\cdots y_q(t-n_q+1) \;\; u_1(t)\cdots u_1(t-m_1+1) \\ &u_2(t)\cdots u_2(t-m_2+1)\cdots u_p(t)\cdots u_p(t-m_p+1)]^{\mathrm{T}} \end{aligned} \tag{3}$$

The system output column vector $Y(t)$ depends (in the sense defined by the nonlinear column vector map F) on the past n_i, $i=1,2,\ldots,q$ output values, where, q is the number of outputs and on the past m_i, $i=1,2,\ldots,p$ input values, where, p is the number of inputs.

Multi-input multi-output feed-forward artificial neural networks can be used to model forecasting systems of the form given in Equation (1) or (2). Knowledge about the system dynamics and mapping characteristics is implicitly stored within the network that is trained using historical time series input-output process data. The ANN model is a nonlinear functional approximation of the real system. Useful information and theory about ANN's can be found in [22]. These networks are used as models for processes that have input-output data available. The historical observations allow the neural network to be trained such that the errors between the real outputs and the estimated (neural net) outputs are minimized. The model is then used for different purposes among which are estimation and control.

It is satisfactory to consider neural networks with one hidden layer as explained in [1]. An artificial neural net mathematical model is written as

$$Y_{nn}(t+1) = W_o * \tanh(W_i * U(t) + B_i) + B_o \tag{4}$$

where, Y_{nn} $(t+1)$ is a column vector that contains the q outputs of the neural net model, $U(t)$ is given in Equation (3) and is a column vector of size $N=n_1+\ldots+n_q+m_1+\ldots+m_p$ that contains the inputs to the ANN, W_o is a matrix of size qxh that contains the output weights from the hidden layer to the outputs with h being the number of hidden neurons, W_i is a matrix of size hxN that contains the input weights from the inputs to the hidden layer, B_i is a column vector of size h that contains the input biases and B_o is a column vector of size q that contains the output biases. Note that $\tanh(W_i * U(t) + B_i)$ is the activation function of the hidden layer. It is a column vector of size h.

For a given number of hidden neurons the network is trained to calculate the optimum values of the weights and biases that minimize the errors between the real and ANN outputs. We assume that after an appropriate choice of the number of hidden neurons and a suitable training period, the network gives a good representation of the forecasting system given in Equation (1) or (2). The resultant model is of the form given in Equation (4) and can be rewritten as

$$y_{nni}(t+1) = \beta(U(t))\alpha_i, i = 1,2,\cdots,q \tag{5}$$

where, y_{nni} is the ith output of the ANN model,

$$\beta(U(t)) = \begin{bmatrix} \tanh(W_i * U(t) + B_i) \\ 1 \end{bmatrix}^{\mathrm{T}}, \tag{6}$$

$$\alpha_i = \begin{bmatrix} W_{oi}^{\mathrm{T}} \\ b_{oi} \end{bmatrix}, \tag{7}$$

W_{oi} is the ith row of the matrix W_o, and b_{oi} is the ith element of the vector B_o.

It is clear that the estimated (neural network) future value of the ith output is linear in the parameter α_i that is a column vector composed of the ith output weight vector W_{oi} and bias b_{oi}. On the other hand, the dependence on the input weight matrix and bias vector is nonlinear. The linearity in the parameter α_i will enable adapting the ANN using the Kaczmarz's projection algorithm as we will see in the next section.

3. Neural Network Adaptation

The ANN is used to predict the real system outputs in the future. Due to possible gradual degradation of the underlying process, the short-term predictions will be more accurate than the long-term ones. Thus, there is a need to adapt the neural net in order to better accommodate the changing environment and improve the net's prediction accuracy over the long run. The linearity property of the network outputs y_{nni} in the parameters α_i, $i = 1,2,\ldots,q$ (see Equation (5)) allows the use of available adaptation schemes for updating these parameters. One of those simplified adaptation algorithms is the Kaczmarz's algorithm [23]. It leads to less computational complexity and quick convergence. The real-time parameter estimation Kaczmarz's algorithm when applied to the ANN model described by Equation (5) gives the following recursive adaptation scheme for α_i

$$\alpha_i(t+1) = \alpha_i(t) + \frac{\beta(U(t-1))^{\mathrm{T}}}{\beta(U(t-1))\beta(U(t-1))^{\mathrm{T}}} e_i \tag{8}$$

where the error, e_i, is the difference between the ith real and estimated outputs at the time instant t, that is

$$e_i = y_i(t) - y_{nn_i}(t) \tag{9}$$

We can rewrite Equation (8) as

$$[\alpha_1(t+1)\,\alpha_2(t+1)\cdots\alpha_q(t+1)] = [\alpha_1(t)\,\alpha_2(t)\cdots\alpha_q(t)] + \frac{\beta(U(t-1))^{\mathrm{T}}}{\beta(U(t-1))\beta(U(t-1))^{\mathrm{T}}}[e_1\,e_2\cdots e_q] \tag{10}$$

We use Equation (7) to simplify Equation (10) to the following form

$$\alpha(t+1) = \alpha(t) + \frac{\beta(U(t-1))^{\mathrm{T}}}{\beta(U(t-1))\beta(U(t-1))^{\mathrm{T}}} E^{\mathrm{T}}, \tag{11}$$

where,

$$\alpha = \begin{bmatrix} W_o^T \\ B_o^T \end{bmatrix} \tag{12}$$

and

$$E = Y(t) - Y_{nn}(t). \tag{13}$$

Note that the on-line adaptive algorithm described in Equations (11), (12) and (13) assumes that the outputs measurements at the current time t are available. This adaptation scheme is a generalization of the technique developed in [1] for single-output feed-forward neural networks. The estimated outputs of the ANN model are compared with the measured real outputs of the system. If there is a difference between any two of the values the parameter estimation Kaczmarz's algorithm updates α_i (according to Equation (8)) to be used in the subsequent predictions. The parameter estimator or adaptive algorithm compensates for process degradation and modelling errors and provides more accurate long-term predictions.

4. Application to Camless Engines

An interesting application where the proposed adaptation algorithm can be applied is the cylinder air charge/energy loss prediction during the breathing process in camless internal combustion engines. In this kind of engines the intake valve motion that allows air into the cylinder is controlled by an electrohydraulic actuator. The actuator controls the intake valve lift and intake valve closing timing. During breathing air enters into the engine cylinders while the intake valve is open. The mass of air that enters the cylinder during the breathing process is called cylinder air charge (*CAC*) whereas the energy loss is called the pumping loss (*PL*). These two quantities depend on the intake valve lift (*IVL*), intake valve closing timing (*IVC*) and engine speed (*S*). We will consider *IVL* and *IVC* as inputs whereas *S* will be considered as a system parameter. The outputs are the cylinder air charge and pumping loss. A model based on thermodynamics laws was developed for the breathing process in [24]. The cylinder air charge and pumping loss onc time unit ahead are written as

$$\begin{bmatrix} CAC(t+1) \\ PL(t+1) \end{bmatrix} = \begin{bmatrix} f_1(IVL(t), IVC(t), S(t)) \\ f_2(IVL(t), IVC(t), S(t)) \end{bmatrix}. \tag{14}$$

Based on the model developed in [24] we generate input-output data at a speed $S = 1500$ RPM. The data contains 84 patterns. Each pattern includes data about the two inputs and the corresponding two outputs. The intake valve lift ranges from 1 mm to 7 mm with a 1 mm increment, whereas the intake valve closing timing ranges from 60 to 280 degrees with a 20 degree increment of crank angle.

Seventy two of the eighty four data patterns are used to train an artificial neural net model for the breathing process, whereas the remaining twelve data patterns are used to test the performance of the model. The training was done with the software package Matlab. We ran experiments for different numbers of hidden neurons. It was observed that the quality of the results depends on the number of hidden neurons. We choose the neural net with seven hidden neurons. This is the smallest number of hidden neurons with acceptable least square errors over the training and test regions. The real and predicted (ANN) values of cylinder air charge and pumping loss over the training and test regions are plotted in Figure 1 as functions of the pattern index. Note that the predicted values are very close to the real ones which indicates that the ANN model is accurate.

The ANN is used to predict the *CAC* and *PL*. It is shown in [25] that the cylinder air charge and pumping loss can be calculated from available engine sensors. Thus, the ANN can be adapted as described in this research to compensate for variations in speed *S* (from 1500 RPM), actuator characteristics, geometric parameters of the engine and intake valve effective area which depends on deposits and wear. We will test the ability of the adaptive ANN in capturing speed variations. Let us suppose that we have the following scenario:

Time	*IVL*	*IVC*	*S*
0-9	4	180	1500
10-19	4	180	3000
20-29	5	180	3000
30-39	6	200	3000

The corresponding simulation results are shown in Figures 2 and 3. Note that adapting the ANN produces accurate *CAC* and *PL* predictions. If the net is not updated it fails to predict the *CAC* and *PL* accurately. The speed changes from 1500 RPM to 3000 RPM at $t = 10$. The *CAC* and *PL* measurements at $t = 11$ are compared with the corresponding ANN values and the errors are used to update the ANN parameters. Prediction accuracy is achieved at $t = 12$ and therefore the adaptive ANN compensates for changes due to speed. In addition, the adaptive ANN corrects for errors as the inputs (*IVL* and *IVC*) change at $t = 20$ and 30. Therefore, the adaptive algorithm achieves its objective.

The adaptive ANN algorithm can be connected to the ANN based camless engine inverse controller developed in [14] whose performance depends highly on the accuracy of the ANN. This leads to a better and more robust controller performance and hence a more satisfactory driver's torque demand and better fuel economy. In addition, the adaptive ANN eliminates the need to produce engine data at speeds other than 1500 RPM which reduces the memory usage in the engine computer.

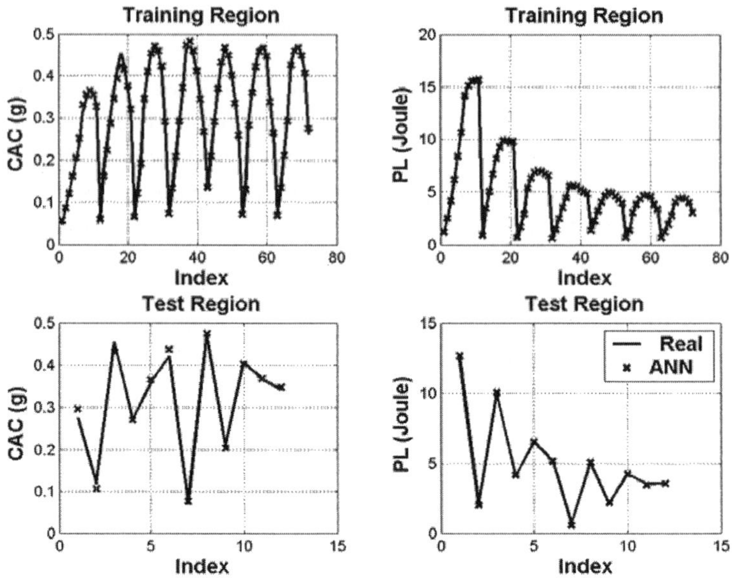

Figure 1 Camless engine ANN modelling results.

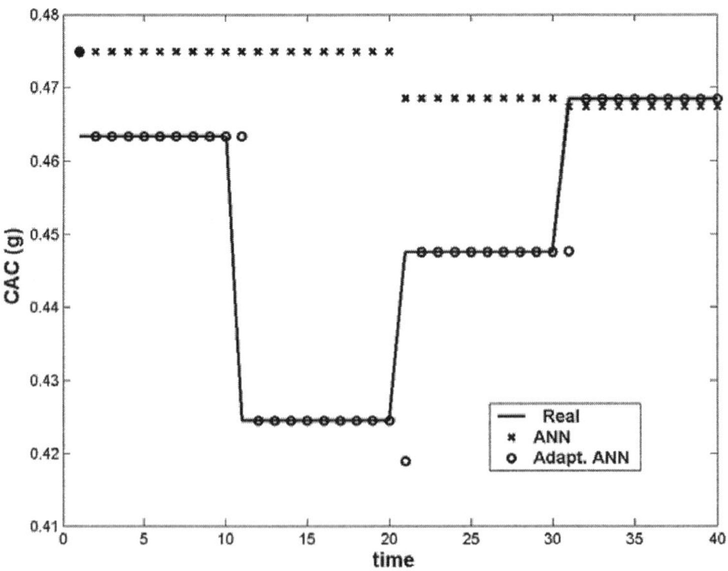

Figure 2 Camless engine simulation results for *CAC*.

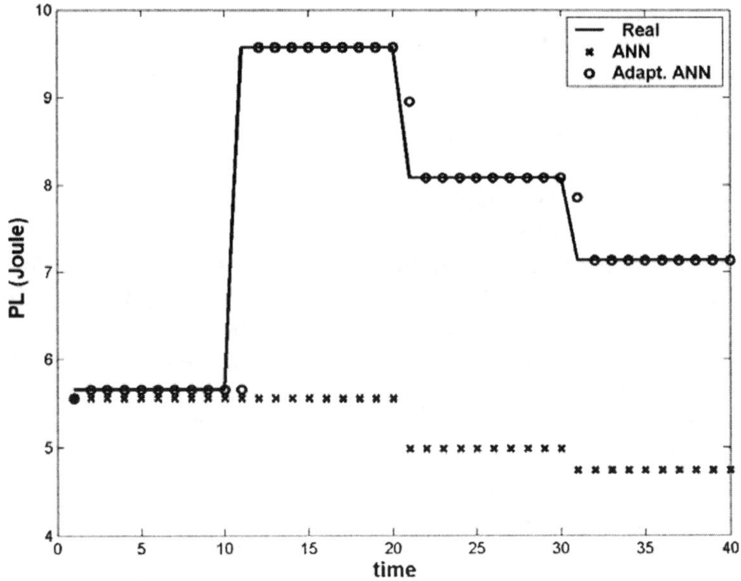

Figure 3 Camless engine simulation results for *PL*.

5. Conclusion

An algorithm for adapting multi-input multi-output feed-forward artificial neural networks was developed. The presented technique is useful for forecasting applications and is a generalization of previous research results presented in [1] where adaptation of single output feed-forward neural networks was developed. The adaptation scheme is based on the Kaczmarz's projection algorithm that offers less computational complexity and quick convergence. It uses the real outputs measurements to correct for errors in the predicted (neural net) outputs by updating the linear parameters of the network on-line. The developed algorithm was tested on cylinder air charge/energy loss prediction in camless engines. The simulation results showed that the algorithm is effective. The adaptive ANN can improve the performance and accuracy of a previously developed ANN based camless engine inverse controller.

References

1. M. S. Ashhab, *The 5th IASTED International Conference on Modelling, Simulation and Optimization*, Oranjestad-Aruba, 114 (2005).
2. I. Vajk and J. Hetthessy, *Control Engineering Practice* **13**, 895 (2005).

3. K. Kalaitzakis, G. Stavrakakis and E. Anagnostakis, *Electric Power Systems Research* **63**, 185 (2002).
4. G. Nasr, E. Badr and C. Joun, *Energy Conversion and Management* **44**, 893 (2003).
5. K. Reddy and M. Ranjan, *Energy Conversion and Management* **44**, 2519 (2003).
6. J. Vandegriff, K. Wagstaff, G. Ho and J. Plauger, *Advances in Space Research* **36**, 2323 (2005).
7. M. Mungiole and D. Keith Wilson, *Applied Acoustics* **67**, 324 (2006).
8. Y. Chiang, L. Chang and F. Chang, *Journal of Hydrology* **290**, 297 (2004).
9. O. Makarynskyy, *Ocean Engineering* **31**, 709 (2004).
10. S. Heravi, D. R. Osborn and C. R. Birchenhall, *International Journal of Forecasting* **20**, 435 (2004).
11. A. Chen, M. Leung and H. Daouk, *Computers & Operations Research* **30**, 901 (2003).
12. G. Tkacz, *International Journal of Forecasting* **17**, 57 (2001).
13. M. S. Ashhab, *The 4th IASTED International Conference on Modelling, Simulation and Optimization*, Hawaii-U.S.A, 66 (2004).
14. M. Hafner, M. Schüler, O. Nelles and R. Isermann, *Control Engineering Practice* **8**, 1211 (2000).
15. T. Slanvetpan, R. Barat and J. Stevens, *Computers & Chemical Engineering* **27**, 1605 (2003).
16. A. Datta, M. Hareesh, P. Kalra, B. Deo and R. Boom, *Steel Research* **65**, 466 (1994).
17. G. Zhang, B. Patuwo and M. Hu, *International Journal of Forecasting* **14**, 35 (1998).
18. M. Ghiassi, H. Saidane and D. K. Zimbra, *International Journal of Forecasting* **21**, 341 (2005).
19. Y. Chen, B. Yang, J. Dong and A. Abraham, *Information Sciences* **174**, 219 (2005).
20. R. Boné, M. Crucianu and J. Asselin de Beauville, *Neurocomputing* **48**, 251 (2002).
21. S. Kim, *Neurocomputing* **20**, 253 (1998).
22. S. Haykin, *Neural networks: a comprehensive foundation*, Englewood Cliffs, NJ: Prentice-Hall (1998).
23. K. Astrom and B. Wittenmark, *Adaptive control*, Paramus, NJ: Prentice-Hall (1994).
24. M. Ashhab, A. Stefanopoulou, J. Cook and M. Levin, *Journal of Dynamic Systems, Measurement, and Control* **122**, 122 (2000).
25. M. Ashhab, A. Stefanopoulou, J. Cook and M. Levin, *Journal of Dynamic Systems, Measurement, and Control* **122**, 131 (2000).

DYNAMIC MODELING OF AN ELECTROSTATIC ACTUATED CANTILEVERED MICROMIRROR

JIANLIANG YOU
MUTHUKUMARAN PACKIRISAMY
ION STIHARU

Optical Microsystems Laboratory, Concave Research Center
Department of Mechanical & Industrial Engineering, Concordia University
1455 de Maisonneuve Blvd. West, Montreal, Quebec H3M 1G8, Canada

This paper investigates dynamic performance of a rectangular micromirror supported by a straight cantilevered beam and actuated by either one of the two bottom electrodes. The established analytical model is based on assumptions that bending motion and torsion motion of the actuator are independent with each other, and bending motion of the microplate is decomposed as a translation at the cantilever tip and a rotation about the lateral axis of the point, as well as that the microplate is assumed rigid. The model has demonstrated the independence between torsion and bending motions. Hence a uniform beam cross-section that leads to stiff bending and soft torsion of the structure can be used to form the cantilevered hinge for a simple torsional micromirror. This new concept is then verified through case study of both analytical model and the corresponding FEA simulation. Dynamic performance of the actuator was analyzed through dynamic responses of the structure under various applied voltages. Comparisons of the calculated resonant frequencies and responses with simulated results also proved the accuracy of the analytical model.

1. Introduction

Micro electrostatic actuators that consist of a microplate and the corresponding suspension have vast applications in optical cross-connects (OXC), scanners, optical switches, digital displays, accelerometers, etc. Though electrostatic actuation exhibits binary unstable deflection when the drive voltage reaches at the pull-in voltage of the structure, it shows many advantages that have been proved in recent years, such as, simple structure, easy to fabricate, out-of-plane motion, quasi-linearity before pull-in, tunable working frequency, and controllable static and dynamic performance. The electrostatic actuators fabricated are usually designed to realize either a translation or a rotation with one DOF (degree of freedom) of motion. Coupled motion and pull-in phenomenon of a structure by electrostatic actuation with two degrees of freedom has been modeled and validated through FEA simulations [1-4].

However, there are few articles discussing on the dynamic performance of electrostatic actuators. As it is common to deal with problems of oscillation, dynamic analysis has become an essential issue. This characterization helps predict the possible response of the actuator under excitation [5].

Several torsional structures with electrostatic actuation have appeared in recent years. A rectangular micromirror suspended symmetrically by two hinges is one of the structures [6]. Another structure for torsional actuators is composed of a rectangular micromirror that is supported by two straight beams attached to an edge of the mirror [1]. In order to soften the torsional stiffness, hinges of straight beams are replaced by planar torsional serpentine springs [7], thereby reducing the drive voltage of actuation and achieving a compatibility of integration with CMOS circuits. It is noted that torsional spring constant is usually much smaller than bending stiffness of the beams. Based on this knowledge, the microplate suspended by a single cantilever instead of by two symmetric cantilevers is modeled for applications such as spatial optical switching or steering. Although nonlinear electrostatics around pull-in status deviates a bit from linear assumption, for a wide range of drive voltage the calculated static bending, torsion and rotation angles according to the derived analytical model are still within an acceptable range of error, as compared with the FEA results. The investigation thus has verified the applicability of the deduced model with assumption of small deflection [8].

With only one support cantilever the proposed torsional micromirror is shown in Fig. 1. Dynamic characterization of the actuator is presented in this paper. Spring constants of the cantilevered microplate are less than half of the originally symmetric beam-plate structure. Investigation of dynamic responses of the actuator to step DC voltage excitation is the focus of the work. Characteristic analysis of the dynamic performance is performed through a lumped model with three degrees of freedom. This simple prismatic cantilever and microplate structure can be easily fabricated by Micragem™ technology (**Micra**lyne **GE**neralized **M**EMS: An SOI-based MEMS fabrication process) [9]. Verification of dynamic performance is implemented by FEA simulation using ANSYS. The torsional micromirror with very soft torsion but stiff bending (negligible bending displacement) can thus be realized through a deliberate selection of cantilever geometry parameters.

2. The Coupled Dynamic Model

The cantilever-plate actuator is composed of an assumed rigid rectangular plate and a straight cantilever with uniform rectangular cross-section. The suspended plate can rotate with respect to the cantilever axis as well as to the conjunction of the cantilever and the plate when an electrical bias is applied between the plate and one of the two bottom electrodes. The two isolated, equally sized bottom electrodes are symmetrically arranged with respect to the axis of the cantilever and have an isolated distance of a between them. Fig. 1(a) shows the 3D view of the actuator, and Fig. 1(b), (c) indicates geometry parameters of the actuator.

Due to highly uniform surface and relatively large size, the SOI (silicon on insulator) fabricated microplate is assumed to be rigid and thus can be used to reflect the incident light without divergence or diffusion. The thin cantilever beam is deliberately designed so that the resulted slope angle at the cantilever tip due to a drive voltage can be negligible as compared with that of the torsion angle. It has been noticed in [8] that for a short cantilever the torsion angle is much larger than the possible bending slope at the tip, or the torsional spring constant of a cantilever is several orders smaller than the bending stiffness. However, as is shown in Fig. 1 (d), the equivalent force diagram for the actuator considers the rigid plate as a lumped mass and the electrostatic loads as vertical bending force F_e, torque T_e and bending moment M_e that correspond to deflections z, α and β at the cantilever tip, respectively.

Though the ratios between spring constants of the actuator determine the coupling effect among the assumed three DOFs, the actuator can be specifically designed for a customized OXC (optical cross connection) with simple control algorithm if the designed structure has a negligible bending deflection and no coupling effect between torsion and bending. As a matter of fact, two of the assumed 3-DOF model are actually belonged to the same DOF, i.e., both vertical translation and bending slope at the tip together constitute the bending deflection of the actuator. Nevertheless, the decoupled 3D analytical model is adopted in this paper for convenience.

The applied electrostatic load on the cantilever tip of the actuator has been defined in [8] and is described as follows, where ε is the air permittivity, V the bias, g_0 the gap between the plate and the bottom electrode, W the width, and L the length of the plate, a the isolated distance between two bottom electrodes.

$$F_e = \frac{\varepsilon V^2}{2\alpha\beta}\left[\ln\frac{g_0 - z - \frac{W}{2}\alpha}{g_0 - z - L\beta - \frac{W}{2}\alpha} - \ln\frac{g_0 - z - \frac{a}{2}\alpha}{g_0 - z - L\beta - \frac{a}{2}\alpha}\right] \tag{1}$$

$$T_e = \frac{\varepsilon V^2}{2\alpha^2\beta}\left[(g_0 - z - L\beta)\ln\frac{g_0 - z - L\beta - \frac{a}{2}\alpha}{g_0 - z - L\beta - \frac{W}{2}\alpha} - (g_0 - z)\ln\frac{g_0 - z - \frac{a}{2}\alpha}{g_0 - z - \frac{W}{2}\alpha}\right] \quad (2)$$

$$M_e = \frac{\varepsilon V^2}{2\alpha\beta^2}\left[\begin{array}{l}(g_0 - z - \frac{W}{2}\alpha)\ln(g_0 - z - \frac{W}{2}\alpha) - (g_0 - z - \frac{a}{2}\alpha)\ln(g_0 - z - \frac{a}{2}\alpha) \\ + (g_0 - z - L\beta - \frac{a}{2}\alpha)\ln(g_0 - z - L\beta - \frac{a}{2}\alpha) \\ - (g_0 - z - L\beta - \frac{W}{2}\alpha)\ln(g_0 - z - L\beta - \frac{W}{2}\alpha) + L\beta\ln\frac{g_0 - z - L\beta - \frac{a}{2}\alpha}{g_0 - z - L\beta - \frac{W}{2}\alpha}\end{array}\right] \quad (3)$$

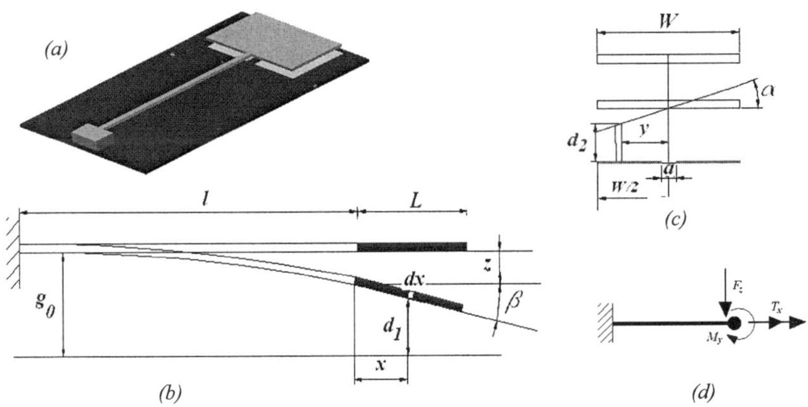

Fig. 1 The schematic view, geometry parameters and the equivalent force diagram of the actuator.

Distributed mass of the lengthy cantilever beam has to be included when dealing with dynamic performance. The equivalent mass of the cantilever beam with respect to the tip motions is deduced through assumption of third-order polynomial mode shapes of the beam and their integration along the total length is given with the following formula:

$$m_{ij} = \int_0^l m\varphi_i\varphi_j dx \quad (4)$$

where m, l are the mass per unit length and length of the beam. The equivalent mass is thus derived as

$$m_e = \frac{ml}{210}\begin{bmatrix} 78 & 0 & -11l \\ 0 & 70 & 0 \\ -11l & 0 & 2l^2 \end{bmatrix} \quad (5)$$

According to Lagrange's equations,

$$\frac{d}{dt}\left(\frac{\partial(T-U)}{\partial \dot{q}_i}\right)-\frac{\partial(T-U)}{\partial q_i}=Q_i \tag{6}$$

where T and U are the system's kinetic and potential energies; q_i is a deflection variable, and Q_i is the external load. The matrix dynamic equation of the actuator when including the effect from the cantilever beam can then be given

$$\begin{bmatrix} m_{11} & 0 & m_{13} \\ 0 & m_{22} & 0 \\ m_{31} & 0 & m_{33} \end{bmatrix}\begin{bmatrix} \ddot{z} \\ \ddot{\alpha} \\ \ddot{\beta} \end{bmatrix}+\begin{bmatrix} k_{11} & 0 & k_{13} \\ 0 & k_{22} & 0 \\ k_{31} & 0 & k_{33} \end{bmatrix}\begin{Bmatrix} z \\ \alpha \\ \beta \end{Bmatrix}=\begin{Bmatrix} F_e \\ T_e \\ M_e \end{Bmatrix} \tag{7}$$

where

$$m_{11}=M+\frac{13}{35}ml \qquad m_{13}=\frac{ML}{2}-\frac{11}{210}ml^2$$

$$m_{22}=\frac{MW^2}{12}+\frac{1}{3}ml \qquad m_{31}=m_{13} \qquad m_{33}=\frac{ML^2}{4}+\frac{1}{105}ml^3$$

and M is the mass of the plate.

The stiffness matrix in the equation corresponds to

$$k_{11}=\frac{12EI}{l^3} \qquad k_{13}=-\frac{6EI}{l^2} \qquad k_{22}=\frac{GJ}{l} \qquad k_{31}=k_{13} \qquad k_{33}=\frac{4EI}{l}$$

where I is the cross-sectional area moment of inertia of the beam and J the cross-sectional polar moment of inertia of the beam, which are given by

$$I=\frac{wt^3}{12} \qquad J=\frac{wt^3}{16}\left[\frac{16}{3}-3.36\frac{t}{w}\left(1-\frac{t^4}{12w^4}\right)\right] \quad (t\leq w) \tag{8}$$

E is Young's modulus, G is the shear modulus. The polar moment of inertia of the beam can also be defined more precisely as

$$J=\frac{wt^3}{3}\left[1-\frac{192}{\pi^5}\frac{t}{w}\sum_{k=0}^{\infty}\frac{1}{(2k+1)^5}\tanh\left(\frac{(2k+1)\pi w}{2t}\right)\right] \quad (t\leq w) \tag{9}$$

For $t\geq w$, w and t should be replaced between each other in the formulae.

It is noticed that the second order differential equations of the actuator in (7) are nonlinear and coupled among the three variables due to the nonlinear electrostatics. Resonant frequencies of the structure can be estimated by solving the determinant equation:

$$(k_{22}-m_{22}\omega^2)\left((k_{11}-m_{11}\omega^2)(k_{33}-m_{33}\omega^2)-(k_{13}-m_{13}\omega^2)^2\right)=0 \tag{10}$$

It can be seen from the equation that the torsional motion of the actuator is independent from other motions, that is, the natural frequency for torsion is deduced from letting the first parenthesis term equal to zero as

$$\omega_2 = \sqrt{k_{22} / m_{22}} \tag{11}$$

whereas the resonant frequency of bending motion is resolved from the second parenthesis term.

3. Simulations and Discussions

3.1. *Electrostatic Equivalent Stiffness Matrix*

Generally eigen-frequencies of a structure mainly depend on the stiffness term and mass term of its governing equation. Nonlinear electrostatic loads imposed on the actuator can be approximated through Taylor series expansion. The electrostatic load has been decoupled with respect to the three coordinates as defined in the previous section. Developing expressions for F_e, T_e and M_e about an initial static point (z_0, α_0, β_0) of the actuator under an applied voltage and discarding terms of the second and higher orders, yield the following expressions for equivalent spring constants:

$$k_{e11} = \frac{\partial F_e(z_0,\alpha_0,\beta_0)}{\partial z} \quad k_{e12} = \frac{\partial F_e(z_0,\alpha_0,\beta_0)}{\partial \alpha} \quad k_{e13} = \frac{\partial F_e(z_0,\alpha_0,\beta_0)}{\partial \beta}$$

$$k_{e21} = \frac{\partial T_e(z_0,\alpha_0,\beta_0)}{\partial z} \quad k_{e22} = \frac{\partial T_e(z_0,\alpha_0,\beta_0)}{\partial \alpha} \quad k_{e23} = \frac{\partial T_e(z_0,\alpha_0,\beta_0)}{\partial \beta}$$

$$k_{e31} = \frac{\partial M_e(z_0,\alpha_0,\beta_0)}{\partial z} \quad k_{e32} = \frac{\partial M_e(z_0,\alpha_0,\beta_0)}{\partial \alpha} \quad k_{e33} = \frac{\partial M_e(z_0,\alpha_0,\beta_0)}{\partial \beta}$$

where k_{e11}, k_{e12}, k_{e13} mean the equivalent translational bending, rotational bending and torsional spring constants due to the vertical electrostatic load F_e. The other equivalent stiffness coefficients are defined similarly. Substitution of these coefficients into the governing matrix equation (7) gives

$$\begin{bmatrix} m_{11} & 0 & m_{13} \\ 0 & m_{22} & 0 \\ m_{31} & 0 & m_{33} \end{bmatrix} \begin{bmatrix} \ddot{z} \\ \ddot{\alpha} \\ \ddot{\beta} \end{bmatrix} + \begin{bmatrix} k_{11} - k_{e11} & -k_{e12} & k_{13} - k_{e13} \\ -k_{e21} & k_{22} - k_{e22} & -k_{e23} \\ k_{31} - k_{e31} & -k_{e32} & k_{33} - k_{e33} \end{bmatrix} \begin{Bmatrix} z \\ \alpha \\ \beta \end{Bmatrix} = \begin{Bmatrix} F_{e0} \\ T_{e0} \\ M_{e0} \end{Bmatrix} \tag{12}$$

The right side vector represents the electrostatic force, torque and moment that are resulted from the electrical load at an equilibrium point. It is clear from the stiffness matrix that each individual member in the stiffness matrix is subtracted by a positive value, indicating the softening effect of the electrostatics on the structure. Three fundamental eigen-frequencies can thus be calculated by solving the corresponding deterministic equation.

3.2. *Frequency Spectrum*

The specified parameters of the actuator for analysis are shown in Table 1. Simulation and calculation for free vibration of the actuator without applied voltage give a slight difference of the fundamental frequency. As shown in Table 2, the first bending mode simulated from ANSYS model has a slight higher frequency than that of analytical model and with an error of 1.8% between them. There even exists a small difference when SOLID45 and SOLID95 element meshing are involved respectively in the same model. Still, it should be noticed that the first torsional mode frequency is disappeared in ANSYS simulation, which has not been mentioned in literature so far, although there apparently exists this mode due to the relatively large lateral dimension of the plate and much more flexibleness of torsion motion than that of bending.

Table 1 Parameters of An Actuator for Analysis

microplate $L\times W$ (μm^2)	cantilever $l\times w$ (μm^2)	thickness t (μm)	gap g_0 (μm)	isolated distance a (μm)	Young modulus E (GPa)	Poisson ratio ν	density ρ (kg/m^3)
200×400	600×20	10	12	40	129.5	0.21	2320

Table 2 The fundamental frequency without applied voltage

Frequency	MATLAB	ANSYS (solid45)	ANSYS (solid95)	error
1st bending	5029.8 Hz	5122.5	5120.8	1.8%
1st torsion	10.56 Hz	Not shown	Not shown	---

With different voltages applied to the actuator the resonant frequencies are shown in Table 3 for comparison. Using different element types in ANSYS model for meshing of the structure has resulted in a very small difference of the values when the applied voltage is much lower than pull-in value. As the voltage increases, nonlinearity of electrostatics becomes more dominant and this difference becomes larger. The results show that Solid95 meshing has much closer values than Solid45 meshing to the calculated results. However, for an actuator of this type used for oscillation purposes the applied voltage is usually within an almost linear range of the electrostatics, thus either element types can be used for simulation though preference should be given to Solid95.

Table 3 Variation of the first bending mode frequency with applied voltage

Voltage (V)	Analyzed (MATLAB)	Simulated (ANSYS, solid45)	Difference (%)	Simulated (ANSYS, solid95)	Difference (%)
0	5029.8	5122.5	1.8	5120.8	1.8
10	5024.2	4970.0	1.1	5097.7	1.5
20	5006.1	4888.5	2.3	5023.4	0.3
30	4970.7	4725.1	4.9	4878.4	1.9
40	4902.6	4401.9	10.2	4607.7	6.0

Table 3 also validates that a linear working range exists since the difference between analysis and simulation is within an acceptable range. However, when dealing with pull-in or near pull-in performance of the actuator, analytical results again shows a large deviation from simulated results, which demonstrated that electrostatic non-linearity must be considered within this range. The first bending frequency versus applied voltage from the three models is shown in Fig. 2 (*a*).

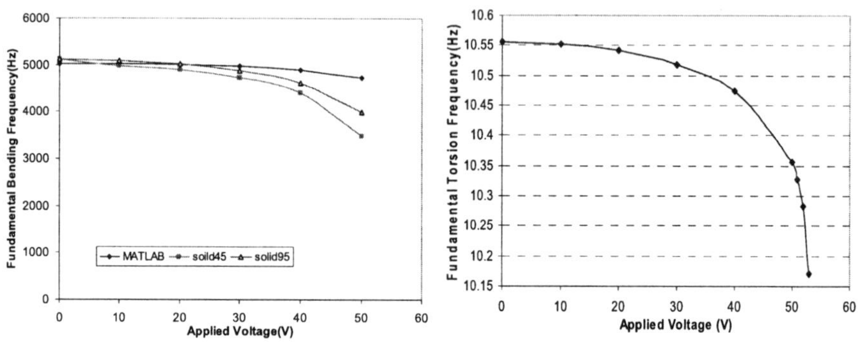

(*a*) (b)

Fig. 2 Resonant frequencies against the applied voltage: (*a*) the first bending frequency versus the applied voltage. (*b*) the first torsion frequency versus the applied voltage.

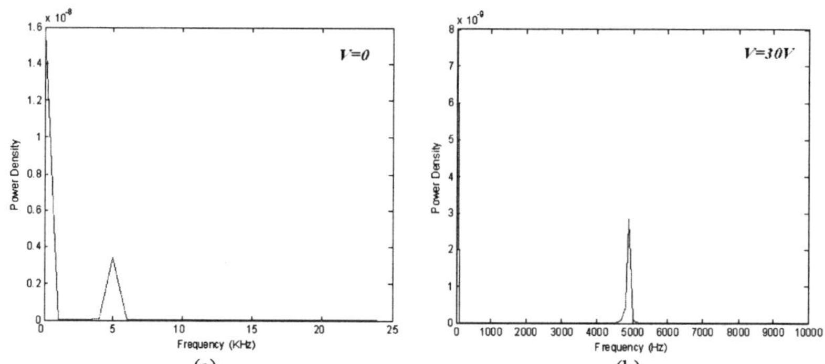

(a) (b)

Fig. 3 Frequency spectrums with (*a*) no voltage and (*b*) applied voltage of 30 V.

It is interesting to note that the first torsional mode frequency varies only within a 1Hz range when the applied voltage varies from zero to a maximum value close to pull-in (Fig. 2 (*b*)). Moreover, there is a huge difference in frequency between torsion and bending (10Hz and 5000Hz, respectively). This reveals a fact that the cantilevered microplates can be applied for torsional micromirrors through proper selection of the cantilever dimensions. As long as the bending deflection induced is kept within an acceptable range of errors, the cantilevered micromirrors can replace the current torsional micromirrors in the applications for precise optical switching. By Runge-Kutta method, frequency spectrum of the actuator corresponding to a stepped applied voltage is solved and shown in Fig. 3. There are two frequencies, one with 10Hz frequency which has a much larger power density and the other with around 5000Hz that has a much lower power of vibration.

3.3. *Dynamic Responses*

Dynamic performance of the actuator here only considered deflection responses to a suddenly applied constant voltage. Fig. 4 shows the torsion responses of the actuator with three applied voltages. Lower applied voltage leads to smaller amplitude of response and the torsion angle is always kept at a range from zero to a positive or a negative value. The average torsion angle or the amplitude of torsion motion can be estimated through the response. The period or frequency of the torsion motion is read from the responses, which is approximately 0.95s or 10.5Hz, being conformant with the previous calculation.

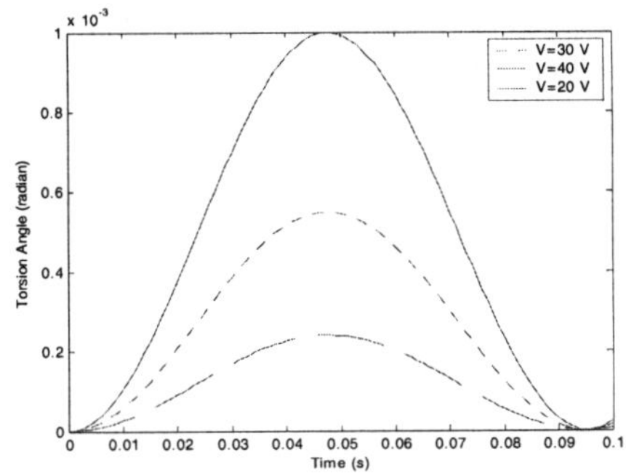

Fig. 4 Response of torsional mode to the step voltage applied.

Comparatively, vertical or translational bending and rotational bending motions have shown a much higher frequency. Fig. 5 lists the translational bending responses at tip of the actuator to different DC voltages applied. The rotational bending responses at the tip have a little bigger amplitude than that of the translational bending but share the same frequency. This phenomenon agrees with the actual situation since two of the assumed three DOFs are coupled and thus actually belong to the same one. It is again clear from the row of figures that amplitude of the first bending mode decreases as the applied voltage increases, whereas the second bending mode that has a much denser oscillation varies inversely. However the second bending mode is always smaller in amplitude than that of the first mode except at pull-in status. It can also be noticed that the bending motion oscillates about the free-load position when voltage is small. And it has been proved that the bending motion can be in the same order in amplitude with that of the torsion motion if a lengthy cantilever is used. It can be used for spatial steering or switching of light beam if dimensions of the actuator are properly selected. An example of the applications is a cantilever suspended torsional micromirror that has a stiffened bending motion and negligible bending deflection.

4. Conclusions

The dynamic model for electrostatic actuated cantilevered micromirrors is established. The accuracy of the analytical model for dynamic performance of the actuator is verified through FEA simulations. The simulation models from two types of element meshing have resulted in different eigen-frequencies and deflections especially when applied loads are close to pull-in. Element SOLID95 instead of SOLID45 is suggested for dynamic simulations for such electrostatic actuators. Dynamic responses of the actuator to the constant step voltage are solved and analyzed. Some promising applications of the actuator are proposed as long as the dimensions of both the cantilever beam and the plate are properly selected.

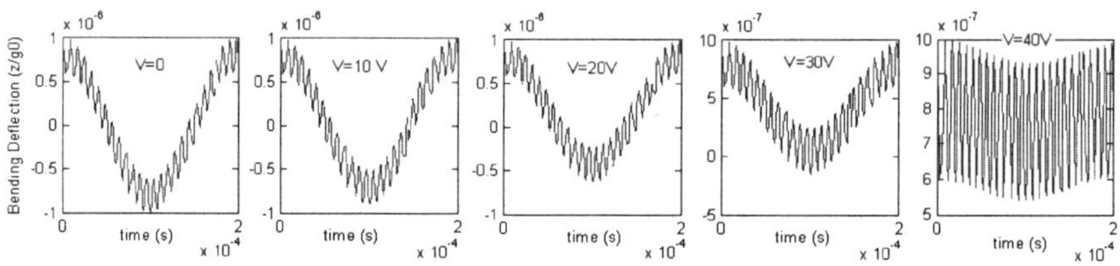

Fig. 5 Responses of bending motion to the step voltage of 0, 10, 20, 30 and 40V.

References

1. Ofir B. Degani and Yael Nemirovsky, *J. of Microelectromechanical systems* **11**, 20(2002).
2. J.-M. Huang, A.Q. Liu, et al, *Sensors and Actuators* **A115**, 36 (2004).
3. Ofir B. Degani, Yael Nemirosky, *Sensors and Actuators*, **A97-98**, 569 (2002).
4. Z. Xiao, W. Peng and K. R. Farmer, *J. of Microelectromechanical systems* **12**, 929(2003).
5. Jian Ping Zhao, Hua Ling Chen, *Microsystem Technologies* **11**, 1301 (2005).
6. Avinash K. Bhaskar, Muthukumaran Packirisamy, Rama B. Bhat, *Mechanism and Machine Theory* **39**, 1399 (2004).

7. Jianliang You, Muthukumaran Packirisamy and Ion Stiharu, *The 3rd Int. Conf. on Intelligent Sensing and Information Processing*, Bangalore, India (2005).
8. Jianliang You, Muthukumaran Packirisamy and Ion Stiharu, *IEEE Int. Symposium on Industrial Electronics (ISIE'06)*, Montreal, Canada (2006). (Accepted)
9. www.micralyne.com, Micralyne Inc. 1911-94 Street, Edmonton, Alberta, Canada, T6N 1E6.

AUTOTUNING OF A DC SERVOMECHANISM USING CLOSED LOOP IDENTIFICATION

RUBÉN GARRIDO, ROGER MIRANDA

Departamento de Control Automático, CINVESTAV
México, D.F

In this paper a methodology for automatic tuning of position controlled servomotors is presented. The key element is a novel closed loop identification technique where the loop is closed using a standard Proportional Derivative controller. Once the servomotors parameters are estimated they are employed for setting up a controller using Linear Quadratic Regulator techniques. It is experimentally shown that within a few seconds the closed loop system is tuned.

1. Introduction

In the early days of motion control selecting the parameters of a compensation filter for servomotor control was a time consuming and laborious task. Today, much of the task can be automated depending on what is known about the servo. For the user, it would be desirable to have an automatic tuning procedure allowing for a fast commissioning of the motion system. Automatic tuning or autotuning for short is composed of two sequential parts. In the first part parameters of the servo are identified. In the second a control law is computed using the parameters obtained in the first part. Regarding the second part there exists a great number of designs including linear and nonlinear techniques. Among the linear techniques Proportional Derivative (PD) and Proportional Integral Derivative (PID) controllers are well established in the realm of motion control. On the other hand, regarding the identification part situation is not as good. Even if there exist a lot of work concerning parameter identification [14], [15] most of the proposed algorithms deal with open loop stable systems. Note that for a linear model of position controlled servomotor its transfer function has a pole on the imaginary axis then making the open loop system marginally stable, then, most of the identification methods do not apply for this case. Moreover, if the identification is performed when the servomotor is coupled to a mechanical load, for example to a robot arm, closed loop identification with the loop closed around a position sensor would be desirable for security reasons since open loop techniques would lead to unbounded motor behavior. From the above situation it would be desirable to develop algorithms for fast identification and tuning of position controlled servos. In the following paragraphs a brief literature review concerning identification and autotunig is given.

There are several works dealing with parameter estimation of servomotors and among them two classes may be distinguished: open loop based and closed loop based. In the open loop case it is assumed than the servomotor works in velocity mode, i.e., the variable of interest is the servomotor velocity. An advantage of this approach is the fact that in this case the servomotor is open loop stable and then a standard off-line or on-line identification technique may be applied. Works in this vein are [1], [2], [3], [4], [5], [6] and [7]. It is interesting to mention that in [6] and [7] closed loop identification techniques are reported for velocity controlled servos, in particular, in [6] a Proportional Integral controller (PI) is employed for identifying a discrete-time model of a three-mass electromechanical system. Closed loop identification of position controlled servos is studied in [9], [10], [11]. In all of these works a technique used to obtain a stable movement is to close the feedback loop using a relay in the same way as in industrial processes [13] for controller tuning purposes. A drawback of the relay technique is the fact that tuning of the relay controller can be cumbersome, further, it is not clear that the signal generated by the relay have an adequate frequency spectrum to identify the servomotor parameters, specially if the parameters of non linear terms such that nonlinear friction and nonlinear position depending loads need to be identified. Finally, works presenting identification and tuning results simultaneously are [8] using relay techniques and [12] where a robust controller is designed from results using classic identification algorithms.

In this work a new automatic tuning methodology is presented. The loop around the servo is closed using a PD controller, in this way closed loop stability is easily ensured. The key ingredient of this methodology is a novel closed loop identification technique. Once the servomotor parameters are identified, a PID controller is computed using Linear Quadratic Regulator (LQR) techniques. It is experimentally shown that after several seconds the servo motor is tuned. Other controller techniques may be accommodated under this approach, moreover, the methodology may be applied to DC brush and AC brushless servos provided that an inner current loop is closed, i.e., the servo-amplifier feeding the servomotors is working in current mode.

The paper is organized as follows. Section 2 is devoted to the identification algorithm. Section 3 deals with the LQR controller design. Experimental results for identification and tuning are shown in Section 4 and some concluding remarks are given in Section 5.

2. Closed Loop Parameter Identification

The idea of the closed loop identification algorithm is as follows (see Fig.1). Two PD controllers are applied to the real servomotor and to a servomotor model. Note that the same controller gains are used in both controllers. A reference consisting of a signal fulfilling the persistence of excitation condition [16] is fed to both closed loop systems. The error between the output of these systems and its time derivative feeds an identification algorithm and the model parameters are adjusted using the parameters obtained from the identification algorithm. It is clear that even if the closed loop system associated to the real servo is stable, the same conclusion can not be stated for the closed loop corresponding to the identified model then making necessary to analyze the stability of the later.

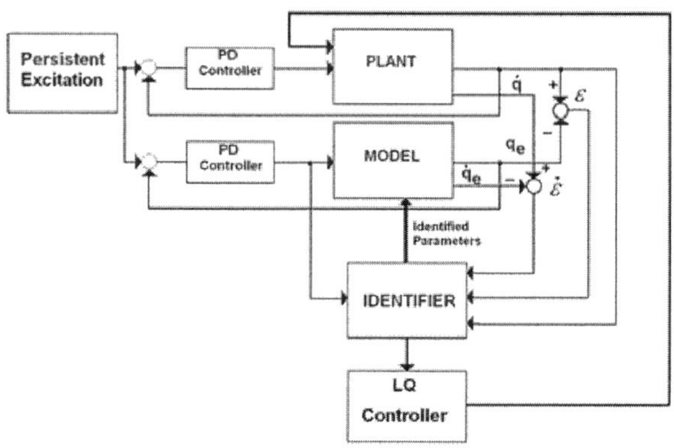

Figure 1. Blocks diagram of the autotuning process

The mathematical description of the DC servomechanism is given by

$$J\ddot{q} + f\dot{q} = \tau = ku \tag{1}$$

where J and f are the servo inertia and viscous friction respectively, $\tau = ku$ the input torque, u the control input voltage and k the amplifier gain. Model (1) assumes that the servo electric time constant is small. The above is reasonably for small servomotors; however, in large servos this condition may be not fulfilled. However, by closing a current loop around the amplifier feeding the servomotor, a common industrial practice, allows to effectively reduce the electric time constant. We have from (1)

$$\ddot{q} = -a\dot{q} + bu \tag{2}$$

where $a=f/J, b=k/J$. Let the PD control law

$$u = k_p e - k_d \dot{q} \tag{3}$$

Where $e = q_d - q$ is the position error and q_d a reference. Consider now the following estimated model, with $\hat{a}$ and $\hat{b}$ being estimates of a and b

$$\ddot{q}_e = -\hat{a}\dot{q}_e + \hat{b}u_e \tag{4}$$

to which is applied a PD control law

$$u_e = k_p e_e - k_d \dot{q}_e \tag{5}$$

With $e_e = q_d - q_e$. Note that the same gains are used in (3) and (5). Substituying (3) into (2) and (5) into (4) we have

$$\ddot{q} = -a\dot{q} + bk_p e - bk_d \dot{q} \tag{6}$$

$$\ddot{q}_e = -\hat{a}\dot{q}_e + \hat{b}k_p e_e - \hat{b}k_d \dot{q}_e \tag{7}$$

Let define the error between the outputs of the plant and the model $\varepsilon = q - q_e$ whose second time derivative is

$$\ddot{\varepsilon} = (-a - bk_d)\dot{\varepsilon} - bk_p\varepsilon + \left(\hat{a} - a\right)\dot{q}_e + \left(\hat{b} - b\right)\left[k_d \dot{q}_e - k_p e_e\right] \tag{8}$$

where (6) and (7) were employed. Let define $c = a + bk_d > 0$ and consider the parameter vector $d\theta/dt$ and vector ϕ given by $\theta^T = \begin{bmatrix} \hat{a} - a & \hat{b} - b \end{bmatrix}$ and $\phi^T = [dq_e/dt \;\; (k_d\, dq_e/dt - k_p e_e)]$, then, (8) can be written as

$$\ddot{\varepsilon} = -c\dot{\varepsilon} - bk_p\varepsilon + \tilde{\theta}^T \phi \tag{9}$$

2.1. *Stability*

In the following stability analysis for differential equation (9) we propose the Lyapunov function candidate

$$V\left(\varepsilon,\dot{\varepsilon},\tilde{\theta}\right)=\frac{1}{2}\dot{\varepsilon}^{2}+\frac{1}{2}bk_{p}\varepsilon^{2}+\mu\varepsilon\,\dot{\varepsilon}+\frac{1}{2}\tilde{\theta}^{T}\,\Gamma^{-1}\,\tilde{\theta} \tag{10}$$

and in order to have $V>0$ the inequality $0<\mu<(bk_p)^{1/2}$ should be fulfilled and under this condition we take the time derivate of (10)

$$\begin{aligned}\dot{V}\left(\varepsilon,\dot{\varepsilon},\tilde{\theta}\right)&=-\alpha\,\dot{\varepsilon}^{2}-\mu bk_{p}\varepsilon^{2}-\mu c\varepsilon\,\dot{\varepsilon}\\&+\tilde{\theta}^{T}\left[\phi\,\dot{\varepsilon}+\mu\phi\varepsilon+\Gamma^{-1}\dot{\tilde{\theta}}\right]\end{aligned} \tag{11}$$

Where $\alpha=c-\mu>0$. From this last equation it is clear that

$$\dot{\tilde{\theta}}=\dot{\hat{\theta}}=-\Gamma\phi\left(\mu\varepsilon+\dot{\varepsilon}\right) \tag{12}$$

then, (11) becomes

$$\dot{V}\left(\varepsilon,\dot{\varepsilon},\tilde{\theta}\right)=-\alpha\,\dot{\varepsilon}^{2}-\mu bk_{p}\varepsilon^{2}-\mu c\varepsilon\,\dot{\varepsilon} \tag{13}$$

Completing squares the last equation we have that a condition for negative definiteness of dV/dt is $\beta = [\mu bk_p/\alpha - \mu^2c^2/(4\alpha^2)]>0$ which is equivalent to the inequality $\mu \leq (4bk_pc)/(4bk_p+c^2)$. From the above, boundedness of ε, $d\varepsilon/dt$ and the estimated of θ is concluded. To show that ε and $d\varepsilon/dt$ converge to zero Barbalat lemma [16] is employed. To this end note that $dV/dt \leq -\alpha\beta\varepsilon^2$, then integrating this inequality it can be shown that

$$\int_0^t \varepsilon^2 \leq \frac{V(0)}{\alpha\beta}<\infty \tag{14}$$

and then ε belongs to the space L^2. From the above and since it was proved that ε and $d\varepsilon/dt$ are bounded it follows that ε and $d\varepsilon/dt$ converge to zero. According to the preceding analysis, constant μ must fulfill the following inequality

$$\mu\leq\min\left[\frac{4bk_{p}c}{4bk_{p}+c^{2}},\sqrt{bk_{p}}\right] \tag{15}$$

Finally, boundedness of the model output q_e is concluded from boundedness of ε and q. Parameter convergence is obtained if the following persistence of excitation condition is fulfilled [16]

Definition. A vector $\varphi{:}R_{+} \rightarrow R^{2n}$ is persistently exciting (PE) if there exist α_1, α_2, $\delta>0$ such that for all $t_0\geq 0$ the next equation is fulfilled

$$\alpha_2 I \geq \int_{t_0}^{t_0+\delta} \varphi(\tau)\varphi^T(\tau)d\tau \geq \alpha_1 I \tag{16}$$

3. LQR Controller Design

Because the parameter identification is performed on-line it is possible to carry out on-line the design of a LQR controller for the DC servomotor. Let define

$$e_1(t) = q(t) - q_d, e_2(t) = \dot{e}_1(t) = \dot{q}(t), e_3(t) = \int_0^t e_1(\tau)d\tau \tag{17}$$

then we get the next state space representation of (17)

$$\dot{e}(t) = \begin{bmatrix} \dot{e}_1 \\ \dot{e}_2 \\ \dot{e}_3 \end{bmatrix} = \begin{bmatrix} 0 & 1 & 0 \\ 0 & -a & 0 \\ 1 & 0 & 0 \end{bmatrix} e + \begin{bmatrix} 0 \\ b \\ 0 \end{bmatrix} u, \quad y(t) = \begin{bmatrix} 1 & 0 & 0 \end{bmatrix} e \tag{18}$$

The quadratic performance index is given by

$$J = \frac{1}{2}\int_0^T \left\{ e^T \begin{bmatrix} \gamma & 0 & 0 \\ 0 & \delta & 0 \\ 0 & 0 & \eta \end{bmatrix} e + \kappa u^T u \right\} dt \tag{19}$$

and γ, δ, η are used to weight the value of the states e_1 ,e_2 ,e_3 respectively, while κ is used to weight the magnitude of the control signal u. The steady state solution $S=0$ is considered, then the algebraic Riccati equation yields

$$\begin{aligned} S_{11} &= -S_{23} + aS_{12} + \frac{b^2}{\kappa} S_{12}S_{22}, S_{12} = \sqrt{\frac{\kappa}{b^2}(\gamma + 2S_{13})} \\ S_{13} &= aS_{23} + \frac{b^2}{\kappa} S_{22}S_{23}, S_{23} = \frac{\sqrt{\eta\kappa}}{b}, S_{33} = \frac{b^2}{\kappa} S_{12}S_{23} \\ S_{22} &= \frac{-2a\kappa + \sqrt{4a^2\kappa^2 + 4b^2\kappa(\delta + 2S_{12})}}{2b^2} \end{aligned} \tag{20}$$

Then, the corresponding control law is given as follows

$$u = -\frac{b}{\kappa}\left[e_1 S_{12} + e_2 S_{22} + e_3 S_{23}\right] \tag{21}$$

4. Experimental Results

The servomechanism employed for the experiments (Fig. 2) is controlled by a Copley Controls power amplifier, model 413, configured in current mode. Angular position of the motor is measured by a BEI optical encoder. Resolution of the optical encoder is 2500 pulses per revolution and is directly coupled to the motor shaft.

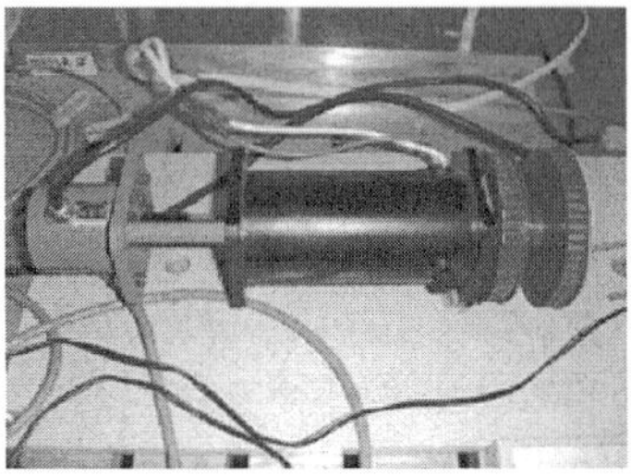

Figure 2. DC motor used in the laboratory test

Data adquisition is performed by a MultiQ 3 card from Quanser Consulting endowed with inputs for optical encoders. The electronics associated to these inputs multiply by 4 the encoder resolution. Angular position is scaled down by a factor of 10,000 corresponding to one motor shaft turn. The card has 12 bits for digital-analog conversion with a output voltage range of ±5V (volts).

Programming was performed using the Matlab-Simulink software operating with the WINCON software from Quanser Consulting. Sampling period was 1 ms. Figure 3 shows the platform used for experiments.

Figure 3. Experimental platform

Velocity estimates were obtained from position measurements through a high-pass filter. As it was indicate before, parameter identification and controller tuning are perfomed sequentially. In the first part of the experiment the loop is closed using a PD controller. It corresponds to the first 5s of the experiment where the values Γ=500 and μ=50 were used for the identification algorithm. In the second part the LQR controller is computed using the estimates obtained from the first part using the values γ=1.5,δ=0.015,η=0.001,κ=0.06.

Figure 4 shows the reference signal employed for the experiments. This signal is composed of two parts. The first part was generated using the block Band Limited White Noise and it is used for parameter identification during the first 5s. The second part is a square wave with an amplitude of 1 motor turn and a period of 5s.

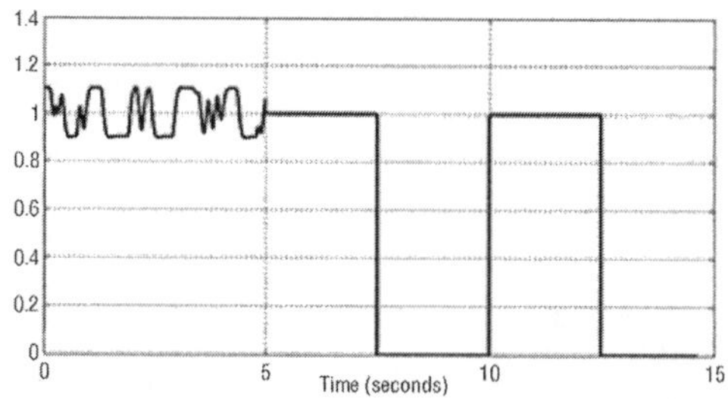

Figure 4. Reference signal q_d

The identified parameters are shown in the Figures 5 and 6, where the average values are a=0.2,b=120. Note that the servo parameters are identified within the first 5 seconds.

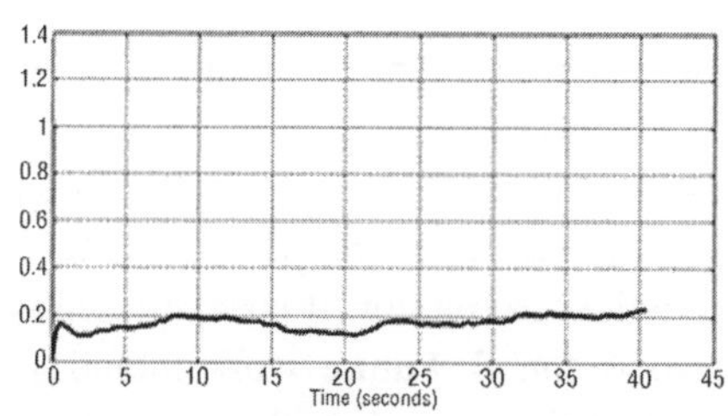

Figure 5. Parameter estimated a

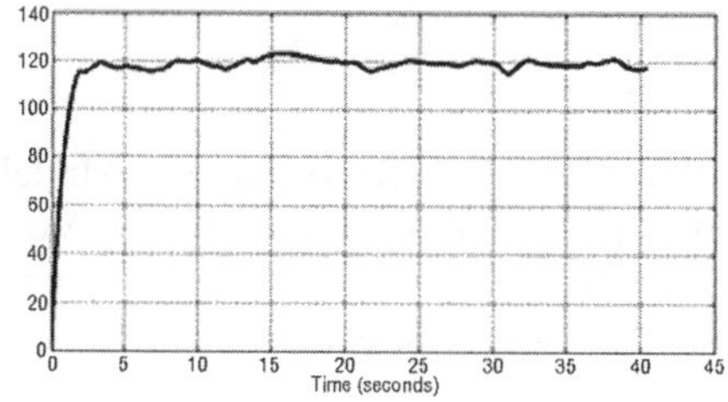

Figure 6. Parameter estimated b

Figures 7 and 8 show the response of the servomotor. The first 5 seconds correspond to the servo response to the persisting exciting reference. The next seconds correspond to the servo response using the LQR controller.

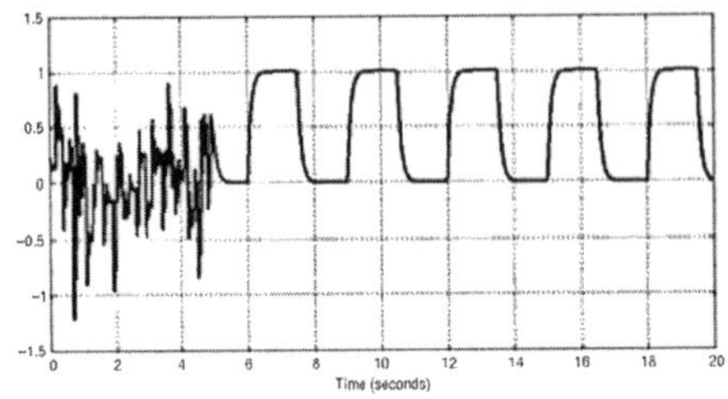

Figure 7. Response of the system with autotuning

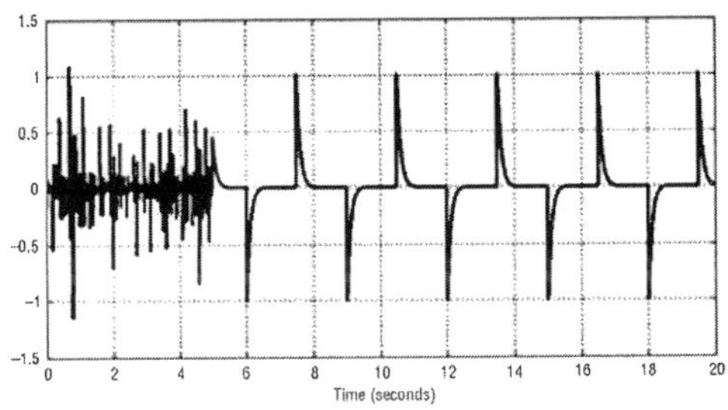

Figure 8. Position error

5. Conclusion

In this paper a methodology for automatic tuning of servomotors is presented. The proposed approach is not based on relay techniques and its key feature is a novel closed loop identification technique where the loop is closed under PD control. Once the servo parameters are identified an LQR control law is computed then applied to the servomotor. It is experimentally shown that within a few seconds the servo is tuned. Future work includes to dispense velocity measurements in the identification algorithm and to use more elaborated controllers.

References

1. W. Lord, J. H. Hwang, "DC Servomotors: Modeling and parameter determination", IEEE Trans. Industrial Electronics, Vol. 1A-13, N° 3,1977
2. O. S. Turkay, H. Yalcinkaya, B. Turkay, "Implementation of frecuency domain and time domain Least-Squares parameter estimation methods on

DC motors", Proc. of 3° European Control Conference, Rome Italy, September, 1995

3. S. Daniel-Berhe, H. Unbehauen, "Experimental physical parameter estimation of a thyristor driven DC motor using the HMF method", Control Engineering Practice, Vol. 6, pp. 615-626, 1998
4. S. A. Soliman, A. M. Al-Kandari, M. E. El-Hawari, "Parameter identification method of a separately excited DC motor for speed control", Electric Machines and Power Systems, Vol. 26, N° 8, 1998
5. Y. Y. Chen, P. Y. Huang, J. Y. Yen, "Frecuency domain identification algorithms for servo systems with friction", IEEE Trans. Control Systems Technology, Vol. 10, N° 5, 2002
6. I. Eker, "Open loop and closed loop experimental on-line identification of a three mass electromechanical system", Mecatronics, vol. 14, pp. 549-565, 2004
7. T. Kara, I. Eker, "Non linear closed loop identification of DC motor with load for low speed two directional operation", Electrical Engineering, Vol. 86, N° 2, 2004
8. K. K. Tan, T. H. Lee, P. Vadakkepat, F. M. Leu, "Automatic tuning of two degree of fredom control for DC servomotor system", Intelligent Automation and Soft Computing, Vol. 6, N° 4, 2000
9. K. K. Tan, Y. Xie, T. H. Lee, "Automatic friction identification and compensation with a self adapting dual relay", Intelligent Automation and Soft Computing, Vol. 9, N° 2, 2003
10. A. Besancon-Voda, G. Besancon, "Analysis of a two relay systems configuration", Automatica, Vol. 35, N° 8, 1999
11. K. K. Tan, T. H. Lee, S. N. Huang, X. J. Jiang, "Friction Modelling and adaptive compensation using a relay feedback approach", IEEE Trans. Industrial Electronics, Vol. 48, N° 1, 2001
12. E. J. Adam, E.D. Guestrin, "Identification and robust control for an experimental servomotor", ISA Transactions, Vol. 41, N° 2, 2002
13. K. J. Astrom, T. Hagglund, "PID controllers: theory, design and tuning, Second edition", Instruments Society of America, 1994
14. Nelles, Oliver, "Nonlinear System Identification", Ed. Springer, 2001
15. Ljung, Lennart, "System Identification", Ed. Prentice Hall, 1987
16. Shankar Sastry, Marc Bodson, "Adaptive Control, Stability, Convergence and Robustness", Ed. Prentice Hall, 1989